The Circadian Clock

Protein Reviews

Series Editor:

M. Zouhair Atassi
Baylor College of Medicine
Houston, Texas

For other titles published in this series, go to
www.springer.com/series/6876

Urs Albrecht
Editor

The Circadian Clock

Springer

Editor
Urs Albrecht
Unit of Biochemistry
Department of Medicine
University of Fribourg
1700 Fribourg
Switzerland
urs.albrecht@unifr.ch

ISBN 978-1-4419-1261-9 e-ISBN 978-1-4419-1262-6
DOI 10.1007/978-1-4419-1262-6
Springer New York Dordrecht Heidelberg London

Library of Congress Control Number: 2009932755

Printed on acid-free paper

Springer is part of Springer Science+Business Media (www.springer.com)

Preface

With the invitation to edit this volume, I wanted to take the opportunity to assemble reviews on different aspects of circadian clocks and rhythms. Although most contributions in this volume focus on mammalian circadian clocks, the historical introduction and comparative clocks section illustrate the importance of various other organisms in deciphering the mechanisms and principles of circadian biology.

Circadian rhythms have been studied for centuries, but only recently, a molecular understanding of this process has emerged. This has taken research on circadian clocks from mystic phenomenology to a mechanistic level; chains of molecular events can describe phenomena with remarkable accuracy. Nevertheless, current models of the functioning of circadian clocks are still rudimentary. This is not due to the faultiness of discovered mechanisms, but due to the lack of undiscovered processes involved in contributing to circadian rhythmicity. We know for example, that the general circadian mechanism is not regulated equally in all tissues of mammals. Hence, a lot still needs to be discovered to get a full understanding of circadian rhythms at the systems level. In this respect, technology has advanced at high speed in the last years and provided us with data illustrating the sheer complexity of regulation of physiological processes in organisms. To handle this information, computer aided integration of the results is of utmost importance in order to discover novel concepts that ultimately need to be tested experimentally. In this development of new concepts lies the chance to understand living organisms better and to develop strategies to apply these new concepts for the benefit of society. A decade ago, the influence of changing day-night cycles (e.g. jet-lag and shift work) on health was intuitively possible but not taken very seriously due to lack of convincing scientific evidence. This has changed in the last years due to discoveries illustrating the involvement of clock components in the development of cancer, obesity and mood disorders.

The main task of the circadian clock is to optimize an organism's performance and tune it with temporal changes in the environment. In that sense, the clock links the genetic setting of the organism with its environment. A better understanding of the circadian clock, therefore, will open avenues for treatment of diseases via environmental stimuli such as light. The importance of this approach is evident in the treatment of seasonal affective disorder and probably could be extended to the treatment of other neuropsychiatric disorders in the future.

However, this non-pharmacological approach is strongly underrated since no direct economic benefit is evident, except for reducing costs for health care. On the other hand, pharmacological approaches to influence biochemical mechanisms via the circadian clock might emerge. First attempts to optimize available cancer treatments taking the temporal dimension into account, have indicated to be beneficial for patients. However, the awareness of the importance for the timing of medical treatment is still very rudimentary in the medical community.

I hope that the different topics described in this book illustrate the importance of the circadian clock for health, although the evidence presented is certainly not complete. The selected topics are thought as starting points for discussions and hopefully ignite new ideas and hypotheses.

I thank the authors for their contributions. I also thank the series editor Dr. Zou Atassi, the publisher, and the various funding agencies including the EUCLOCK project for making this work possible. Finally, I thank my wife Andrea and my four children for their patience and support.

Fribourg, Switzerland

Urs Albrecht
October 2009

Contents

Contributors

Urs Albrecht
Unit of Biochemistry, Department of Medicine, University of Fribourg,
1700 Fribourg, Switzerland
urs.albrecht@unifr.ch

Steven A. Brown
Institute of Pharmacology and Toxicology, University of Zürich, 8057 Zürich,
Switzerland
Steven.brown@pharma.unizh.ch

Christian Cajochen
Centre for Chronobiology, Psychiatric University Clinics,
Wilhelm Kleinstrasse 27, CH-4025 Basel, Switzerland
Christian.Cajochen@upkbs.ch

Sarah Laxhmi Chellappa
The CAPES Foundation/Ministry of Education of Brazil, Caixa Postal 365,
CEP. 70359-970, Brasilia - DF, Brazil;
Centre for Chronobiology, Psychiatric University Clinics,
Wilhelm Kleinstrasse 27, CH-4025 Basel, Switzerland
Sarah.Chellappa@upkbs.ch

Serge Daan
Unit of Chronobiology, Centre for Life Sciences,
University of Groningen, The Netherlands
s.daan@rug.nl

Thomas d'Eysmond
School of Life Sciences, Ecole Polytechnique Fédérale de Lausanne (EPFL),
and Swiss Institute of Bioinformatics (SIB), CH-1015 Lausanne, Switzerland
thomas.deysmond@epfl.ch

Céline Feillet
Unit of Biochemistry, Department of Medicine, University of Fribourg,
1700 Fribourg, Switzerland
celine.feillet@unifr.ch

Russell G. Foster
Nuffield Laboratory of Ophthalmology, The John Radcliffe Hospital, University of Oxford, Level 5 and 6 West Wing, Headley Way, Headington, Oxford OX3 9DU, UK
russell.foster@eye.ox.ac.uk

Erik D. Herzog
Department of Biology, Washington University, St. Louis, MO 63130, USA
herzog@wustl.edu

Achim Kramer
Laboratory of Chronobiology, Charité - Universitätsmedizin Berlin, Germany
achim.kramer@charite.de

Cheng Chi Lee
Department of Biochemistry and Molecular Biology, University of Texas Health Science Center at Houston, 6431 Fannin Street, MSB 6.200, Houston, TX 77030
Cheng.C.Lee@uth.tmc.edu

David Lenssen
Unit of Chronobiology, The University of Groningen, Haren, The Netherlands

Martha Merrow
Unit of Chronobiology, The University of Groningen, Haren, The Netherlands
m.merrow@rug.nl

Felix Naef
School of Life Sciences, Ecole Polytechnique Fédérale de Lausanne (EPFL), and Swiss Institute of Bioinformatics (SIB), CH-1015 Lausanne, Switzerland
felix.naef@epfl.ch

Henrik Oster
Circadian Rhythms Group, Max Planck Institute for Biophysical Chemistry, 37077 Göttingen, Germany
henrik.oster@mpibpc.mpg.de

Stuart N. Peirson
Nuffield Laboratory of Ophthalmology, The John Radcliffe Hospital, University of Oxford, Level 5 and 6 West Wing, Headley Way, Headington, Oxford OX3 9DU, UK

Jürgen A. Ripperger
Unit of Biochemistry, Department of Medicine, University of Fribourg, 1700 Fribourg, Switzerland
juergenalexandereduard.ripperger@unifr.ch

Till Roenneberg
Institute of Medical Psychology, Ludwig-Maximilians-University Munich, Munich, Germany
till.roenneberg@med.uni-muenchen.de

Paul H. Taghert
Department of Anatomy and Neurobiology, Washington University Medical School, St. Louis, MO 63110, USA
taghertp@pcg.wustl.edu

Jens T. Vanselow
Laboratory of Chronobiology, Charité - Universitätsmedizin Berlin, Germany
Center for Experimental BioInformatics (CEBI), University of Southern Denmark, Odense

Zhaoyang Zhao
Department of Biochemistry and Molecular Biology, University of Texas Health Science Center at Houston, 6431 Fannin Street, MSB 6.200, Houston, TX 77030, USA
Zhaoyang.Zhao@uth.tmc.edu

Chapter 1
A History of Chronobiological Concepts

Serge Daan

1.1 Prelude

The perpetual alternation of night and day could not escape being noticed by the earliest humans. It must have marked to them the passage of time. Tilling their fertile soil, the early Sumerians needed precise knowledge of time. When should the wheat be sown, when be harvested? How many days to wait till the great floods in spring? How allocate their stores of grain such that daily rations would last till the new harvest? Scores of practical questions. The Sumerians went to their temples to seek the answers. Much more than the nomads' opportunistic way of life, early agricultural civilisation relied on planning, on anticipation, on keeping track of time. There was a growing caste of those who assembled the information: the priests. They noticed the rigid patterns of annual change in the sun's position. They started to observe the movements of other celestial bodies. They constructed their temples almost as astronomical observatories, with the major axes aligned to the stellar constellation on days of importance. The priests in Sumer were the first, but not the last in this respect. Most of the Mesopotamian temples, such as the Ziggurath in Babylon had a long East–West axis. In Egypt, temples were often oriented towards the direction where the sun rises on the longest day. Once per year, at dawn of the summer solstice, the first rays would illuminate the god-statue at the end of a narrow pillar gallery. Much later, the Incas in South-, the Anasazi in North America again dramatized special days in the annual cycle by the architecture and orientation of their places of worship. It was the place where priests engaged both in scientific observation and in religious duties. It also became the great storage room of past observations and events, of accumulating knowledge. The observation of stellar constellations as a means of measuring time became the first scientific activity in most early settlements, long before the cause of their movements was evident.

S. Daan
Unit of Chronobiology, University of Groningen, Groningen, The Netherlands
e-mail: s.daan@rug.nl

U. Albrecht (ed.), *The Circadian Clock*, Protein Reviews 12,
DOI 10.1007/978-1-4419-1262-6_1,

Early humans must also have been thoroughly aware of the periodic internal changes within themselves, and in the living world around them. Their own patterns of sleep and wakefulness were obvious, and so were the daily opening and closing of flowers, the seasonal up and down of their crops. Perhaps so obvious that this attracted virtually no scientific attention, until Aristotle wrote extensively about sleep and wake behaviour in animals. He jotted down the observation:

> *In the bee the fact of its being asleep is very obvious; for at night-time bees are at rest and cease to hum. But the fact that insects sleep may be very well seen in the case of common every-day creatures; for not only do they rest at night-time from dimness of vision (and, by the way, all hard-eyed creatures see but indistinctly), but even if a lighted candle be presented they continue sleeping quite as soundly.* (History of Animals; 350 B.C.; translation from Greek by D'Arcy Wentworth Thompson)

1.2 Persistence

It took more than 2000 years until the first experimental attempt was made to attack the true nature of Aristotle's phenomenon. In 1729, the French Astronomer Jean Jacques d'Ortous de Mairan wondered whether the daily leaf movements known from some plants are generated by the alternation of light and dark. He hid a *Mimosa* plant in a dark cupboard and saw that its leaves assumed their open day position in daytime even when there was no light [1].

De Mairan's observation is the classic start of every textbook on the subject of biological rhythms. Quite justly so, since it both was the first known experiment in the field, and it also revealed the first basic concept, which has turned out to have ubiquitous validity: Persistence in the absence of an external cue. De Mairan was well aware of this potential general meaning, since he speculated right away on the importance his experiment on 'sleep movements' may have for human sleep. His experiment is often cited as demonstrating the endogenous, internal generation of the leaf movements in plants. It could not be more than a suggestion since only the light conditions were changed. And De Mairan himself did not draw that conclusion. He concluded that '*the sensitive plant thus senses the sun without seeing it in any way*'. There are other variations resulting from the earth's rotation, such as in temperature and humidity, and only after these were all excluded the endogenous nature might be considered proven. Indeed Duhamel in 1758 reported leaf movements in a dark cave with constant temperature, and probably humidity [2]. But if not light or temperature, yet other variables in the environment that are correlated with time of day might be instrumental in eliciting biological periodicity. This possibility was still favoured by the authoritative plant physiologists of the nineteenth century, such as Julius Sachs, and Wilhelm Pfeffer [3] in his early writings.

Much of the research on rhythms early in the twentieth century was a hunt for the unknown correlate of the earth's axial rotation ('factor X') that drove daily cycles. Attempts to eliminate factor X included experiments in the arctic continuous daylight [4] and deep in the earth in abandoned saltmines. The last ardent defendant of such an unknown factor was the American endocrinologist Frank Brown, who at the end of his career, though not convinced, admitted that *'living things may actually have an internal timer as well, for redundancies are far from uncommon'* [5]. Probably by Brown's perseverance, several remarkable attempts were made in the second half of the twentieth century to definitively settle the issue by studies in the absence of geophysical cycles. Karl Hamner placed hamsters on a turntable on the South pole that rotated against the earth each 24 h, and saw their activity rhythm persisting [6]. In the 1980s, Sulzman and colleagues ran an experiment in the Spacelab away from the earth's influence and found the rhythm of *Neurospora* conidiation persisting indistinguishably from that in a simultaneous batch studied at Cape Kennedy on earth [7].

For nearly two centuries, the study of biological rhythms was fully dominated by the studies on the 'sleep movements' of plants. It was only at the close of the nineteenth century that De Mairan's phenomenon was confirmed in animals, in the daily migration of retinal pigment in moths by the Viennese physiologist Kiesel [8] and the body temperature rhythm of squirrel monkeys by Simpson and Galbraith [9].

1.3 Endogenous Generation

The searchers for factor X could have saved themselves the effort, had they better interpreted a finding going back to the early nineteenth century. In 1832, Auguste de Candolle, a Swiss botanist, had made a critical observation. He noted that the daily rhythm of leaf movements of *Mimosa* plants housed in continuous illumination had a cycle length up to 2 h shorter than the exact 24 h periodicity seen in natural daylight. The plants gradually drifted out of synchrony with the cycle of day and night when deprived of light cues, even though their behaviour remained rhythmic. It would be impossible to claim that the continued rhythm is a reaction to any cue associated with the alternation of day and night, if the movements were no longer in synchrony with it. The only alternative conclusion can be that the rhythm is generated internally by the plant.

Why did it take the botanists, who were a century ahead of the animal physiologists, so long to accept this conclusion? The leaf movements of *Mimosa*, and the other plants, observed in constant light tended to get smaller (damp out), much as the oscillations of a simple pendulum do after the stimulus which has set it in motion is removed. This observation may have suggested that no endogenous mechanism was involved. Hence leaf movements were at the time generally discussed in terms of residual 'Nachwirkungen' [3] or 'Nachschwingungen' (after-oscillations) that survived briefly after removal of the driving light cues. Charles Darwin had challenged this idea. He disagreed with Pfeffer, and became convinced of the endogenous nature of the sleep movements since they continued in darkness [10]. Darwin added the insightful hypothesis that these movements are adaptive in the sense of minimizing heat loss by radiation at night, while fully exposing their surface to the light in daytime. In his opening address to the Cold Spring Harbor Symposium on Biological Clocks, Erwin Bünning [11] mistakenly wrote: 'It never occurred to Darwin that the endogenous diurnal rhythm might have a selective value'. Darwins hypothesis was supported by experiments carried out a century later by James Enright [12].

A problem with the persistence argument is that it depends on conditions. In constant darkness photosynthesis is gone and the plant in trouble, in constant light the oscillation damps out. The difficulty was eventually resolved by the influential German plant physiologist Wilhelm Pfeffer who had dismissed the idea of endogenous generation of rhythms early in his career [3]. Forty years later, after having developed a mechanism by which leaf movements were automatically recorded continuously on a smoked-drum kymograph, he reversed his position. He then produced a lengthy account with careful documentation to demonstrate that plant leaf movements are not dependent on external cues. One of Pfeffer's crucial observations was that the leaf movements in bean plants *(Phaseolus vitellinus)* did not damp out when only light-sensing organs at the base of each leaf were covered with black cotton to shield them from the light. Pfeffer published his findings during World War I in an obscure German-language journal [13], where they long remained unnoticed, even by German readers [14]. During the 1920s, two people definitively settled the issue of the endogenous nature of daily plant movements. Anthonia Kleinhoonte, a Dutch student in Delft, and Erwin Bünning, a German botanist. Kleinhoonte did extensive measurements on the leaf movements of large bean plant *Canavalia ensiformis*. She became convinced of the endogenous nature because after a shift in the light–dark (LD) cycle, the phase of the rhythm in constant conditions had no longer any relationship to the day and night outside [15]. Bünning confirmed that plants housed in constant light showed sustained rhythmic movements with a cycle length that deviated from 24 h [16]. Both drew the conclusion that the rhythm could not be attributed to any 'factor X' related to the earth's rotation, but had to come from within the organism.

By that time, zoologists also had been forced by experimental data to the same conclusion. The first animal rhythm yielding evidence of the deviation from 24 h was that of pigment change in the crustacean *Hippolyte varians* [17]. The American psychologist Curt Richter extensively studied the activity rhythms of rats and observed the persistence of rhythmicity in constant environmental conditions with a period shorter than 24 h [18]. A freerunning rhythm, with a cycle deviant from 24 h in constant conditions is today considered the true proof of endogenous generation, a proof now experimentally available for hundreds of species of plants as well as animals. The conclusion gradually dawned on biology: virtually all living species, all evolved on a planet in continuous rotation, have developed an internal oscillating system to match the external variations. This conclusion was first drawn in the landmark Cold Spring Harbor Symposium on Biological Clocks [19]. By then, the word *circadian* had been coined by Franz Halberg [20] to emphasize that 'about a day' rather than precisely 24 h is the true hallmark of these periodicities.

Had humans been the object of experimental study, the conclusion of endogenous generation might have much sooner been accepted. There was intense awareness of the pervasive nature of daily periodicity in human behaviour and physiology [21], and little inclination to attribute this to simple unconscious responses to light. But it took until Jürgen Aschoff in the 1960s started to systematically exclude the external cues to human subjects that De Candolle's phenomenon was immediately observed in the first experiment and interpreted correctly [22].

1.4 Innateness

The conclusion that rhythms are generated endogenously in adult organisms does not necessarily imply that the underlying mechanism is innate or genetically determined. The possibility that the rhythm was imprinted on the organism by the light–dark cycle prevailing during development was advocated by Richard Semon early in the century, who still described the periodicity as '*Nachschwingungen*' or after-oscillations [23]. Semon soon changed his position when he observed that exposure to LD 6:6, and LD 24:24 also led to a subsequent 24 h rhythm in constant light. He then suggested that the LD rhythm was imposed on the ancestors and inherited to the seedlings [24].

In the 1950s, Semon's original proposition of a role for 'learning' the LD cycle was laid to rest definitively with experimental evidence. By then Jürgen Aschoff, a german medical doctor by training, had entered the scene. Aschoff was a thermophysiologist who had developed the Core-Shell model of thermoregulation, and described the day–night periodicity of heat loss from his hand in cold water. He started his fundamental research on daily rhythms by attacking Semon's original conjecture. He raised mice for seven generations in constant light, and found that the daily rhythms of the last generation were as robust as those seen earlier, despite the persistent absence of a lighting cycle [25]. This absence did not establish an arrhythmic environment for the rats, since each embryo would be exposed to its mother's daily rhythms in utero – and indeed we now know that the mother does exert a prenatal synchronizing influence [26]. But Aschoff also raised chickens (developing inside the egg under constant conditions), and found that they expressed a normal daily rhythm [27], indicating that no experience of a light–dark cycle was needed for the development of rhythmicity. Similar experiments have more recently been done by Vijay Sharma in Bangalore, who bred *Drosophila* in constant light over 600 generations, and still found no suppression of circadian rhythmicity [28].

While elsewhere in the developing science of behaviour the *nature vs. nurture* debate was raging, the issue was settled early on in biological rhythms research, in favour of nature: the periodic behavioural program was not only endogenously generated, it was apparently also transmitted through the genetic make-up of the individual animal. This conclusion does not imply that experience plays no role in shaping the expression of rhythmicity, only that experience of a rhythmic environment is not essential for the organism to develop a fully functional rhythm generating system. It also implies that the rhythm is self-excited, a question later addressed by Rütger Wever, Aschoff's longtime collaborator. Wever established that songbirds would spontaneously display circadian rhythmicity in low intensity illumination after rhythmicity had been damped out in high intensity constant light [29]. Thus, they had the capacity for self-excitation, a non-trivial aspect if one thinks of physical oscillators like a pendulum, which need an external stimulus before starting to oscillate.

1.5 Clock and Frequency

To serve its 'purpose', a circadian rhythm in whatever function must run at the frequency of the earth's rotation (one cycle per 24 h), as well as maintain a specific phase relationship to the day–night cycle. The word purpose is used here in the loose sense of a property that contributes to the individual's Darwinian fitness, i.e., the expected rate of propagation of its genes to future generations. That rate is determined by its life span and by its rate of reproduction of surviving offspring. This purpose and function have rarely been exactly established. Nonetheless, specific functional considerations have greatly inspired the scientists to gain insight in the mechanisms of frequency and phase control, and I briefly digress into one of the functional landmark studies.

Ingeborg Beling [30], a student of the later Nobel prize winning biologist Karl von Frisch, trained honeybees to approach a particular food source at a restricted time of day. The bees returned to that location on subsequent days *in anticipation* of the training time. It seemed that the bees were aware of the time of day, and anticipated finding food again in the same location at the particular daily phase of training. The knowledge of time of day did not depend on the lighting cycle, since it persisted in constant darkness, temperature and humidity. The source of this timing information was inside the bees. Twenty years later, Von Frisch made another crucial discovery. He had meanwhile unravelled the meaning of the 'dance language', by which honeybees transmit information about particularly rich food sources to fellow-workers in the hive. During the dance on the honeycomb the bee informs her hive-mates of both the direction, relative to the sun's azimuth, and the distance to the food source. The use of such a *'sun compass'* for orientation had been known since the beginning of the century in ants [31], and would by itself not have stirred much excitement. However, von Frisch reported in 1950 that honeybees which had not seen the sun for some time since finding a food source, compensated for the time elapsed by adjusting the angle of their dance. Since the sun's azimuth moves across the sky at roughly 15°/h, directing other bees to the food source after a lapse of time requires a compensatory adjustment in the angle indicated by the dance. This requires knowledge of the passage of time to determine how far the sun would have moved since the food was found. Somehow, the bees must use a clock system to measure the time of day and to compensate for the sun's movement [32]. Simultaneously, Gustav Kramer, an ornithologist in Wilhelmshaven on the German north coast, reached exactly to the same conclusion for starlings. He demonstrated that in order to navigate during their seasonal migrations, starlings also use a *time-compensated sun-compass* [33].

These lines of work together provided compelling evidence that these species have an internal clock that is used much as humans use artificial chronometers: to provide time-of-day information that is essential for orientation or navigation with respect to the movements of celestial bodies. In order to permit sun-compass orientation, it must also have the property of a continuously consultable clock: one that, like a watch, can be accessed at any time to measure the passage of time and determine the time of day. Any model of this timing system would have to incorporate its ability to persist with an altered period in constant conditions, its regulation of a variety of overt behaviours, and its availability as a consultable clock.

These observations had a deep impact by inspiring other researchers to consider circadian rhythms as clocks, and to dig into the properties of these clocks which take care of sufficient precision in the face of external variability. Foremost among them was Colin S. Pittendrigh, a British scientist who became one of the founding fathers of modern biological rhythms research. Pittendrigh was deeply impressed with the results of Kramer and von Frisch, and systematically began to use the word biological clocks. But he was not the first. Already in 1814, the French pharmacist Julien-Jospeh Virey , the first to report the daily cycle in human mortality , had used the word *horloge vivante* or living clock [34]. Again, it was easier to appreciate that the humans had a large degree of independence from the environment than it was to accept that in plants or animals.

The simultaneous frequency and phase control needed for a proper clock to function in nature poses a dilemma. For phase control (see Sect. 1.7) sensitivity to a reliable external signal is needed. For frequency control, the system should be shielded as much as possible against the environmental influences. One of the first who systematically investigated environmental effects on frequency in constant conditions was Maynard Johnson, who studied white-footed mice (*Peromyscus leucopus*) kept in constant light in the laboratory. Johnson [35] concluded his report with the insightful statement that '*the animal has an exceptionally substantial and durable self-winding and self-regulating physiological clock, the mechanism of which remains to be worked out*'. Johnson also made the important observation that the periodicity expressed in constant environmental conditions changed depending on the intensity of the constant light to which the mice were exposed [35]. Higher intensity of illumination caused the rhythm to slow down, while constant darkness speeded them up. How could a system with such properties maintain stable phase under the caprices of natural conditions? Others were to follow. Hans Kalmus, for instance, reported that the daily rhythm of pupal eclosion of *Drosophila* was strongly dependent on the ambient temperature [36].

Pittendrigh took up the challenge. He had worked during the war on the daily cycles of biting by mosquitoes transmitting malaria on Trinidad. Trained as an evolutionary geneticist with Theodosius Dobzhansky at Columbia University, he then moved to Princeton University and focussed his research on the system of Kalmus, the eclosion rhythm in *Drosophila*. He resolved the paradox by demonstrating that Kalmus had looked at a transient phenomenon in the first cycle after a temperature change, and that the frequency of the freerunning rhythm was in fact hardly affected by temperature [37]. The phenomenon of relative temperature independence became known as *temperature compensation* was soon confirmed in other systems [38] and is today considered one of the functional hallmarks of circadian clocks. Pittendrigh later went a step further. He suggested on the basis of experiments in cockroaches, combining temperature with other variables, that temperature is only one aspect of a general homeostatic mechanism buffering circadian frequency against external influences [39].

Johnson's observation of the effect of light intensity on circadian frequency was meanwhile repeated in scores of animals. In the 1960 Cold Spring Harbor Symposium, Aschoff put forward the generalization that diurnal animals have rhythms that shorten their period with an increasing light intensity, while those of nocturnal animals lengthen it [40]. This generalization became known as *Aschoff's rule*, and was expected to reveal an ecologically functional property of the circadian mechanism. When more data became available, Aschoff himself realized that the difference was rather taxonomic in origin, and had resulted from the fact that most diurnal animals studied initially were birds, most nocturnal species were mammals [41].

Another crucial aspect of the frequency of freerunning circadian rhythms were the *after-effects* of prior conditions, first observed by Pittendrigh [19], and later demonstrated extensively in mice [42]. Mice kept in light–dark cycles of 20 or 28 h for several weeks, and then released into constant darkness had initially a circadian cycle length (or period, τ, the inverse of frequency) that was considerably shorter than 23 h, or longer than 25 h, respectively. The prior LD cycle was not needed for a subsequent oscillation, as in the *Nachschwingungen* of Pfeffer and Semon, but it did affect its expression. The difference between the two groups of mice gradaually disappeared and after 100 days the average periods of both had become indistinguishable. Lifelong aftereffects have subsequently been shown in other creatures, notably in cockroaches [43]. Aftereffects reveal a property with substantial functional meaning since they bring the frequency of the endogenously generated rhythm close to that of the environmental cycle. The rhythm, while innate and endogenously generated, at least to some extent 'learns' to fine-tune its frequency to the external world [44]. The importance of a close match between the endogenous circadian period and that of the environment has been most impressively documented by the group of Carl Johnson, working with circadian clockmutants of the cyanobacterian *Synechococcus elongatus.* They mixed mutants with different circadian periods and studied their fate in LD cycles with different periods. The result was that always the mutant with the closest match between internal and external time outcompeted the others and survived [45].

1.6 Oscillator

Until 1960, research on biological rhythms was a scattered enterprise. Scientists worked on a wide variety of systems, but except in a few minds, such as of the German botanist Erwin Bünning, there was no general awareness that circadian rhythmicity is a fundamental property of life. There was an international society

(*'Internationale Gesellschaft für Biologische Rhythmusforschung'*) with periodic meetings, starting in Ronneby, Sweden, in 1937. These meetings were dominated by studies on human physiology and did not succeed in bringing conceptual unity. The field of circadian rhythms research received its unifying theoretical framework when Pittendrigh introduced the concept of Endogenous Self-Sustained Oscillators (ESSOs) [46]. This made explicit the analogy of periodic biological phenomena with the action of physical oscillators, which were well understood mathematically. The physical theory dates back to the seventeenth century. The Dutch scholar Christiaan Huygens derived the period (τ) of the movement of a pendulum from its length (l): $\tau = \pi V(l/g)$, where g is the acceleration of gravity. Huygens invented the first pendulum clock (and patented it), which meant a leap forward in the precision of time measurement. He also made a key observation. He saw that two pendulum clocks attached to the same wooden beam tended to synchronize each other, i.e., to tick in a constant phase relationship, to which they gradually returned (see [47]). A physical oscillator can be synchronized by another oscillator if there is a coupling force, and if their frequencies are not too far remote from each other.

The analogy of biological rhythms with such physical ocillators provided ageneral framework for discussing a wide variety of periodic biological phenomena. This insight became the focus and success of a key event in the history of our field, the Cold Spring Harbor Symposium on Biological Clocks. Jürgen Aschoff, who by then had been appointed as one of the directors of the Max Planck Institute for Behavioural Physiology, had visited Pittendrigh for the first time in 1958 in Princeton. There was immediate friendship. They were two visionary people. Pittendrigh was an evolutionary biologist with a marvellous intuition for experimental design and theoretical analysis. He was continuously inspired by functional considerations on what the timing system was selected to do for an organism. Aschoff was the scholar who devoured every study he could lay his hands on, who kept a wonderfully rich card catalogue of his readings and developed empirical generalizations in many insightful reviews. Pittendrigh made a return visit to Heidelberg the next year. Together they planned the 1960 meeting. They assembled biologists and mathematicians from a wide variety of fields. In Cold Spring Harbor both Aschoff and Pittendrigh presented their broad generalizing views, and provided a strong functional and formal analytical context to analyse behavioural rhythmicity. Aschoff and Pittendrigh, aided by the force of their flamboyant and charismatic personalities, thereby laid the foundations of the field. The view that circadian rhythms are ubiquitous, and follow general rules, made it possible and worthwhile for any system to make a contribution to the field, and inspired the researchers from many different disciplines. It provided a common language of description of the phenomena, even though there was not yet a glimpse of understanding on the underlying physiological mechanisms.

In Cold Spring Harbor, the remarkable precision of some circadian rhythms was also documented, especially by Patricia DeCoursey, who showed that her flying squirrels time their daily onsets of activity with a cycle-to-cycle variation of only a few minutes, a coefficient of variance smaller than 5‰ [48]. This must have impressed those attending as evidence of the almost physical nature of this biological pendulum.

The oscillator analogy inspired many theoreticians to design specific mathematical models to bring together a wide array of hitherto unconnected data. This included Rütger Wever [49], Aschoff's longtime collaborator in Andechs, and Theodosius Pavlidis, who worked out a model for the *Drosophila* eclosion rhythm [50]. Sometimes such models led to experimental predictions that could be verified experimentally. A prominent example was the work of Arthur Winfree, originally one of Pittendrigh's pupils in Princeton. Winfree argued that biological oscillators could be described as moving in a limit cycle to which they would return after any brief disturbance (see Sect. 1.8) [51].

Pittendrigh himself emphasized the distinction between the fundamental or master oscillator and the overt rhythm observed, behavioural or physiological. He pointed out that the temperature dependence that Kalmus [36] had observed in the phase of the *Drosophila* pupal eclosion rhythm was the consequence of transient in the overt rhythm rather than evidence for temperature effects on the master oscillator.

1.7 Entrainment

As we now know, biological daily rhythms are synchronized by external periodic signals emerging from the rotation of the earth, most notably the alternation of light and darkness. This concept of what is now called *entrainment* only gradually dawned upon the researchers of biological rhythms. The word itself is old enough. Already Virey talked about '*... une sorte d'horloge vivante, montée par la nature, entrainée par le mouvement rapide du soleil et de notre sphere*'. It later became the common word in physical oscillator theory to indicate synchronization of a self-sustaining oscillator by an external signal [52]. For this signal Aschoff added the concept of a 'Zeitgeber' [53]. This word, meaning 'time giver' in German, has remained in the chronobiological jargon. It indicates any periodic signal that is able to elicit such influence on the rhythm that it can be entrained to it, i.e., assume a stable phase relationship with the zeitgeber. Long before these concepts were common, Kleinhoonte exposed her Canavalia plants to LD cycles with different duration. She observed that the leaf movements could be entrained to a range of periods around 24 h. A zeitgeber with twice the frequency of the endogenous rhythm, i.e., a period of 12 h, was also able to entrain the rhythm but now with a period twice that of the zeitgeber [15]. This phenomenon is now known as *frequency demultiplication.*

The way in which endogenous rhythms are entrained by light and darkness became the focus of much of the work in the 1950s and 1960s. It was then one of the few experimental handles one had on the analysis of biological clocks. This was essentially a black box analysis, since one did not have a shimmer of an idea what these clocks were. Their mechanism might be approached by varying the cues they were sensitive to and observe their responses. The leading theories of entrainment were elaborated in the labs of Aschoff and Pittendrigh in the decade following Cold Spring Harbor.

Pittendrigh's approach emphasized entrainment by discrete instantaneous phase resetting once or twice per day, which circadian systems employ to correct for the innate deviation of their cycle length from 24 h. We return to this in Sect. 1.8. Aschoff initially pursued his own views on the basis of the effects of illumination intensity on the spontaneous frequency. He and Rütger Wever designed a model that used continuous influences of light on the velocity of the oscillator [49]. It was in essence based on parametric effects of light, i.e., on an effect of light on a parameter in the basic equation. This contrasted with the instantaneous shift to a different phase that Pittendrigh assumed, which was called non-parametric because it involved no alteration of the generating oscillation itself. The model of Wever was able to bring a vast amount of data that had gradually accumulated in a general theory, but it yielded only qualitative predictions. It was gradually forgotten when Pittendrigh's competing theory of non-parametric entrainment by daily phase shifts did produce quantitative predictions.

Entrainment by light turned out to be the dominant pattern in circadian rhythms across the animal and plant kingdom. Indeed the alternation of light and darkness provides the most precise signal in the environment as a cue to time of day, and it now seems trivial that organisms have been equipped by natural selection with pathways sensitive to this signal. Nonetheless, entrainment by other modalities was also established early on. Entrainment by temperature cycles became known in plants and poikilotherms, but eventually also in mammalian homeotherms such as the Squirrel monkey [54]. Other periodicities capable of entraining the circadian systems are for instance barometric pressure in pocket mice, *Perognathus longimembris* [55], periodic conspefic song in songbirds [56] and social interaction in bats [57]. In the first attempts to establish entrainment in humans, it was thought that cognitive information on time of day played a major role [58]. This was under conditions where the subjects themselves were free to choose their lighting cycle, and the cognitive information (gong signals) may have secondarily generated a light–dark cycle. A few years later, the same researchers demonstrated clear entrainment by light also in humans [59]. Recently, entrainment by the solar cycle rather than by the social environment has been shown to be the prevalent pattern in the large-scale countrywide 'chronotype questionnaire' research in Germany [60].

Till Roenneberg, who pioneered these large-scale internet based human studies, was one of Aschoff's students, and is currently one of the leading rhythms researchers in Europe. He contributed an important conceptual feature to the issue of circadian entrainment: the '*Zeitnehmer*' [61]. He realized that the circadian clocks, while controlling behaviour and physiology, also modulate the sensitivity of sensory pathways to external entraining factors. This concept is of significance when we try to separate, as Pittendrigh did, the internal clock from the processes it steers. Roenneberg has further persuasively argued that self-sustainment is not an essential prerequisite for entrainment from a functional perspective, but rather a reflection of the internal circadian machinery [62]. Self-sustainment is often restricted to a limited set of ambient conditions.

1.8 Phase Response

That single stimuli of light can phase shift a circadian rhythm turned out to have much greater predictive power than the effects of constant illumination intensity on its frequency. Early in the twentieth century, Antonia Kleinhoonte had observed that stimuli as brief as 1 min of light can shift the rhythm of the large leaves of the bean plant *Canavalia ensiformis*. She noted that the shift may be either forward or backward, depending on the phase of the rhythm at which the stimulus was applied [15]. A more complete study of such stimuli covering a whole cycle of the rhythm was for the first time made by John Burchard, an undergraduate student in Pittendrigh's lab in 1956. This was the first Phase response curve (PRC), although it was based on only two individual Syrian hamsters, and was never published. Shortly afterwards, Patricia DeCoursey published a more detailed rodent PRC for two Flying squirrels, *Glaucomys volans* [63].

Pittendrigh himself soon came up with his PRC for the *Drosophila* pupal eclosion rhythm in 1958 [46], describing the phase shifts elicited by a standard 4 h light pulse as dependent on the phase of the circadian cycle. In the same year, Hastings and Sweeney published their PRC for 3 h light pulses in the luminescence rhythm of the marine dinoflagellate *Gonyaulax polyedra* [64]. At the Cold Spring Harbor Symposium, Pittendrigh included the phase response to single perturbations as one of his empirical generalizations about circadian rhythms [19]. From then on, the assessment of PRCs became one of the standard assays of circadian systems. A then complete Atlas containing over 200 PRCs was assembled by Johnson [65]. Also human circadian systems share the property of a PRC for light [66]. The first demonstration in humans was achieved by Ken-Ichi Honma and collaborators in Sapporo [67], and later confirmed in the lab of Charles Czeisler in Harvard [68].

PRCs for brief light pulses are ubiquitously characterized by phase delays during part of the circadian cycle, followed by a part where phase advance responses are observed. Between the advance and the delay part often only small or insignificant phase shifts are seen. During entrainment to the day–night cycle in the natural environment, light will shift the delay part to the early night and the advance part to the late night, such that a stable phase relationship emerges. The PRC for light thus reflects an excellently functional property, common to all circadian systems, which keeps life in a synchrony with the rotation of the earth.

An important conceptual step in the understanding of PRCs was made by Arthur Winfree, who distinguished type-1 and type-0 PRCs [51]. These names derive from the slopes of the corresponding phase transition curves (PTCs). The PTC is a plot of the new phase to which an oscillation is reset as a function of the old phase. A low intensity pulse will have small effects and lead to only small deviations from a straight line with slope 1.0, hence type-1. A high intensity pulse, or a system highly sensitive to the perturbation will cause the system to be reset on average to the same new phase with little dependence on the old phase, hence type-0. This insight brought together PRCs with seemingly different waveforms. Winfree also predicted the *singularity*, i.e. the combination of phase and stimulus strength, as the latter is gradually increased from low (type-1) to high (type-0) intensity, that stops the oscillation. In a careful series of experiments on *Drosophila* he indeed found this singularity [51]. Others were to follow his exercise in other systems.

Pittendrigh had the vision to exploit the PRC for brief light pulses to quantitatively predict entrainment under a wide variety of exotic experimental light cycles. These consisted of one or two pulses every T hours, where T=the period of the zeitgeber. Under the assumption that each pulse given at a certain phase φ resets the oscillator instantaneously to a phase φ' as dictated by the empirically measured PRC ($\Delta\varphi(\varphi)$) for that pulse. One could thus derive the phase of stable entrainment by solving the simple equation $\Delta\varphi(\varphi)=\tau-T$. An extensive set of studies testing these and related predictions was made both for the pupal eclosion rhythm in Drosophila populations and for the individual rodents. The predictions were very well matched by experiments in the highly sensitive (type-0) Drosophila system [69]. In the rodents, the success was more limited, and it became clear that a light pulse itself also affected both τ and the PRC, so that a second light pulse, if administered soon after could not be precisely predicted [70].

Pittendrigh's extended search for the entrainment mechanism was inspired by the idea that circadian clocks are somehow involved in measuring daylength (Bünning 1936; see Sect. 14). He proposed that the transitions from light to darkness and back were sensed by the circadian oscillator. The PRC somehow reflected the response of the oscillator to these transitions, by instantaneous, non-parametric phase shifts. The light in between dawn and dusk would not contribute much. This was appealing in its simplicity. The rotation of the earth produces two clear and precise signals, dawn and dusk. Resetting in response to both provided a mechanism for entrainment as well as the proper cues for daylength and season. Two brief light pulses in a 'skeleton photoperiod' could readily replace a long day [71]. The success of the model let the field convinced that the deviation of τ from 24 h is corrected every day in response to the light transitions. However, transitions by themselves are not instrumental in the phase shift response as demonstrated experimentally in hamsters by Nelson and Takahashi [72]. A recent extensive analysis in mice has shown that a continuous parametric effect of light on the angular velocity of the oscillator can fully explain the responses to different daylengths [73]. The parametric effect is suppressed by an ongoing illumination, and slowly restored by prior darkness. This view integrates the different approaches of Aschoff and Pittendrigh. It accounts for the fact that some burying animals never see other light–dark transitions than those generated by their own behaviour [74]. It also explains the PRC as a reflection of the dynamic properties of the system rather than as a mechanism designed for daily corrections of τ [73].

1.9 Pacemaker

The early students of circadian rhythms were not yet much concerned about the question,where in the body the rhythm is generated or controlled. It was treated usually as a diffuse capacity of the organism as a whole. This view is plausible in plants, which dominated the experimental scene for the first 200 years.

Plants have no central nervous system controlling the whole organism. In animals, despite the notion of a 'clock', the widely accepted view was one of a multioscillatory system [19]. Yet Pittendrigh developed the idea of a distinct *'light-sensitive (A-) oscillation that serves as a pacemaker for the organism'* [19, 75]. This notion was inspired by the apparent temperature dependence of the overt rhythm of Drosophila eclosion, contrasting with the temperature independence of the underlying oscillator.

From then on, the search for such a pacemaker was high on the agenda. An early claim of a hormonal central clock was made for the suboesophageal ganglion in the cockroach *Periplaneta americana* by Janet Harker in Cambridge [76]. Her claim was based on transplantation experiments which were poorly substantiated and widely criticized (see [77]). Taking up the challenge, Junko Nishiitsutsuji-Uwo, a postdoc in Pittendrigh's lab, found positive evidence against Harker's claim and located a potential pacemaker in the optic lobe of the cockroach [78]. She severed the neural connection from the left optic lobe to the brain, and from the right eye to the optic lobe, and observed a freerunning rhythm in a 24 h light–dark cycle. She was careful enough to realize that true evidence of a pacemaker requires more. The putative pacemaker must be rhythmic itself, resettable by appropriate light stimuli, and – most crucially – must upon transplantation into an arrhythmic animal be able to restore overt rhythmicity with phase or period determined by the transplant, not the host. This proof was provided later for the cockroach optic lobe – or more specifically the medulla of the optic lobe – by Terry Page in a series of elegant experiments [79].

Meanwhile, progress had been made in other systems. The pineal organ was shown, by the group of Michael Menaker, to be required for circadian rhythmicity of sparrows [80]. The pineal also restored rhythmicity when transplanted into a pinealectomized host, with a phase reflecting the light–dark cycle imposed on the donor [81]. This important breakthrough demonstrated that the pacemaker concept was correct. It could not be extended to all birds, since many species, such as the quail, were found to retain rhythmicity after pinealectomy [82]. This is probably related to the fact that in quail, as well as in pigeons, the retina itself produces melatonin rhythmically, and may well act as a pacemaker itself.

In mammals, progress was slower. In Baltimore Curt Richter did a series of brain lesions in rats in the 1960s and suggested that the hypothalamic area was a candidate for harbouring a circadian pacemaker [83]. 1972 saw the simultaneous appearance of two publications narrowing down this suggestion to the suprachiasmatic nuclei (SCN) [84, 85]. These are two bilateral nuclei sitting at the lower boundary of the brain just above the optic chiasm. Ablation of the SCN caused rhythmicity to disappear. Electrophysiological recordings showed the SCN to be rhythmic, in antiphase with the surrounding tissue. More importantly, Inouye and Kawamura demonstrated that the SCN could be isolated from the surrounding brain tissue and then retained its rhythmicity in multiple unit activity, while the rest of the brain and the animal became arrhythmic [86]. The definitive proof was given when the first mammalian circadian mutant was exploited for the crucial transplantation experiment. Martin Ralph, who had detected this mutant as a student in Michael

Menaker's lab [87], transplanted the SCN from homozygous mutant embryos into SCN-lesioned wildtype hosts and vice versa. He and his colleagues made the eminent discovery that the donor always determined the period of the restored circadian rhythm [88].

The field was initially not easily convinced of the existence of circadian pacemakers. I remember well how the publication of the key paper of Stephan and Zucker was received in Aschoff's Max Planck Institute, where I was a postdoc at the time. Aschoff, whose door was always open, saw me passing through the corridor, called me in, pushed the PNAS issue into my hands and said: 'Have a look, do you believe this?'. One had lived with a diffuse view of the whole body as a complex oscillator for too long to readily accept the speculation that rhythmicity came from a single tiny piece of tissue.

Nearly 40 years later, we know that in all model systems where circadian pacemakers have been identified, they are located in the central nervous system. They are either themselves photosensitive as the sparrow pineal, or strategically located in immediate contact with the photoreceptive organs: the cockroach optic lobe medulla just behind the compound eyes [78], the basal neurons in the eyes of the mollusc *Bulla gouldiana* [89], the SCN with their monosynaptic connections to the mammalian retina [90, 91]. Thus, they are excellently suited to entrain to the world outside, convey the external time to the internal organization of the animal, and to let it perform its functions properly timed.

The identification of pacemakers kindled a tremendous new enthusiasm in the field. Pacemakers were real and tangible. That they existed removed the last suspicions that some people had about rhythms research as a slightly esoteric and sometimes even superstitious endeavour. The SCN became the focus of many first rate laboratories worldwide in the last two decades of the twentieth century.

1.10 Photoreception

The notion of central circadian pacemakers emerging in the 1960s initiated simultaneously a search for the input pathways by which zeitgebers, the light–dark cycle in particular, are perceived. In animals, the first attempt to identify the photoreceptors involved had already been made in 1953 by Cloudsley-Thomson [92], who painted a black cover over both the eyes and the ocelli. His conclusions from these experiments were later found flawed, but the technique was followed by others. S.K. de F. Roberts, doing the same thing in another species, established that the roaches would display a freerunning rhythm in an LD cycle. He also found that removing the ocelli, or painting the head except the ocelli, did not affect entrainment [93]. Junko Nishiitsutsuji-Uwo confirmed these observations and definitively demonstrated by optic tract sections that the compound eyes mediate the light information to the pacemaker [94]. Where exactly photoreception takes place, inside the ommatidia or behind them remained uncertain.

But retinal entrainment was not the general pattern. Around 1970 it became clear that there must be photoreceptors in the brain, both in insects and vertebrates, that could be sufficient for circadian entrainment. Jim Truman transplanted brains of silkmoth (*Antherya pernyi*) pupae into the abdomen or the head of brainless pupae and exposed the front and back parts of the pupae to light cycles in antiphase. The result was that the timing of pupal eclosion was determined by the LD cycle to which the 'loose brain' was exposed [95]. This was preceded by Michael Menaker's demonstration of extraretinal photoreception in sparrows: the eyes are not needed for circadian entrainment, but light penetrating through the skull into the brain is, even at low intensities of light [96]. This was later confirmed for several species of fish, reptiles and birds. Over the years it has become clear that there are multiple photoreceptive inputs into their complex circadian systems, including the eyes, the pineal and other light-sensitive elements [97].

The situation turned out to be quite different in mammals. Here all circadian entrainment by light employs a retino-hypothalamic pathway. As Curt Richter had extensively demonstrated, blinding of rats and monkeys abolishes entrainment by light while leaving the rhythm intact [98]. Groos & Mason then went on to assess the visual fields in the retina that connected to single cells in the SCN. They observed that the electrical activity of single SCN neurons would tonically respond to illumination over large receptive fields in the retina, quite distinct from the classical visual light perception [99]. Two decades later, a specific network of intrinsically photosensitive ganglia was found in the inner retina, after Russell Foster and his group had shown normal entrainment in a strain of mice completely missing both rods and cones, the classical visual photoreceptors [100]. These ganglia were then found to contain the pigment melanopsin [101], to connect directly to the SCN and to be sensitive to light [102].

In humans, the role of the eyes in circadian entrainment took longer to be unequivocally established. In an early study, Lewy and Newsome reported anomalies in the melatonin rhythms of blind people [103]. A recent more extensive analysis makes clear that this depends on the kind of blindness: when both eyes are absent there is usually a freerunning circadian rhythm, when there is retinal degeneration with total visual loss the rhythm may stay entrained (103a), most likely via an intact melanopsin ganglion network.

The physiological details of circadian light perception are as diverse as life itself. Yet their functional relevance is more general, and a few generalizations have emerged. Menaker [104] has suggested that pacemakers are quite often themselves photosensitive, while separate photoreceptors when involved in entrainment frequently themselves have oscillating capacity. Indeed, the isolated retinas of hamsters have shown a self-sustained rhythm in melatonin production [105]. This is probably related to the circadian variation in light sensitivity of the photoreceptive system. Many arthropods have retinal pigment migration shading the eyes from direct sunlight. An innate system 'putting on sunglasses' when their side of the globe is illuminated. The circadian photoreceptors are often separate from the visual system. They are wired to perceive global illumination intensity rather than acquire visual images, and often sensitive in the blue of the scattered sky rather than the white of the direct sunlight.

1.11 Rheostasis

How central pacemakers convey their message over the physiology and behaviour of animals has remained an elusive problem in circadian rhythm research. The SCN has a multitude of neuronal connections to different brain areas that are probably functional in part of the process. Yet transplantation of encapsuled SCN tissue allowing only humoral, no neuronal transmission has been shown by Rae Silver to restore circadian rhythmicity after SCN lesions [106]. Truman's early experiments of pupal brain transplants had demonstrated the same in silkmoths [95]. The pineal in songbirds also employs a humoral signal, the hormone melatonin. Evolution will surely have exploited numerous types of pathways in different animals, and it is too early to identify what the unifying steps have been in unravelling their nature.

I make an exception for the idea which has reconciled rhythmic organization with the old notion of homeostasis as a fundamental principle in physiology. The term homeostasis was coined by Walter Cannon [107], to summarize in one word Claude Bernard's insight that *'all the vital mechanisms, however varied they may be, have but one end, that of preserving constancy in the internal environment'* [108]. The principle has enjoyed nearly universal acceptance. Yet, the field of circadian rhythms documented impressively that in almost every organism there is endogenously generated variance in almost every physiological parameter measured. There is no constancy, even in a perfectly constant environment where this would be most easily maintained. Circadian physiologists have been shy to attack the concept of homeostasis. It took until 1990 when Nicholas Mrosovsky published his little gem *Rheostasis. The physiology of change* [109], that the concepts were reconciled. Mrosovsky's idea of rheostasis in essence is the defense of different values of a variable at different times or in different circumstances. There is great evolutionary advantage of changing homeostatic setpoints over time, in particular when the environmental changes are predictable, as is the case with cosmic cycles such as the day or the year. By changing physiological setpoints over the daily cycle, animals and humans actively prepare for such changes instead of passively responding.

Sleep is a special case of a homeostatic response, that is tightly connected with the circadian cycle. Sleep research has long flourished as a huge field separate from biological rhythms. Some sleep researchers, such as Nathaniel Kleitman [110], a Russian émigré to the USA, had an active interest in the circadian rhythms. Yet for a long time, there was little overlap between the visitors of conferences in the two areas. Nonetheless, the control of daily sleep, arguably the most dominant among human behaviours, is one of the most prominent tasks of our circadian system. Sleep is also one of the most interesting of behaviours, since its true function remains to be unravelled.

The marriage of sleep and rhythms research, as Alex Borbély has called it, took place in 1980, in the romantic setting of the Ringberg castle in the foothills of the Alps. Sleep research had focused on the electrophysiological phenomena at the brain surface. It was vastly inspired by the discovery by Aserinsky and Kleitman [111] of two different types of sleep: Rapid Eye Movement or REM sleep, and Non-REM sleep. The field focussed on the regularities of these phenomena more than on their arrangement within the circadian cycle. The research on human rhythms in Andechs had uncovered the conundrum of *internal desynchronization* [112]. In some of the subjects studied in the bunker of the Max Planck Institute, the rhythms of body temperature and of sleep and wakefulness gradually drifted apart over the weeks. Quite often, the daily up and down of temperature followed a cycle in the neighbourhood of 24.5 h, while the sleep-wake alternation occurred once per up to 33 h. In a few cases, this was even 50 h, or twice the circadian temperature cycle. This '*circabidian*' rhythm had also been observed in prolonged isolation in human subjects in isolation for several months in a French cave by Siffre [113]. The leading authorities in Andechs and in Harvard considered this proof of two separate pacemakers one controlling temperature, the other sleep and wakefulness [59, 114]. There were suggestions about mismatches of these two pacemakers as instrumental in causing depressive illness, even though only a single pacemaker was known. At the conference in Ringberg, Charmane Eastman pointed out the weakness of the argument [115]. Two further ingredients then led to a wholly different, integrative view: Alex Borbély showed that the slow-wave activity (SWA) in the Non-REM EEG signal was homeostatically controlled and reflected an intensity measure of sleep [116]. SWA systematically decreased over sleep time and increased after sleep deprivation. Jürgen Zulley demonstrated that the duration of sleep during internal desynchronization depended on the timing of sleep onset in the body temperature cycle [117]. This led Borbély and Daan to develop the 'two process model' of sleep [118, 119], with a single circadian pacemaker that gates a homeostatically controlled sleep process. The new theory accounted for human behaviour under many conditions, such as internal desynchronization, shift-work and continuous bedrest [120]. It incited a wave of new research, and brought the two fields closer together. By extensively using forced desynchrony protocols, the leading human rhythms laboratory of Charles Czeisler was able to disentangle circadian variations in a host of physiological variations into circadian and homeostatic, sleep-wake driven components [121]. The nature of the sleep process itself has now been tentatively identified as 'synaptic downloading' [122], or the clearance of synaptic overload in the brain as a consequence of prolonged wakefulness. By organizing sleep in the proper time of the day–night cycle, animals rheostatically prepare their brains for optimal functioning during wakefulness.

1.12 Clock Genes

The notion that circadian rhythms are innate (Sect. 1.4) implies that the underlying mechanism is somehow encoded in the genetic material. Already Charles Darwin concluded in his book on the Power of Movements of Plants that '*the periodicity ...*

is to a certain extent inherited' [10], even though he did not yet have any evidence that variations in this periodicity are heritable indeed. Such evidence was obtained later by Erwin Bünning, who saw that the duration of a circadian cycle in an individual plant was inherited to its offspring across many generations [123]. Bünning also crossed two lines of the beanplant *Phaseolus coccineus* which had mean circadian periods of their leaf movements in constant darkness around 23 and 26 h, and observed intermediate periods in plants grown from the seedlings [123].

Further analysis of the role of any genes in the circadian clock had to await first the unravelling of the structure of DNA by Jim Watson, Francis Crick and Rosalind Franklin, and then of the genetic code by a series of other Nobel Prize winners in the 1950s and 1960s. Then Seymour Benzer embarked on the forbidding task to apply forward genetics to identify genes involved in the control of behaviour in *Drosophila*. He exposed flies to mutagenic agents in the hope to find phenotypes with aberrations that could lead the way towards such genes, and eventually allow us to address the question how they are involved. Benzer's Ph.D. student Ronald Konopka took up the task of screening the offspring for aberrations in the circadian system controlling daily pupal eclosion. The approach had unexpected and overwhelming success. Konopka identified three mutants: one with a long period of 28 h (*perL*), one with 20 h (*perS*) and one which had no rhythm (*per0*) [124]. The mutations turned out to be retraceable to the same gene (*per* for period). They laid the basis for the intense search for the molecular genetic foundations of circadian rhythmicity that dominated rhythms research in the two recent decades. The crucial step in that effort was the finding by a research group at Brandeis University that the PER protein was rhythmically produced in the fly brain with a peak in the early night, while the *per* RNA showed a similar expression pattern, about 6 h in advance of the protein. This led Paul Hardin, Jeff Hall and Michael Rosbash in 1990 to propose the idea of the transcription–translation feedback loop [125]. The proposition was that the *per* gene is transcribed in the nucleus of certain cells, leading to messenger RNA leaving the nucleus into the cytoplasm, and to translation into PER protein at the ribosomes, followed by return of the protein into the nucleus to suppress further transcription of the gene.

The transcription–translation loop became the leading theory in the immense effort of the 1990s and 2000s unravelling the – first simple and then ever more complex – ideas on how circadian rhythms are generated at the molecular level. This effort included several model species. In the mold *Neurospora crassa* Jerry Feldman used mutagenesis to isolate and later identify mutants of a circadian clock gene named *frq* (for frequency), in much the same way as Konopka had done it in *Drosophila* [126]. In mammals, the first mutation involved in the clock mechanism was detected by accident in Syrian hamsters, *Mesocricetus auratus*. The circadian period in DD in hamsters varies between 23 and 25 h. In 1986 Martin Ralph, a postdoc in Mike Menaker's lab, then in Eugene, Oregon, saw the remarkable actogram of one hamster displaying a circadian rhythm of activity with a period of 22 h. His insight to breed this individual led to the finding that the mutation followed simple Mendelian inheritance [87]: A single gene (*tau*) must be involved. Homozygous mutants had a period around 20 h, heterozygotes around 22. The gene was later cloned by the group of Joe Takahashi at Northwestern University and found to be equivalent to that coding for a casein kinase enzyme CK1ε [127]. This group had already discovered the first mammalian clock gene in 1994. This was based upon a mutation, again produced by a mutagenesis screen, causing a long period, and baptized *clock* [128].

Since then, a dozen or so genes have been shown to be involved in the complex feedback loops currently held responsible for the molecular generation of circadian rhythmicity in the different organisms and their pacemakers. The many steps involved are beyond the scope of this chapter to review these. The reader is referred here to chapters 2 and 3 and for a well-written narrative of the history of the model's development to the book by Russell Foster and Leon Kreitzman [129]. The development of these models has demonstrated an intriguing correspondence in widely different organisms despite the great diversity in the genes and proteins involved. The mammalian model has also yielded the basis for the first attempts at understanding the genetic variation in human circadian rhythms [130].

1.13 Multioscillatory System

Despite the huge success of the modern molecular genetic approaches in unravelling the clock mechanisms, there are some caveats that make clear that history has by no means ended for this exciting field of research. In the first place, Takao Kondo at the University of Nagoya has recently shown that the clock in Cyanobacteria (*Synechococcus elongatus*) can tick in the proteins involved (KAI-A, -B and -C) alone in vitro, without any of the corresponding genes present [131]. Secondly, Drs. Sato and Ken-Ichi at Hokkaido University have found a circadian oscillation in the activity of rats and mice that is expressed when metamphetamine is supplied in their drinking water and that does not require the presence of SCN [132]. This oscillator has recently been shown to maintain its rhythm in all of the mutants where one or more of the canonical clock genes have been made dysfunctional [133].

These caveats set the stage for a brief discussion of the multioscillatory nature of circadian phenomena. In 1960, Pittendrigh dedicated a 4-page section of his seminal paper to '*Circadian organization: a multioscillator system*' [19]. His arguments for such a system were based: on the persistence of oscillations in excised tissue culture, such as in the hamster intestine [134]; on the fact that single cells can display circadian rhythms as in unicellulars, on the multiple frequencies observed simultaneously in long records [19], and finally on his own explanation of an A- and B-oscillator for the transients observed in the overt eclosion rhythm in *Drosophila* after an instantaneous phase shifts [135]. He proposed that these multiple oscillators were in fact slaves of a master pacemaker. His own model system later beautifully demonstrated rhythmicity all over the body when clock genes had become known and their expression could be visualized with luciferase-promotors [136].

In the post Cold Spring Harbor years, the emphasis of the field was on the central pacemakers, and many researchers focussed on their properties. This was in spite of Pittendrigh's warning that "*we are forced to abandon the common current view that our problem is to isolate…'the internal clock'*"[19]. After central pacemakers were found and silenced, the first endogenous overt rhythm that was found to persist was the so-called food-entrainable oscillator. It had long been known that rodents fed at a regular time of day would develop anticipatory activity for a brief episode preceding this time, a phenomenon not unlike Beling's 'time memory' in honeybees. Fred Stephan, after identifying the SCN as the mammalian pacemaker, applied such food schedules to SCN-lesioned rats and observed that they still display anticipatory activity [137]. In a series of subsequent papers, he demonstrated that these oscillations follow all the rules of circadian systems except that they damp out when food is presented continuously again [138]. Although the location of food-entrainable oscillators remains elusive, a direct synchronizing role of feeding schedules on gene expression rhythms in the liver was demonstrated by the labs of Ueli Schibler in Geneva [139] and of Mike Menaker in Virginia [140]. Schibler followed this up with an elegant series of studies to work out the pathways from the SCN and feeding onto the liver tissue clocks (review: [141]).

With the advent of modern luminescence techniques to monitor clock gene mRNA expression, it has now become possible to firmly assert Pittendrigh's vision that '… *the multicellular system is – literally – a population of autonomous oscillators*' [19]. Indeed many tissues in the mammalian body turn out to harbour self-sustained circadian clocks as demonstrated beautifully by Joe Takahashi's group at Northwestern University [142]. Perhaps even more remarkable for those of us who are still used to talk about 'the clock' of an organism is the existence of multiple oscillators in single cells. In 1993, Roenneberg and Morse reported that populations of the unicellular marine dinoflagellate *Gonyaulax polyedra* can show circadian rhythms in bioluminescence and in behavioural aggregation which simultaneously run free with different frequencies and must hence be attributed to two different circadian systems [143]. The biochemical basis of these oscillators remains to be unravelled.

1.14 Photoperiodism

The North-South axis of rotation of the earth is not perpendicular to the plane of its elliptic movement around the sun. This inclination is the primary cause of the seasons. The alternation of winter and summer leads to massive annual changes in virtually all life on earth, especially in the temperate zones. Growth and flowering in plants are restricted to specific times of year, and so is the reproduction of most animals. Both in plants and animals, there is a bewildering diversity of species, all occupying different niches. We find such diversity also in the temporal dimension of the niche. The snowdrop flowers as the snow melts, the summer snowflake in summer, the autumnal crocus in the fall. Many small mammals breed in the spring, the larger ones often mate in autumn such that after prolonged pregnancy the litters are born in spring at the beginning of the season of food abundance. Even humans used to have pronounced seasonality in birth rates, with peak conceptions in late spring, peak birth rates in late winter [144]. There is a sharp natural selection against the wasting of resources in reproduction at times of year when breeding would lead to offspring without a chance to survive. This is the ultimate cause of rather strict annual timing of a kaleidoscope of specific behaviours in the annual cycle. Among the proximate causes, daylength – the most precise and reliable signal predicting seasonal conditions – is the pre-eminent cue to which life responds.

Before there was any scientific basis for this, the Japanese already for centuries had the tradition of Yogai: extending the daylength artificially brought captive songbirds into reproductive condition so that they would sing. But the first solid documentation that daylength triggers biological processes goes back to two botanists in 1920. Garner and Allard studied flowering of a variety of plants and described that varieties of tobacco and soybean flowered only when the days were shorter than a certain '*critical daylength*'. Others were 'long-day plants' flowering only in longer days [145]. Garner and Allard also introduced the terms '*photoperiod*' and '*photoperiodism*' for such responses. Not long afterwards similar phenomena were detected in animals. Markovitch in 1923 demonstrated control by daylength of the annual transition from asexual to sexual forms in aphids [146]. Rowan in 1926 extended this to gonadal growth and migration in the snowbirds (*Junco hyemalis*) in Canada [147]. Evidence of photoperiodism in mammals was first found by Baker and Ransom in 1932 [148]. Like in plants there are long-day breeders, where reproduction in spring is stimulated by the lengthening days, and short-day breeders – usually the larger mammals such as sheep – where the breeding condition responds to shortening days.

Photoperiodism implies that organisms must somehow be able to measure the length of the daylight – or the night. It was Erwin Bünning who in 1936 voiced the profound intuition that they may use an endogenous circadian timing system to perform this task [149]. This suggestion remained pretty impopular for two decades, until Karl Hamner at UCLA made extensive investigations on flowering in Biloxi soybean plants. Hamners student Nanda exposed this short-day flowering plant to cycles with 8 h of light per cycle followed by 8–64 h. They observed that the same photoperiod induced flowering in cycles of 24, 48 and 72 h, but suppressed flowering in 16, 36 and 60 h cycles. The photoperiodic response followed an endogenous rhythmic pattern. Only when the system was in resonance with the external cycle the response was observed [150]. This 'resonance' experiment or 'Nanda–Hamner protocol' has later been applied to many systems, and in most, the involvement of a circadian system was proven. Brian Follett demonstrated the circadian rhythm in photoperiodic sensitivity in a vertebrate, the white-crowned sparrow, when a single 8 h of light induced gonadal growth when presented in cycles of 24, 48, 72, 96 or 120 h [151].

In his original proposition, Bünning [149] had added the specific suggestion that by extending the daylength beyond a specific critical daylength, light would fall upon a particular phase of the circadian cycle, and there either close or open a switch to a metabolic pathway leading to flowering or gonadal growth or whatever is the appropriate seasonal response. Pittendrigh later called this the '*external coincidence model*' (coincidence of light and circadian phase) as distinct from the *'internal coincidence'* of different circadian oscillators responding differentially to lights-on and lights-off and thus generate an internal reflection of the external day and night [152]. Internal coincidence implies that light is not specifically needed to trigger the response. To distinguish between the two models David Saunders set out to use *thermoperiods* (day of high temperature followed by a night of low temperature). He found that the clock controlling diapause induction in the parasitic wasp *Nasonia vitripennis* responded to daily temperature cycles in a manner similar to its response to light. The critical thermoperiod above which the diapause response was suppressed was 13 h, only slightly shorter than the critical photoperiod (15¼ h) [153]. This gave credence to the internal coincidence model at least for the wasp.

A specific version of an internal coincidence model – with two oscillators, one locking on to dawn, one to dusk, had been suggested originally for insect photoperiodism by the Russian researcher Tyshchenko in 1966 [154]. It was later more explicitly formulated as the hypothesis of a *'morning'* and an *'evening oscillator'* for mammalian systems by Pittendrigh & Daan [155]. The idea was based on the presence of two oscillations in hamster activity patterns. In constant, light two components of the activity rhythm can run free from each other with different frequencies for some time – a process called 'splitting' – and often stably couple in antiphase, with the two bouts of activity spaced about 12 h apart. In a theoretical analysis, Daan and Berde [156] found that stable coupling in two modes – in phase or in antiphase – requires that the two oscillators have nearly exactly the same properties. They suggested that splitting may reflect the left and right SCN drifting apart. Indeed, the split components in another mammal, the tree shrew *Tupaja belangeri*, have indistinguishable light resetting [157]. Bilateral antiphase coupling of the hamster SCNs was later experimentally confirmed in an ingenious study by Horacio De la Iglesia in Bill Schwartz's lab in Boston [158]. The splitting phenomenon has therefore probably nothing to do with photoperiodic time measurement. Nonetheless, morning- and evening oscillators have recently been identified by the Rosbash and Rouyer labs in the *Drosophila* brain [159, 160].

The circadian pacemaker in the SCN is very likely involved in the internal coding for daylength. Its electrophysiological activity in vitro can display two peaks with a phase angle difference that reflects the prior daylength and have indeed been interpreted as morning and evening oscillators [161]. The SCN of rats has a daytime phase of high spontaneous expression of the cFOS protein, which again reflects the duration of prior photoperiod [162].

The pathways downstream from the circadian pacemakers capturing the information on daylength to switch on or off the proper seasonal responses have been subject of thousands of publications in the circadian literature. Many breakthroughs have been made in unravelling these pathways, but they are physiologically as diverse as life itself. It serves no purpose to try and list the major steps in their discovery. An exception is made for melatonin, the nocturnal hormone. This substance was discovered by Alan Lerner in 1958 [163], and it now appears to occur throughout the animal and plant kingdoms. In most vertebrates, it is produced by the pineal organ in the centre of the brain. The involvement of the pineal in photoperiodism became clear in the 1960s when a French group removed the pineal of

Syrian hamsters and found that this blocked the suppression of gonadal regression by short days [164]. Bob Moore and Russ Reiter unravelled the neuronal pathway from the SCN to the pineal [165, 166]. Since then, there has been a vast literature on the pineal and on the hormone melatonin it produces. Melatonin is the internal signal that reflects night in most vertebrates investigated and somehow conveys the duration of external nightlength onto the reproductive system [167]. Melatonin receptors are found in the pituitary where the hormonal responses leading to seasonal reproductive responses are controlled. They are also present in the SCN, possibly to allow feedback onto the central pacemaker and enhancing its stability, as Eberhard Gwinner has suggested. Indeed, the circadian rhythm of activity of birds and mammals can be phase shifted and synchronized by melatonin. In birds, the pineal has been shown in vitro to retain the pattern induced in sparrows by daylength [168], and thus have a 'memory' for night and day. In plants, less is known about the production and actions of melatonin, but evidence is available for nocturnal presence and diurnal absence of melatonin and for melatonin influences on flowering responses. The hormone may indeed be a very old night signalling substance. It moreover has powerful free radical scavenging properties and speculations have been advanced that it originally was a protection system against metabolic damage which by its sensitivity to light evolved into a ubiquitous day–night signalling substance [169].

1.15 Circannual Rhythm

We have seen that seasonal rhythms of flowering, of reproductive activity – and also of moult, migration and other behaviours are often controlled exogenously by a response to daylength. In some situations this simply can not be the case. Mammalian hibernators go through a long seasonal episode of torpor, often for half a year. They do so in a protected situation often in deep burrows where no light penetrates and daily temperature cycles, if present at all, are of minute amplitude. They emerge from their hides in spring. The suggestion that they use an endogenous system to properly time this act was made as early as 1896 by Dubois [170]. Hidden internal drives were already assumed as the basis for many other behaviours, in particular migration, but these suggestions did not articulate that the 'drive' was not triggered by an external stimulus. The first author to propose an underlying internal physiological rhythm was Rowan [147]. After demonstrating the photoperiodic response to daylength in songbirds, Rowan argued that such responses would not work around the equator where daylength is seasonally nearly constant. Hence he postulated an endogenous rhythm to control long-distance migration from the tropics. Suggestions of endogenous annual rhythmicity were voiced repeatedly in subsequent years. Marshall even suggested in 1951 that endogenous gonadal cycles are the main driving force of seasonal breeding cycles [171].

In 1957, the first experimental demonstration was made of an endogenous annual rhythm. This work was done in golden-mantled ground squirrels

(*Spermophilus lateralis*) by Eric Pengelley and Karl Fisher in Toronto. They observed that the seasonal alternation of hibernation and active state, including the associated food intake and body weight, persisted for 2 years in constant conditions of temperature and daylength, but with a period close to 300 days rather than 365 [172, 173]. Clearly, the deviation from 365 days was the necessary proof that the rhythm is indeed fully endogenous, as it is in circadian systems. The authors called this a circannian rhythm, later renamed *circannual.* Some of the ground squirrels were kept all this time in an ambient temperature of 0°C, meaning that their body in the hibernation episodes was just above freezing, and yet their circannual clocks kept ticking. Shortly afterwards, a circannual rhythm of diapause was observed in an insect, the Carpet *beetle (Anthrenus verbasci)* [174]. Eberhard Gwinner was the first to demonstrate the circannual rhythms of gonadal growth and nocturnal migratory restlessness ('*Zugunruhe*') in migrant songbirds [175]. Since then examples of circannual rhythmicity have been found in many different animal taxa [176]. There are even cases of freerunning circannual rhythms in nature in tropical conditions where seasonal rhythmicity in the environment is little pronounced – the rainforest of Borneo [177] and the famous case of the 'wideawake fair'. This is the colony of the Sooty tern (*Sterna fuscata*) on the oceanic island Ascension which return there once in every 10 months as carefully documented over 11 years by James Chapin [178]. This may actually be synchronized by frequency demultiplication by the lunar months, but in any case not by the annual cycle.

The pursuit of the nature of circannual systems has necessarily been much slower than that of the circadian clock. Few laboratories have had sufficient resources – and patience – to maintain active programs unravelling the nature of these extremely low frequencies in the individual pace of life. The outstanding exception was the department of the Max Planck Institute headed by Gwinner in the Bavarian village Erling Andechs. For over three decades Gwinner kept a large-scale program going where thousands of wild song birds were raised by hand from the egg, and their endogenous circannual cycles studied to address dozens of important questions. Gwinner developed the idea that the spontaneous duration of nocturnal restlessness reflects the natural duration of migration. The internal circannual program allows songbirds to find their wintering quarter in Africa without ever having been there [179]. Besides the timing of different events in the cycle (breeding, moult, migration distance) even the direction of migration with respect to the earth's magnetic compass is endogenously programmed with its proper changes in the annual cycle [180]. Gwinner further demonstrated in starlings that persistence of the circannual rhythm in constant daylength is restricted to a narrow range of photoperiods around 12 h. The rhythm can be entrained to cycles of longer and shorter daylengths as short as 2.4 months [181]. From these and many more of Gwinners studies, a concept emerged of the circannual system as a program of slow seasonal changes which cause a sequence of physiological stages each with their adaptive behavioural pattern, which in nature is maintained in precise synchrony with the changing daylength. The system is sufficiently powerful to persist under some constant conditions, though not in other. This is especially the case when the species in nature is deprived of proper timing cues for long episodes in the year.

A key question remaining unanswered in the work of Gwinner and others – notably Irv Zuckers extensive analyses the circannual basis of hibernation [182] – was whether a localized endogenous circannual pacemaker is involved. The claim for such a pacemaker was recently advanced by Gerald Lincoln in Edinburgh [183]. Lincoln's group has made extensive studies of the Soay sheep which undergo a seasonal cycle of gonadal regression and recrudescence under the influence of changing photoperiod. The photoperiod information is relayed by the SCN and pineal as a melatonin signal to the pituitary. In sheep where the neural connections to the pituitary are severed the group found circannual oscillations in prolactin secretion from the pituitary. These persisted for four cycles of around 41 weeks under constant long days. As there was no neural connection and a constant daily melatonin profile, the authors argue in favour of a circannual pacemaker within the pituitary [183].

1.16 Final Remarks

The field of chronobiology has always been focussed on the circadian periodicity. There are many other frequencies in biology. Life itself is periodic. Cells divide, hearts beat, lungs breathe, organisms reproduce, populations often go through cycles of abundance. These phenomena are not generally considered part of chronobiology. The field deals almost exclusively with those rhythms associated with geophysical cycles. I have discussed two: the day and the year. Two other cycles are the tides and the lunar cycle. They have been omitted from this history, but it is fair to mention at least the first researchers to have establish circatidal and circalunar rhythms, using the same criteria as we do for the circadian rhythms. The first true freerunning circatidal rhythm with a period longer than the natural 12.4 h cycle in their seashore habitat was observed in the tidal rhythm of swimming activity of *Synchelidium*, a small amphipod from the sandy beaches in La Jolla, by James Enright in 1963 [184]. Circalunar (ca 28 days) and circasemilunar (ca 14 days between spring tides) were also first observed in the tidal zone, in the emergence rhythm of a coastal midge, *Clunio marinus* [185]. The most elaborate endogenous periodicity, with circatidal and circalunar frequencies in adaptation to the alternating asymmetric neap and high tides on the Californian coast has been found in the isopod, *Excirolana chiltoni* [186]. The marine tidal zone, with its predictable, but complex and altitude depending times of inundation, still offers great challenges in unravelling the physiological basis of such evolutionary achievements.

Still lower frequencies are equally exciting, but forbidding to the experimentalist. The 13- and 17-year periods of synchronous mass emergence of periodical cicadas are more attractive to evolutionary modellers [187] than amenable to analysis. Ultradian rhythms with frequencies higher than the earth rotation have attracted attention especially where they run in synchrony in the population, such as the 2–3 h rhythm of feeding of voles [188], which subserves risk reduction, and is possibly based on a hypothalamic oscillator [189].

These other periodic phenomena have been a sideline to the main thrust of chronobiology. This field has been primarily concerned with circadian organization of behaviour and physiology. It is in the circadian domain that unification was brought suddenly by Pittendrigh and Aschoff, at the Cold Spring Harbor meeting of 1960. Since then the field has exploded, and the post-1960 era has seen a vast diversification of its scopes, methods and systems. To do justice to all the important developments in this era is impossible. What should be related is the punctuation by a series of major conferences following Cold Spring Harbor. There was the meeting in Feldafing, Germany organized by Aschoff in 1964, where a unified terminology was adopted. In 1971, Michael Menaker chaired the Biochronometry meeting in Friday Harbor, Wash. In 1980, the Ringberg conference on Vertebrate Circadian Systems followed. From 1983 onwards, there have been biannual Gordon conferences on Chronobiology, and for 20 years the stimulating biannual Sapporo Symposia on Biological Clocks. In the summer of 2006, a second symposium was held in Cold Spring Harbor to take stock of the impressive progress made in nearly half a century after the first one.

In 1986, Benjamin Rusak started to publish the Journal of Biological Rhythms, which soon became the leading journal of the field, later in the qualified hands of Fred Turek and Marty Zatz. I remember well that Aschoff and Pittendrigh initially resented the idea of a dedicated journal, arguing that biological rhythms should remain in mainstream research and not be separated, but they soon adopted it also as their journal. There was another one on the market already, Chronobiologia. This was initiated in 1974, but initially remained mostly dedicated to human and medical research, but later was renamed Chronobiology International and improved in quality and impact. There are now scores of annual scientific meetings, most prominently that of the Society for Research on Biological Rhythms. There is, since 1986, the biannual Aschoff-Honma prize, established by the late Keizo Honma in Sapporo. The field has branched out into numerous areas of the life sciences. There are subfields named chronoecology, chronoethology, chronopharmacology, *etcetera*. When I studied biology in the 1960s in Amsterdam, I did not hear about rhythms research. Today we have at least five chairs in Chronobiology at Dutch universities. It is exciting to have witnessed the growth of this multidisciplinary field and been part of it.

References

1. De Mairan M (1729) Observation botanique. Hist. de l'Acad. Royal Sciences, Paris, p 1
2. Duhamel du Monceau HL (1758) La physique des arbres, vol 2. H.L.Guerin and L.F. Delatour, Paris
3. Pfeffer W (1875) Die Periodischen Bewegungen der Blattorgane. Wilhelm Engelmann, Leipzig
4. Stoppel R (1926) Die Schlafbewegungen der Blätter von *Phaseolus multiflorus* in Island zur Zeit der Mitternachtsonne. Planta 2:342–355
5. Brown FA (1970) Hypothesis of environmental timing of the clock. In: Brown FA, Hastings JW, Palmer JD (eds) The biological clock: two views. Academic Press, New York, pp 13–59

6. Hamner K, Finn J Jr, Sirohi G, Hoshizaki T, Carpenter B (1962) The biological clock at the South Pole. Nature 195:476–480
7. Sulzman FM, Ellman D, Fuller CA, Moore-Ede MC, Czeisler CA, Wassmer G (1984) *Neurospora* circadian rhythms in space: a reexamination of the endogenous-exogenous question. Science 225:232–234
8. Kiesel A (1894) Untersuchungen zur Physiologie des facettierten Auges. Sitzungsber Akad Wiss Wien 103: 97–139.
9. Simpson S, Galbraith JJ (1905) An investigation into the diurnal variation of the body temperature of nocturnal and other birds and a few mammals. J Physiol 33:225–238
10. Darwin CR, Darwin F (1880) The power of movement in plants. John Murray, London
11. Bünning E (1960) Opening address: biological clocks. Cold Spring Harb Symp Quant Biol 25:1–9
12. Enright J (1982) Sleep movements of leaves: in defense of Darwin's interpretation. Oecologia (Berl.) 54:253–259
13. Pfeffer W (1915) Beiträge zur Kenntnis der Entstehung der Schlafbewegungen. Abh. math-phys. Klasse Königl. Sächs. Ges. Wiss 34:1–154
14. Bünning E, Chandrashekaran MK (1975) Pfeffer's views on rhythms. Chronobiologia 2:160–167
15. Kleinhoonte A (1929) Über die durch das Licht regulierten autonomen Bewegungen der Canavalia-blätter. Arch Neerl Sci Exactes 5:1–110
16. Bünning E, Stern K (1930) Über die tagesperiodischen Bewegungen der Primärblätter von Phaseolus multiflorus. II. Die Bewegungen bei Thermo-konstanz. Ber Deutsche Bot Ges 48:227–252
17. Gamble FW, Keeble F (1900) Hippolyte varians: a study in colour-change. Q J Microsc Sci 43:589–698
18. Richter, CP (1922) A behavioristic study of the activity of the rat. Comparative Psychology Monographs 1:1–54
19. Pittendrigh CS (1960) Circadian rhythms and the circadian organization of living systems. Cold Spring Harb Symp Quant Biol 25:159–184
20. Halberg F (1959) Physiologic 24-hour periodicity in human beings and mice, the lighting regimen and daily routine. In: Withrow E (ed) Photoperiodism and related phenomena in plants and animals. AAAS, Washington, pp 803–878
21. Hufeland CW (1797) Die Kunst das menschliche Leben zu verlängern. Jena: Akademische Buchhandlung
22. Aschoff J, Wever R (1962) Spontanperiodik des Menschen bei Ausschluss aller Zeitgeber. Naturwissenschaften 49:337–342
23. Semon R (1904) Das Mneme als erhaltendes Prinzip im Wechsel des organischen Geschehens. W. Engelmann, Leipzig
24. Semon R (1905) Über die Erblichkeit der Tagesperiode. Biol Zent Bl 15:241–252
25. Aschoff J (1955) Tagesperiodik bei Mäusestämmen unter konstanten Umgebungsbedingungen. Pflügers Arch 262:51–59
26. Davis F, Mannion J (1988) Entrainment of hamster pup circadian rhythms by prenatal melatonin injections to the mother. Am J Physiol 255:R439–R448
27. Aschoff J, Meyer-Lohmann J (1954) Angeborene 24-Stunden-Periodik bei Kücken. Pflügers Arch 260:170–176
28. Sheeba V, Sharma VK, Chandrashekaran MK, Joshi A (1999) Persistence of eclosion rhythm in *Drosophila melanogaster* after 600 generations in an aperiodic environment. Naturwissenschaften 86:448–449
29. Wever R (1980) Circadian rhythms of finches under steadily changing light intensity: are selfsustaining circadian rhythms self-excitatory? J Comp Physiol 140:113–119
30. Beling I (1929) Über das Zeitgedächtnis der Bienen. Z Vgl Physiol 9:259–338
31. Santschi F (1913) A propos de l'orientation virtuelle chez les fourmis. Bull Soc Hist Nat Afr Nord 4–6: 231–235
32. Von Frisch K (1950) Die Sonne als Kompasz im Leben der Bienen. Experientia 6:210–221

33. Kramer G (1950) Weitere Analyse der Faktoren, welche die Zugaktivität des gekäfigten Vogels orientieren. Naturwissenschaften 37:377–378
34. Virey JJ (1814) Ephémerides de la vie humaine, ou recherches sur la révolution journaliere et la periodicité de ses phénomènes dans la santé et les malades. Sorbonne, Paris
35. Johnson MS (1939) Effect of continuous light on periodic spontaneous activity of white-footed mice (*Peromyscus*). J Exp Zool 82:315–328
36. Kalmus H (1940) Diurnal rhythms in the axolotl larva and in *Drosophila*. Nature 145:72–73
37. Pittendrigh CS (1954) On temperature independence of the clock system controlling emergence time in *Drosophila*. Proc Natl Acad Sci U S A 40:1018–1029
38. Hastings WJ, Sweeney BM (1957) On the mechanism of temperature independence in a biological clock. Proc Natl Acad Sci U S A 43:804–811
39. Pittendrigh CS, Caldarola PC (1973) General homeostasis of the frequency of circadian oscillations. Proc Nat Acad Sci U S A 70:2697–2701
40. Aschoff J (1960) Exogenous and endogenous components in circadian rhythms. Cold Spring Harb Symp Quant Biol 25:11–28
41. Aschoff J (1979) Circadian rhythms: influences of internal and external factors on the period measured in constant conditions. Z Tierpsychol 49:225–249
42. Pittendrigh CS, Daan S (1976) A functional analysis of circadian pacemakers in nocturnal rodents I. The stability and lability of spontaneous frequency. J Comp Physiol 106:223–252
43. Page TL, Block GD (1980) Circadian rhythmicity in cockroaches: effects of early post-embryonic development and ageing. Physiol Entomol 5:271–281
44. Daan S (2000) Colin Pittendrigh, Jürgen Aschoff, and the natural entrainment of circadian systems. J Biol Rhythms 15:195–207
45. Ouyang Y, Andersson CR, Kondo T, Golden SS, Johnson CH (1998) Resonating circadian clocks enhance fitness in cyanobacteria. Proc Natl Acad Sci U S A 95:8660–8664
46. Pittendrigh C (1958) Perspectives in the study of biological clocks. In: Buzzati-Traverso AA (ed) Perspectives in marine biology. University of California Press, Berkeley, pp 239–268
47. Bennett M, Schatz MF, Rockwood H, Wiesenfeld H (2002) Huygens's clocks. Proc Math Phys Eng Sci 458:563–579
48. DeCoursey PJ (1960) Phase control of activity in an rodent. Cold Spring Harb Symp Quant Biol 25:49–55
49. Wever R (1966) Ein mathematisches Modell für die circadiane Periodik. Z Angew Math Mech Sonderheft (GAMM-Tagung) 46:148–157
50. Pavlidis T (1973) Biological oscillators: their mathematical analysis. Academic Press, New York
51. Winfree AT (1970) Integrated view of resetting a circadian clock. J Theor Biol 28:327–374
52. Bruce VG (1960) Environmental entrainment of circadian rhythms. Cold Spring Harb Symp Quant Biol 25:29–48
53. Aschoff J (1954) Zeitgeber der tierischen Tagesperiodik. Naturwissenschaften 41:49–56
54. Aschoff J, Tokura H (1986) Circadian activity rhythms in squirrel monkeys: entrainment by temperature cycles. J Biol Rhythms 1:91–99
55. Hayden P, Lindberg RG (1969) Circadian rhythm in mammalian body temperature entrained by cyclic pressure changes. Science 164:1288–1289
56. Gwinner E (1966) Entrainment of a circadian rhythm in birds by species-specific song cycles (Aves, Fringillidae: *Carduelis spinus, Serinus serinus*). Experientia 22:1–3
57. Marimuthu G, Rajan S, Chandrashekaran MK (1981) Social entrainment of the circadian rhythm in the flight activity of the Microchiropteran bat *Hipposideros speoris*. Behav Ecol Sociobiol 8:147–150
58. Wever R (1970) Zur Zeitgeber-Stärke eines Licht-Dunkel-Wechsels für die circadiane periodik des Menschen. Pflügers Arch 321:133–142
59. Wever R (1979) The circadian system of man. Springer, Berlin
60. Roenneberg T, Kumar CJ, Merrow M (2007) The human circadian clock entrains to sun time. Curr Biol 17:R44–R45

61. Roenneberg T, Merrow M (1998) Molecular circadian oscillators: an alternative hypothesis. J Biol Rhythms 13:167–179
62. Roenneberg T, Merrow M (2001) Circadian systems: different levels of complexity. Philos Trans R Soc Lond B Biol Sci 356:1687–1696
63. DeCoursey P (1960) Daily light sensitivity rhythm in a rodent. Science 131:33–35
64. Hastings JW, Sweeney BM (1958) A persistent diurnal rhythm of luminiscence in Gonyaulax polyedra. Biol Bull 115:440–458
65. Johnson CH (1999) Forty years of PRCs – what have we learned? Chronobiol Int 16:711–743
66. Daan S, Lewy A (1984) Scheduled exposure to daylight: a potential strategy to reduce "jet lag" following transmeridian flights. Psychopharmacol Bull 20:566–568
67. Honma K, Honma S (1988) A human phase response curve for bright light pulses. Jpn J Psychiatry Neurol 42:167–168
68. Khalsa SBS, Jewett ME, Cajochen C, Czeisler CA (2003) A phase response curve to single bright light pulses in human subjects. J Physiol 549:945–952
69. Pittendrigh C (1974) Circadian oscillations in cells and the circadian organization of multicellular systems. In: Schmidt FO, Worden FG (eds) The neurosciences IIIrd study program. MIT Press, Cambridge, MA, pp 437–458
70. Pittendrigh CS, Daan S (1976) A functional analysis of circadian pacemakers in nocturnal rodents IV, Entrainment: pacemaker as clock. J Comp Physiol 106:291–331
71. Pittendrigh CS (1964) Entrainment of circadian oscillations by skeleton photoperiods. Science 144:565
72. Nelson DE, Takahashi JS (1999) Integration and saturation within the circadian photic entrainment pathway of hamsters. Am J Physiol 277:R1351–R1361
73. Comas M, Beersma DGM, Spoelstra K, Daan S (2007) Circadian response reduction in light and response restoration in darkness: a "skeleton" light pulse PRC study in mice (*Mus musculus*). J Biol Rhythms 22:432–444
74. Hut RA, Van Oort BEH, Daan S (1999) Natural entrainment without dawn and dusk: the case of the european ground squirrel (*Spermophilus citellus*). J Biol Rhythms 14:290–299
75. Pittendrigh CS, Bruce VG (1957) An oscillator model for biological clocks. In: Rudnick D (ed) Rhythmic and synthetic processes in growth. Princeton University Press, Princeton, pp 239–268
76. Harker JE (1960) Endocrine and nervous factors in insect circadian rhythms. Cold Spring Harb Symp Quant Biol 25:279–287
77. Page TL (1981) Neural and endocrine control of circadian rhythmicity in invertebrates. In: Aschoff J (ed) Handbook of behavioral neurobiology, vol 4. Plenum, New York, pp 145–172
78. Nishiitsutsuji-Uwo J, Pittendrigh CS (1968) Central nervous system control of circadian rhythmicity in the cockroach III. The optic lobes, locus of the driving oscillation? Z Vgl Physiol 58:14–46
79. Page T (1982) Transplantation of the cockroach circadian pacemaker. Science 216:73–75
80. Gaston S, Menaker M (1968) Pineal function: the biological clock in the sparrow? Science 160:1125–1127
81. Zimmerman NH, Menaker M (1979) Pineal-gland – Pacemaker within the circadian system of the house sparrow. Proc Natl Acad Sci U S A 76:999–1003
82. Simpson SM, Follett BK (1981) Pineal and hypothalamic pacemakers – their role in regulating circadian rhythmicity in Japanese quail. J Comp Physiol 144:381–389
83. Richter CP (1967) Sleep and activity: their relation to the 24-hour clock. Res Publ Assoc Nerv Ment Dis 45:8–27
84. Stephan FK, Zucker I (1972) Circadian rhythms in drinking behavior and locomotor activity of rats are eliminated by hypothalamic lesions. Proc Nat Acad Sci USA 69:1583–1586
85. Moore RY, Eichler VB (1972) Loss of a circadian adrenal corticosterone rhythm following suprachiasmatic lesions in rat. Brain Res 42:201–206
86. Inouye S, Kawamura H (1979) Persistence of circadian rhythmicity in a mammalian hypothalamic "island" containing the suprachiasmatic nucleus. Proc Natl Acad Sci U S A 76:5962–5966

87. Ralph MR, Menaker M (1988) A mutation in the circadian system in golden hamsters. Science 241:1225–1227
88. Ralph MR, Foster RG, Davis FC, Menaker M (1990) Transplanted suprachiasmatic nucleus determines circadian period. Science 247:975–978
89. Block GD, Wallace SF (1982) Localization of a circadian pacemaker in the eye of a mollusk, Bulla. Science 217:155–157
90. Hendrickson AE, Wagoner N, Cowan WM (1972) An autoradiographic and electron microscopic study of retino-hypothalamic connections. Z Zellforsch Mikrosk Anat 135:1–26
91. Moore RY, Lenn NJ (1972) A retinohypothalamic projection in the rat. J Comp Neurol 146:1–14
92. Cloudsley-Thompson JL (1953) Studies on diurnal rhythms – III. Photoperiodism in the cockroach *Periplaneta americana* (L.). Ann Mag Nat Hist 6:705–712
93. Roberts SK (1965) Photoreception and entrainment of cockroach activity rhythms. Science 148:958–959
94. Nishiitsutsuji-Uwo J, Pittendrigh CS (1968) Central nervous system control of circadian rhythmicity in the cockroach II. The pathway of light signals that entrain the rhythm. Z Vgl Physiol 58:1–13
95. Truman JW (1971) Circadian rhythms and physiology with special reference to neuroendocrine processes in insects. In: Proceedings of the International Symposium on Circadian Rhymicity, Pudoc Press, Wageningen, pp 111–135
96. Menaker M (1968) Extraretinal light reception in the sparrow I. Entrainment of the biological clock. Proc Natl Acad Sci U S A 59:414–421
97. Underwood H (2001) Circadian organization in nonmammalian vertebrates. In: Takahashi JS, Turek FW, Moore RY (eds) Handbook of behavioral neurobiology, vol 12, Circadian clocks. Kluwer, New York, pp 111–140
98. Richter C (1967) Psychopathology of periodic behavior in animals and man. In: Zubin J, Hunt HF (eds) Comparative psychopathology. Grune & Stratton, New York, pp 205–227
99. Groos GA, Mason R (1980) The visual properties of rat and cat suprachiasmatic neurones. J Comp Physiol 135:349–356
100. Freedman MS, Lucas RJ, Soni B, von Schantz M, Munoz M, David-Gray Z, Foster R (1999) Regulation of mammalian circadian behavior by non-rod, non-cone, ocular photoreceptors. Science 284:502–504
101. Provencio I, Rodriguez IR, Jiang GS, Hayes WP, Moreira EF, Rollag MD (2000) A novel human opsin in the inner retina. J Neurosci 20:600–605
102. Berson DM, Dunn FA, Takao M (2002) Phototransduction by retinal ganglion cells that set the circadian clock. Science 295:1070–1073
103. Lewy A, Newsome D (1983) Different types of melatonin circadian secretory rhythms in some blind subjects. J Clin Endocrinol Metab 56:1103–1107
103a. Lockley SW, Skene DJ, Arendt J, Tabandeh H, Bird AC, Defrance R (1999) Relationship between melatonin rhythms and visual loss in the blind J.clin. Endocrinol Metab 82:3763–3770
104. Menaker M, Binkley S (1981) Neural and endocrine control of circadian rhythms in the vertebrates. In: Aschoff J (ed) Handbook of behavioral neurobiology. Plenum, New York, pp 243–256
105. Tosini G, Menaker M (1996) Circadian rhythms in cultured mammalian retina. Science 272:419–421
106. Silver R, Lesauter J, Tresco PA, Lehman MN (1996) A diffusible coupling signal from the transplanted suprachiasmatic nucleus controlling circadian locomotor rhythms. Nature 382:810–813
107. Cannon WB (1929) Organization for physiological homeostasis. Physiol Rev 9:399–431
108. Bernard C (1878) Leçons sur les phénomènes de la vie communs aux animaux et aux végétaux. J.-B. Bailliere et fils, Paris
109. Mrosovsky N (1990) Rheostasis. The physiology of change. Oxford University Press, Oxford
110. Kleitman N (1963) Sleep and Wakefulness. Revised and enlarged edition. University of Chicago Press, Chicago
111. Aserinsky E, Kleitman N (1953) Regularly occurring periods of eye motility, and concomitant phenomena, during sleep. Science 118:273–274
112. Aschoff J (1965) Circadian rhythms in man – a self-sustained oscillator with an inherent frequency underlies human 24-hour periodicity. Science 148:1427–1432

113. Jouvet M, Mouret J, Chouvet G, Siffre M (1974) Toward a 48-hour day: experimental bicircadian rhythms in man. In: Schmitt F (ed) The neurosciences: third study program. MIT Press, Cambridge, MA, pp 491–497
114. Moore-Ede M (1983) The circadian timing system in mammals: two pacemakers preside over many secondary oscillators. Fed Proc 42:2802–2808
115. Eastman C (1982) The phase-shift model of spontaneous internal desynchronization in humans. In: Aschoff J, Daan S, Groos G (eds) Vertebrate circadian systems. Springer, Berlin-Heidelberg, pp 262–267
116. Borbély AA (1982) Circadian and sleep-dependent processes in sleep regulation. In: Aschoff J, Daan S, Groos GA (eds) Vertebrate circadian systems. Springer, Berlin-Heidelberg, pp 237–242
117. Zulley J, Wever RA (1982) Interaction between the sleep-wake cycle and the rhythm of rectal temperature. In: Aschoff J, Daan S, Groos GA (eds) Vertebrate circadian systems. Springer, Berlin-Heidelberg, pp 253–261
118. Borbély AA (1982) A two-process model of sleep regulation: I. Physiological basis and outline. Hum Neurobiol 1:195–204
119. Daan S, Beersma DGM (1983) Circadian gating of human sleep-wake cycles. In: Moore-Ede MC, Czeisler CA (eds) Mathematical models of the circadian sleep-wake cycle. Raven Press, New York, pp 129–158
120. Daan S, Beersma DGM, Borbély AA (1984) Timing of human sleep: recovery process gated by a circadian pacemaker. Am J Physiol 246:R161–R178
121. Dijk D-J, Duffy JF, Czeisler CA (1992) Circadian and sleep/wake dependent aspects of subjective alertness and cognitive performance. J Sleep Res 1:112–117
122. Tononi G, Cirelli C (2006) Sleep function and synaptic homeostasis. Sleep Med Rev 10:49–62
123. Bünning E (1935) Zur Kenntnis der erblichen Tagesperiodizitat bei den Primarblattern von Phaseolus multiflorus. Jahrb wiss Bot 81:411–418
124. Konopka RJ, Benzer S (1971) Clock mutants of *Drosophila melanogaster*. Proc Natl Acad Sci U S A 68:2112–2116
125. Hardin PE, Hall JC, Rosbash M (1990) Feedback of the Drosophila period gene product on circadian cycling of its messenger RNA levels. Nature 343:536–540
126. Feldman JF, Hoyle MN (1973) Isolation of circadian clock mutants of *Neurospora crassa*. Genetics 75:605–613
127. Lowrey PL, Shimomura K, Antoch MP, Yamazaki S, Zemenides PD, Ralph MR, Menaker M, Takahashi JS (2000) Positional syntenic cloning and functional characterization of the mammalian circadian mutation tau. Science 288:483–491
128. Hotz Vitaterna M, King D, Chang A, Kornhauser JM, Lowrey PL, McDonald JD, Dove WF, Pinto LH, Turek FW, Takahashi JS (1994) Mutagenesis and mapping of a mouse gene clock, essential for circadian behaviour. Science 264:719–725
129. Foster R, Kreitzman L (2004) Rhythms of life. The biological clocks that control the daily lives of every living thing. Profile Books Ltd., London
130. Toh KL, Jones CR, He Y, Eide EJ, Hinz WA, Virshup DM, Ptaçek LJ, Fu Y-H (2001) An hPer2 phosphorylation site mutation in familial advanced sleep phase syndrome. Science 291:1040–1043
131. Nakajima M, Imai K, Ito H, Nishiwaki T, Murayama Y, Iwasaki H, Oyarna T, Kondo T (2005) Reconstitution of circadian oscillation of cyanobacterial KaiC phosphorylation in vitro. Science 308:414–415
132. Honma K, Honma S, Hiroshige T (1987) Activity rhythms in the circadian domain appear in suprachiasmatic nuclei lesioned rats given methamphetamine. Physiol Behav 40:767–774
133. Mohawk JA, Baer ML, Menaker M (2009) The methamphetamine-sensitive circadian oscillator does not employ canonical clock genes. Proc Natl Acad Sci U S A 106:3519–3524
134. Bünning E (1958) Das Weiterlaufen der "physiologischen Uhr" im Säugerdarm ohne zentrale Steuerung. Naturwissenschaften 45:68
135. Pittendrigh CS, Bruce V, Kaus P (1958) On the significance of transients in daily rhythms. Proc Natl Acad Sci U S A 44:965–973

136. Plautz JD, Kaneko M, Hall JC, Kay SA (1997) Independent photoreceptive circadian clocks throughout Drosophila. Science 278:1632–1635
137. Stephan F, Swann J, Sisk C (1979) Anticipation of 24-hr feeding schedules inrats with lesions of the suprachiasmatic nucleus. Behav Neural Biol 25:346–363
138. Stephan FK (2001) Food-entrainable oscillators in mammals. In: Takahashi JS, Turek FW, Moore RY (eds) Handbook of behavioral neurobiology, vol 12, Circadian clocks. Kluwer/ Plenum, New York, pp 223–246
139. Damiola F, Le Minh N, Preitner N, Kornmann B, Fleury-Olela F, Schibler U (2000) Restricted feeding uncouples circadian oscillators in peripheral tissues from the central pacemaker in the suprachiasmatic nucleus. Genes Dev 14:2950–2961
140. Stokkan K-A, Yamazaki S, Tei H, Sakaki Y, Menaker M (2001) Entrainment of the liver clock by feeding. Science 291:490–493
141. Schibler U (2009) The 2008 Pittendrigh/Aschoff lecture: peripheral phase coordination in the mammalian circadian timing system. J Biol Rhythms 24:3–15
142. Yoo SH, Yamazaki S, Lowrey PL, Shimomura K, Ko CH, Buhr ED, Siepka SM, Hong HK, Oh WJ, Yoo OJ, Menaker M, Takahashi JS (2004) PERIOD2:: LUCIFERASE real-time reporting of circadian dynamics reveals persistent circadian oscillations in mouse peripheral tissues. Proc Natl Acad Sci U S A 101:5339–5346
143. Roenneberg T, Morse D (1993) 2 Circadian oscillators in one cell. Nature 362:362–364
144. Roenneberg T, Aschoff J (1990) Annual rhythm of human-reproduction.1. Biology, Sociology, or Both. J Biol Rhythms 5:195–216
145. Garner WW, Allard HA (1920) Effect of the relative length of day and night and other factors of the environment on growth and reproduction in plants. J Agric Res 18:553–606
146. Marcovitch S (1923) Plant lice and light exposure. Science 58:537–538
147. Rowan W (1926) On photoperiodism, reproductive periodicity and the annual migration of birds and certain fishes. Proc Boston Soc Nat Hist 38:147–189
148. Baker JR, Ransom JR (1932) Factors affecting the breeding of the field mouse (*Microtus agrestis*): 1. Light. Proc R Soc Lond B Biol Sci 110:313–322
149. Bünning E (1936) Die endonome Tagesrhythmik als Grundlage der photoperiodischen Reaktion. Ber Dtsch Bot Ges 54:590–607
150. Nanda KK, Hamner KC (1959) Photoperiodic cycles of different lengths in relation to flowering in Biloxi soybean. Planta 53:45–52
151. Follett BK, Mattocks PW, Farner DS (1974) Circadian function in the photoperiodic induction of gonadotrophin secretion in the white-crowned sparrow. Proc Natl Acad Sci U S A 71:1666–1669
152. Pittendrigh CS (1972) Circadian surfaces and the diversity of possible roles of circadian organization in photoperiodic induction. Proc Natl Acad Sci U S A 69:2734–2737
153. Saunders DS (1973) Thermoperiodic control of diapause in an insect: theory of internal coincidence. Science 181:358–360
154. Tyshchenko VM (1966) Two-oscillatory model of the physiological mechanism of the photoperiodic reaction of insects. Zh Obshch Biol 27:209–222
155. Pittendrigh CS, Daan S (1976) A functional analysis of circadian pacemakers in nocturnal rodents V. Pacemaker structure: a clock for all seasons. J Comp Physiol 106:333–355
156. Daan S, Berde C (1978) Two coupled oscillators: simulations of the circadian pacemaker in mammalian activity rhythms. J Theor Biol 70:297–313
157. Meijer JH, Daan S, Overkamp GJF, Hermann PM (1990) The 2-oscillator circadian system of tree shrews (*Tupaia belangeri*) and its response to light and dark pulses. J Biol Rhythms 5:1–16
158. de la Iglesia HO, Meyer J, Carpino A, Schwartz WJ (2000) Antiphase oscillation of the left and right suprachiasmatic nuclei. Science 290:799–801
159. Stoleru D, Peng Y, Agosto J, Rosbash M (2004) Coupled oscillators control morning and evening locomotor behaviour of *Drosophila*. Nature 431:862–868
160. Grima B, Chelot E, Xia RH, Rouyer F (2004) Morning and evening peaks of activity rely on different clock neurons of the *Drosophila* brain. Nature 431:869–873

161. Jagota A, de la Iglesia HO, Schwartz WJ (2000) Morning and Evening circadian oscillators in the suprachiasmatic nucleus in vitro. Nature 3:372–376
162. Sumová A, Travnicková Z, Peters R, Schwartz WJ, Illnerová H (1995) The rat suprachiasmatic nucleus is a clock for all seasons. Proc Natl Acad Sci U S A 92:7754–7758
163. Lerner AB, Case JD, Takahashi Y, Lee TH, Mori W (1958) Isolation of melatonin, the pineal gland factor that lightens melanocytes. J Am Chem Soc 80:2587
164. Czyba JC, Girod C, Durand N (1964) Sur l' antagonisme épiphysohypophysaire et les variations saisonnieres de la spermatogenèse chez le hamster doré (*Mesocricetus auratus*). C R Soc Biol (Paris) 158:742–745
165. Moore RY, Heller A, Wurtman RJ, Axelrod J (1967) Visual pathway mediating pineal response to environmental light. Science 155:220–223
166. Reiter RJ (1978) Interaction of photoperiod, pineal and seasonal reproduction as exemplified by findings in the hamster. In: Reiter RJ (ed) Progress in reproductive biology, vol 4. Basel, S. Karger AG, pp 169–190
167. Carter DS, Goldman BD (1983) Antigonadal effects of timed melatonin infusion in pinealectomized male Djungarian hamsters (*Phodopus sungorus sungorus*): duration is the critical parameter. Endocrinology 113:1261–1267
168. Brandstätter R, Kumar V, Abraham U, Gwinner E (2000) Photoperiodic information acquired and stored in vivo is retained in vitro by a circadian oscillator, the avian pineal gland. Proc Natl Acad Sci U S A 97:12324–12328
169. Kolar J, Machackova I (2005) Melatonin in higher plants: occurrence and possible functions. J Pineal Res 39:333–341
170. Dubois R (1896) Étude sur le méchanisme de la thermogenèse et du sommeil chez les mammifères. Physiologie comparée de la Marmotte. Annales de l' Université de Lyon 25:1–268
171. Marshall AJ (1951) The refractory period of testis rhythm in birds and its possible bearing on breeding and migration. Wilson Bull 63:238–261
172. Pengelley ET, Fisher KC (1957) Onset and cessation of hibernation under constant temperature and light in the golden-mantled ground squirrel (*Citellus lateralis*). Nature 180:1371–1372
173. Pengelley ET, Fisher KC (1963) The effect of temperature and photoperiod on the yearly hibernating behavior of captive golden-mantled ground squirrels (*Citellus lateralis tescorum*). Can J Zool 41:1103–1120
174. Blake GM (1959) Control of diapause by an "internal clock" in *Anthrenus verbasci* (L.) (*Col. Dermestidae*). Nature 183:126–127
175. Gwinner E (1968) Circannuale Periodik als Grundlage des jahreszeitlichen Funktionswandels bei Zugvögeln: Untersuchungen am Fitis (*Phylloscopus trochilus*) und am Waldlaubsänger (*P. sibilatrix*). J Ornithol 109:70–95
176. Gwinner E (1986) Circannual rhythms. Springer, Berlin
177. Fogden MPL (1972) The seasonality and population dynamics of equatorial forest birds in sarawak. Ibis 114:307–309
178. Chapin JP (1954) The calendar of wideawake fair. Auk 71:1–15
179. Gwinner E (1968) Artspezifische Muster der Zugunruhe bei Laubsängern und ihre mögliche Bedeutung für die Beendigung des Zuges im Winterquartier. Z Tierpsychol 25:843–853
180. Gwinner E, Wiltschko W (1980) Circannual changes in migratory orientation of the garden warbler, Sylvia borin. Behav Ecol Sociobiol 7:73–78
181. Gwinner E (1981) Circannuale Rhytmen bei Tieren und ihre photoperiodische Synchronisation. Naturwissenschaften 68:542–551
182. Zucker I (2001) Circannual rhythms. In: Takahashi JS, Turek FW, Moore RY (eds) Handbook of behavioural neurobiology, vol 12, Circadian clocks. Kluwer/Plenum, New York, pp 511–528
183. Lincoln GA, Clarke IJ, Hut RA, Hazlerigg DG (2006) Characterizing a mammalian circannual pacemaker. Science 314:1941–1944
184. Enright JT (1963) The tidal rhythm of activity of a sand-beach amphipod. Z Vgl Physiol 46:276–313

185. Neumann D (1966) Die Lunare Und Tägliche Schlupfperiodik Der Mücke *Clunio* - Steuerung Und Abstimmung Auf Die Gezeitenperiodik. Z Vgl Physiol 53:1–61
186. Enright JT (1972) A virtuoso isopod – circa-lunar rhythms and their tidal fine-structure. J Comp Physiol 77:141–161
187. Lloyd M, Dybas HS (1966) Periodical cicada problem. 2. Evolution. Evolution 20:466–505
188. Daan S, Slopsema S (1978) Short-term rhythms in foraging behavior of common vole, Microtus arvalis. J Comp Physiol 127:215–227
189. Gerkema MP, Groos GA, Daan S (1990) Differential elimination of circadian and ultradian rhythmicity by hypothalamic lesions in the common vole, Microtus arvalis. J Biol Rhythms 5:81–95

Chapter 2
Transcriptional Regulation of Circadian Clocks

Jürgen A. Ripperger and Steven A. Brown

2.1 Introduction

The first chapter of this book has introduced the historical background of the circadian clock, as well as its anatomical organization. It has described how researchers over the past several decades have grappled with the problem of biological timekeeping: how a constantly-changing organism can measure time, and in particular solar time, accurately in a changing environment? In the case of simpler eukaryotes, the desired metric is longer than the lifespan of the organism, and the mechanism must be cell-autonomous and robust to cellular division. Added to this already-daunting problem is the difficulty of temperature: biochemical reactions occur with greater rapidity as temperature increases, and any timekeeping mechanism must be immune to these changes. In this chapter, we shall consider the molecular mechanisms by which metazoan organisms have organized timekeeping mechanisms that fulfill all of these criteria.

A cell-autonomous circadian system is present in nearly all cells of all metazoans studied so far, from flies to man, and its component proteins share high homology from one organism to the next. In fact, the same general mechanism is even conserved in plants and simpler eukaryotes. Though individual components are no longer precisely homologous, identical general lessons can be drawn. For those interested in these interesting comparisons, Chap. 7 is devoted to comparing clocks among different organisms later in this book. In it, similarities and differences among circadian systems in metazoans, in plants, in simple eukaryotes like the bread mold *Neurospora crassa*, and in the evolutionarily ancient clocks of photosynthetic cyanobacteria are considered. The present chapter, however, considers the basic design principles of metazoan clocks, the ways in which they are controlled

J.A. Ripperger (✉)
Department of Medicine, Unit of Biochemistry, University of Fribourg, 1700, Fribourg, Switzerland
e-mail: juergenalexandereduard.ripperger@unifr.ch

S.A. Brown
Institute of Pharmacology and Toxicology, University of Zürich, 8057, Zürich, Switzerland
e-mail: steven.brown@pharma.uzh.ch

U. Albrecht (ed.), *The Circadian Clock*, Protein Reviews 12,
DOI 10.1007/978-1-4419-1262-6_2,

by the environment, and the ways in which they in turn control the vast spectrum of circadian output processes.

2.2 Basic Design Principles: The Transcriptional Feedback Loop

Transcription is necessary to exploit the genetic information stored in the genome of an organism. This information has to be converted into an mRNA copy before it can be used as template for the synthesis of its corresponding gene product. In principle, regulation of this process can be achieved by two opposing mechanisms: transcriptional activation or repression. In this section, we will elaborate the principal concepts how to build stable circadian oscillators from simple transcriptional regulatory loops. From the observation of Hardin, Hall, and Rosbash in 1990 that the product of the circadian clock protein PERIOD regulates its own transcription, a model was proposed that has become the cornerstone of thinking about the circadian clock for the past 20 years – a transcriptional feedback loop of gene expression [1]. Since its origin, the idea possessed an immediate appeal. Without any consideration for biology, it was mathematically apparent that such auto-repression could explain the oscillatory behavior – of genes, of proteins, or of anything else. (For a basic description of the mathematics, see Appendix 1. For a brief introduction to the biology of transcription and translation, see Appendix 2.)

2.2.1 The Simple Transcriptional Feedback Loop

Plainly stated, for the circadian clock the basic idea of a feedback loop of gene expression is that the transcription of a "clock gene" is repressed indirectly by its product. Although elegantly simple, this idea has two fundamental problems. Most importantly, it does not explain how the circadian oscillator measures daily time. From the moment a eukaryotic gene is "activated" or switched on, the time taken for its transcription and translation is up to 2 h. Thus, in its simplest form, a transcriptional feedback loop would have a period of between 1 and 2 h, and certainly not 24.

This difficulty is best highlighted by "designed" oscillators of gene expression that have been created by multiple groups in an attempt to mimic the functions of the circadian oscillator. For example, Elowitz and Leibler have created a simple oscillator in *E. coli* by introducing synthetic genes that regulate each other, using three known transcriptional repressors from other systems. In their system, the lacI transcriptional repressor inhibited the transcription of the tetR transcriptional repressor, tetR inhibited transcription of the cI transcriptional repressor, and cI inhibited transcription of the original lacI repressor, thereby "closing" the feedback loop. The basic promoters that turned on each gene in the absence of repressor were strong, but were able to be tightly shut off, and the half-life of each protein was short (less than 1 h). The resulting oscillator had a period of around 2.5 h [2] (See Fig. 2.1a and b.). Already, this simple design was robust to cellular division (in *E. coli* every 20–60 min depending upon nutrients). In natural systems, a similarly short period can be seen in the clock that directs somite formation during vertebrate development. Here, the HES-7 gene product directly represses its own transcription, and the resultant oscillatory period is 2 h long [3].

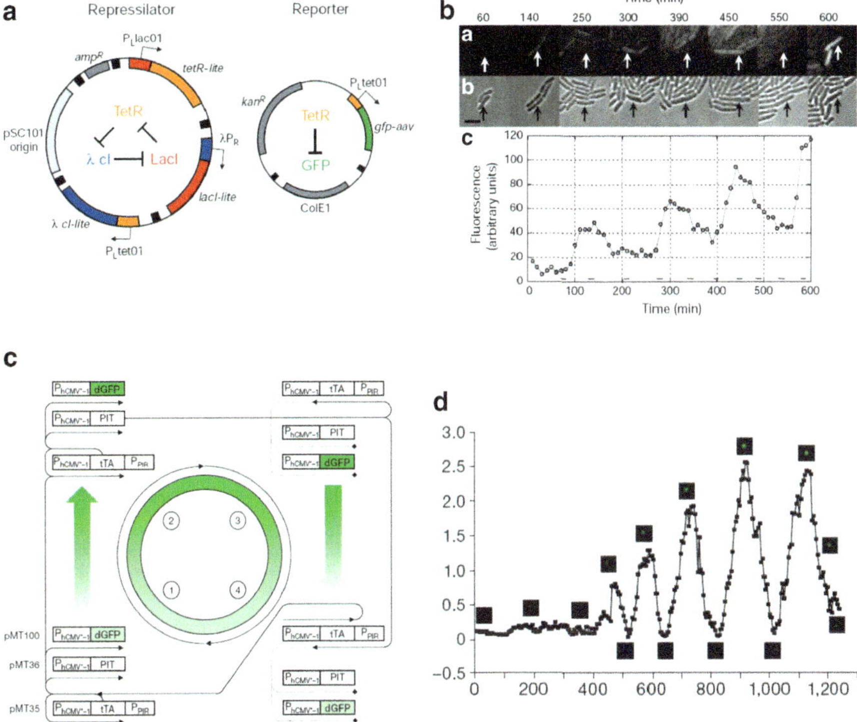

Fig. 2.1 **(a)** The bacterial "repressilator" of Elowitz and Leibler. It is composed of three repressor genes and their corresponding promoters. It uses pllacO1 and pLtetO1, which are strong, tightly repressible promoters containing lac and tet operators, respectively, as well as pR, the right promoter from phage lambda. The compatible reporter plasmid at right expresses an intermediate-stability GFP variant (gfp-aav). **(b)** Growth and time course of GFP expression of a single cell of *E. coli* strain MC4100 **containing** the repressilator plasmids. Fluorescent (top) and brightfield (middle) snapshots are shown, along with quantitation of observed fluorescence. **(c)** The mammalian oscillator of Tigges et al. Autoregulated phCMV-1-driven ttA transcription triggers increasing expression of sense ttA (pMT35), UbV76-GFP (pMT100), and PIT (pMT36) (1). As UbV76-GFP and PIT levels reach a peak (2), PIT steadily induces pPIR-driven tTA antisense expression (3), resulting in a gradual decrease in sense tTA, PIT, and UbV76-GFP (4). **(d)** Sample output from mammalian CHO cells transfected with equimolar ratios of each of the plasmids of the oscillator system. Text and Figure parts **a** and **b** are reproduced from Elowitz and Leibler (2000), parts **c** and **d** are reproduced from Tigges et al. [4] with permission

2.2.2 *Additional Features Stabilizing Transcriptional Feedback Loops*

The second major problem faced by a simple "feedback loop" oscillator is robustness. In the simple form that has been discussed, the period length of the resulting clock – as well as whether it cycled at all – would be highly influenced by the concentration of its components, and could also dampen rapidly. Thus, it would be highly susceptible to "stochastic noise", the variation of transcription or translation rates from one cell to another based upon random availability of components. Here, again, the ramifications are best illustrated synthetically. The *E. coli* oscillatory system described in the previous paragraph showed both rapid damping and relatively unstable period [2]. To achieve a stable period length, more precise control of nonrepressed transcription – i.e. the transcription of feedback loop components in the "on" state – is required. Such an example can be found in a mammalian synthetic feedback loop designed by Tigges et al. [4]. Here, transcription of the ttA tetracycline-mediated activator was driven by a constitutive strong promoter, the CMV promoter. Antisense transcription of the same gene – i.e. transcription of the other strand of DNA – was driven by the pristamycin-dependent transactivator PIT. Negative feedback was provided at two levels. First, transcription of the PIT gene was itself turned on by the ttA activator; and second, antisense transcription of the ttA locus interferes with ttA production. The activation properties of this network can be modulated by antibiotics, because both the ttA activator and the PIT activator can be potentiated by the presence of antibiotic (tetracycline or pristamycin, respectively), thereby controlling the degree of activation. The resultant oscillator displayed a stable period length in individual cells that was tunable from 2 to 6 h in length, but critically dependent upon activator concentrations for its stability. (See Fig. 2.1c and d) In addition, this synthetic system still displayed significant stochastic variation from cell to cell, with period variations of one-third to one-half the average period length [4]. Overall, based upon this experiment and from others like it, it is likely that two design features aid in robust oscillations: a time delay in the negative feedback loop, and the additional input of positive factors [5].

From these examples, one can conclude that a circadian oscillator based upon a simple feedback loop of gene expression would be very imprecise and only a few hours long. Nevertheless, all circadian oscillators studied so far are remarkably reliable daily timekeepers. Thus, other factors must be operational to aid in their stabilization and in the lengthening of their period. A first clue to these "other factors" is offered by the dazing and evergrowing array of genes that have been shown to be important to the circadian oscillator.

2.3 Clock Genes, Clock Gene Functions

Beginning with the discovery of *Drosophila* mutations that changed the period length of fly activity measured in constant environmental conditions, an ever-increasing array of loci has been shown to influence the circadian clock function.

These genes have been discovered in a variety of different organisms using both genetic and biochemical techniques. Most have been shown to be regulated by other clock gene products, or to interact with them. Set out next is a list of these "clock genes" and their demonstrated or presumed functions within the circadian clock. Subsequently, we shall consider their interactions in a feedback loop model of the circadian oscillator. According to their genetic or biochemical activities, these genes have been classed below as "negative" or "positive" depending upon whether they play a repressive or activating role within this feedback loop. For those wishing to see the interactions more globally while reading about the individual genes, the overall network for mammals is diagrammed in Fig. 2.2, and it will be discussed in detail after the individual genes have been introduced.

2.3.1 The Period Genes

These first-discovered of clock genes were initially characterized as mutations of a *Drosophila* gene that affected the period length of fly circadian behavior [6]. All of the mutations cosegregated to the same fly gene, Period (abbreviated Per). Nevertheless, homology-based cloning in mammals has indicated three Period genes, *Per1*, *Per2*, and *Per3* [7]. Because the expression of Per in flies represses its own transcription by direct or indirect means [1], it is traditionally indicated to be at the heart of the circadian "transcriptional feedback loop", generally in a negative or repressive role. It has also been shown to play an activating role for the *Bmal1* gene [8], discussed below, but this interaction is likely indirect (e.g. the repressor of a repressor).

Genetically, hypomorphic mutations (causing reduction of function) or deletions of one or more *Per* genes have resulted in shorter circadian period length or in arrythmicity – i.e. the lack of a functional oscillator. Even in humans, a familial mutation mapped to the *Per2* gene causes Familial Advanced Sleep Phase Syndrome, a disease characterized by short circadian period and early behavioral phase [9]. In *Drosophila* mutations can also be found in the *Per* gene that lengthen circadian period [10]. These map to a particular helix believed to be involved in PER protein homo- or heterodimerisation and in temperature compensation, the mechanism by which the circadian clock succeeds in maintaining the same period length at different temperatures [11, 12].

Structurally, the PER proteins contain two PAS (PER-ARNDT-SIM) protein–protein interaction motifs [13], two other C-terminal alpha helices likely involved in interprotein interactions [12], nuclear localization and export signals [14], and sites for post-translational modifications. Hence, it is not surprising that the PERIOD proteins have been shown to interact biochemically with multiple different dedicated members of the circadian oscillator, including Timeless and Cryptochromes. (For a description of these and other mentioned proteins, as well as cited literature, please see their corresponding rubrics below.) The actions of PER proteins are probably facilitated or hindered by a number of nondedicated

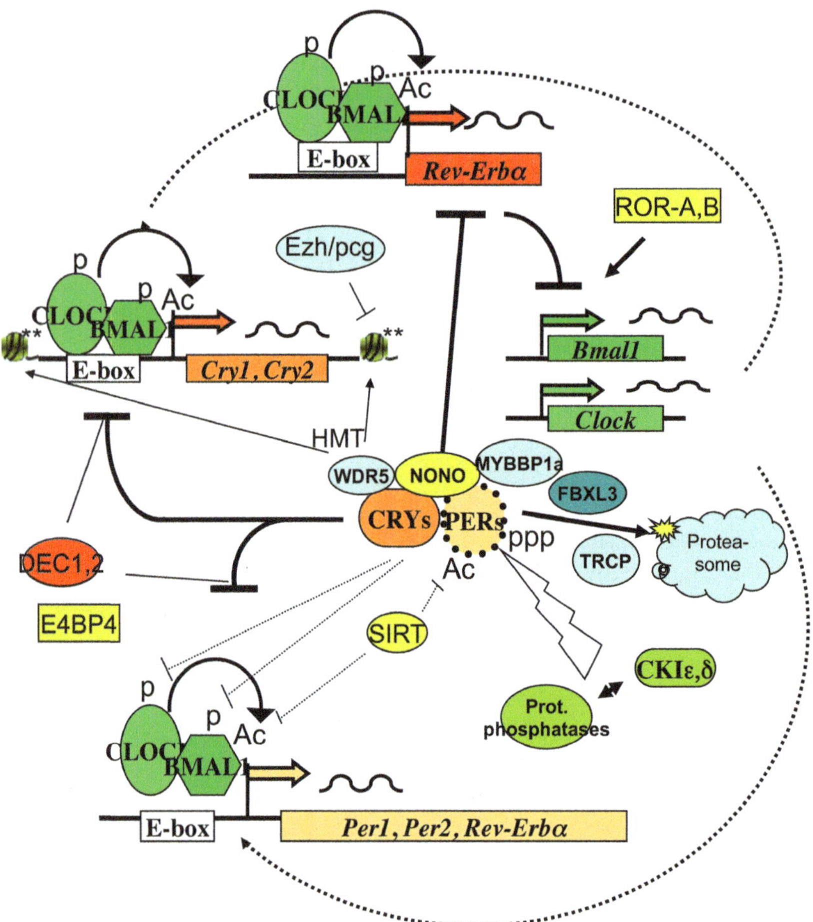

Fig. 2.2 Model of the mammalian circadian oscillator. A pair of transcriptional activators, BMAL1 and CLOCK, activates transcription via E-box motifs of two classes of repressors. In the stabilizing loop, REV-ERBα represses immediately the transcription of the *Bmal1* and *Clock* genes. The transcriptional activators RORα and RORβ can rhythmically compete with the action of REV-ERBα to fine-tune circadian gene expression. In the core loop, BMAL1 and CLOCK activate the transcription of the *Per* and *Cry* genes. Upon reaching a certain threshold concentrations, these factors counteract the positive factors to repress the *Per*, *Cry*, and *Rev-Erbα* genes. This generates two interlocked feedback-loops with their phases separated by about 12 h. Post-translational modifications (p for phosphorylation, e.g. by CKIε,δ, Ac for acetylation) regulate the activity or halves-lives of the different proteins. In particular, SIRT may influence the activity of BMAL1 or the half-life of PER2, FBXL3 determines the half-life of the CRY proteins and TRCP determines the half-life of the PER proteins via proteosome-dependent degradation pathways, and various factors (WDR5, Ezh/pcg, and the HAT activity of CLOCK) may regulate the local chromatin structure. Some factors, like NONO and MYBBP1α, interact with PER or CRY proteins, respectively, but have yet to precise functions. There are additional factors, which are involved in the regulation of circadian genes like the *Dec1* and *Dec2* genes, and E4BP4

proteins – i.e. proteins which play an important circadian function, but additionally play functional roles in other noncircadian systems. These include adaptors for chromatin modifying complexes like WDR5 [15], F-box-containing ubiquitin ligase complex members like β- TRCP in mammals [16] and SLIMB in *Drosophila* [17], corepressors such as MYBBP1a [18] and E4BP4 (a homolog of *Drosophila* Vrille) [19, 20], and RNA-binding proteins such as NONO [15], all of which have been shown to interact with PER protein itself. Another RNA-binding protein, LARK, has been shown to interact with the *Per* mRNA to modulate its stability [21].

Period proteins are modified post-translationally by a number of kinases including casein kinase 1ε, casein kinase 1δ, and casein kinase 2 [22–26]. In *Drosophila*, the same conserved domain phosphorylated by these kinases in the PER protein has been linked to its nuclear localization and transcriptional repression activity, suggesting that many actions of and upon PER may be inter-related [27, 28]. In mammals, different phosphorylation events have been shown to affect the stabilization of PER and its nuclear localization in different ways (see Chap. 3) [29]. PER protein is also acetylated, and its deacetylation by SIRT1 facilitates its degradation and perhaps also connects PER protein function to cellular metabolism [30].

In mammals, the period genes *Per1* (and possibly *Per2*) are also acutely induced by light in the suprachiasmatic nucleus (SCN) (see also Sect. 2.5.1), and probably play a role in the input of light into the circadian molecular circuit [31, 32]. *Per* genes are also induced in cells by a variety of stimuli that reset the circadian oscillator, and therefore are likely to play a role in clock synchronization at all systemic levels [33, 34]. This role is not completely conserved in all metazoans. In zebra fish, at least one of the (multiple) Per genes demonstrates a behavior that is the reverse of the mammalian one, and is repressed by light [35], and in *Drosophila*, the role of PER in light-induced phase shifting is an indirect one: the Timeless and Cryptochrome proteins are likely the direct mediators of light upon the circadian oscillator [36].

2.3.2 The Timeless Gene

This gene was also first isolated in *Drosophila*, where its function was shown to be critical to the circadian oscillator, and its presence necessary for the nuclear localization of Period proteins [37, 38]. Since these two proteins dimerize in the cytoplasm prior to translocating to the nucleus, it was largely assumed that TIM and PER translocated as a complex; however, recent FRET studies have disproved this notion, and instead suggest that the two proteins accumulate as dimers together in the cytoplasm and then enter the nucleus separately within the same approximate temporal window [39]. Consistent with this observation, although PER and TIM are both classed as "negative" factors, PER proteins appear capable of directing transcriptional repression in the absence of TIM [40].

TIM also serves as a central regulation point for the effects of light upon the circadian oscillator via its light-dependent degradation mediated through Cryptochromes

[36], discussed next. This degradation also requires proteasome function, probably recruited via the JETLAG protein [41]. In mammals, however, the role of Timeless is highly controversial. The mammalian TIM protein has been shown to interact with other clock proteins in transfection assays [42, 43], and antisense oligo-based loss-of-function experiments in the SCN also suggest a role in the clockwork [44]. Nevertheless, the mammalian TIM is in fact probably the homolog of the distantly-related *Drosophila* Timeout protein important in development, and not of the Timeless protein itself [45]. A mouse *Timeless* knockout perishes early in development at embryonic day 8 [46]. Hence, its direct role in the mammalian circadian clockwork remains a disputed question, and the Timeless protein itself remains one of the most significant differences between insect and mammalian circadian systems.

In insects, however, the importance of Timeless to the circadian oscillator remains unquestioned, and its interaction with PER is important both for PER nuclear localization as discussed earlier, and for the modification of PER by casein kinase 2 [47]. TIM protein is itself post-translationally modified by another kinase crucial to insect circadian function, Shaggy [48]. Shaggy is the *Drosophila* homolog of the mammalian glycogen synthase kinase 3β kinase, and cellular expression and inhibition studies suggest that this kinase too may play a role in the circadian clockwork [49].

2.3.3 The Cryptochrome Genes

The third major dedicated class of circadian genes that play a repressive role in the circadian oscillator are the Cryptochrome genes. These genes were first identified by their homology to blue-light photoreceptors in plants and bacteria, and their effects upon the circadian oscillator were therefore presumed to be light-driven [50]. In fact, mouse knockout studies and numerous functional ones show that in mammals, cryptochromes play an essential role in the inherent mechanism of the circadian oscillator [51], and specifically in transcriptional repression [52]. Surprisingly, they have little or no circadian photoreceptive role at the whole-organism level [53]. Nevertheless, in *Drosophila*, these proteins clearly carry out both functions: on the one hand, they act as blue-light photoreceptors that mediate the light-dependent degradation of the TIM protein [36, 54]; and on the other, they act as direct or indirect transcriptional repressors that play a necessary light-independent role in the circadian clockwork [55].

Structurally, CRY proteins possess an N-terminal domain homologous to bacterial photolyases which is sufficient for phototransduction and also apparently for transcriptional repression [56], and a carboxy-terminal section that is responsible for interaction with other proteins, including TIM and PER [57]. All cryptochrome proteins also bind two cofactors, a pterin (methenyltetrahydrofolate) and a flavin (FADH). In photolyases, the pterin cofactor harvests light and transfers it to the FADH, which in turn interacts with DNA. Although all important residues for photolyase function appear conserved, no photolyase activity has been detected in vertebrate CRY proteins.

Like PER proteins, CRY proteins are implicated in transcriptional repression within the core circadian clock mechanism. In fact, CRY proteins have transcriptional repressive activity independent of PER [58]. It is perhaps due to this potentially redundant function that deletions of one *Cry* gene in mammals can suppress the effects of deletion of a *Per* gene, a hypothesis discussed further below [59]. Finally, tangential to their clock roles, insect CRY proteins also play an important role in sun-compass navigation and magnetosensitivity [60, 61].

2.3.4 The Clock Gene

The *Clock* (Circadian **L**ocomotor **O**utput **C**ycles **K**aput) gene was first identified via a landmark forward mutagenesis screen in the mouse, followed by positional cloning [62, 63]. A close homolog of similar function exists in *Drosophila* [64]. Together with its partner BMAL1 (described below), CLOCK acts as the principal transcriptional activator of the circadian feedback system. It binds to *cis*-acting elements called E-boxes [65], which are present in the promoter sequences in multiple circadian clock genes of repressive function (including the *Periods* and *Cryptochromes,* and the *Rev-Erbα* repressor gene described below). In some tissues, a second CLOCK-like protein termed NPAS2 is also present [66]. Probably for this reason, the *Clock* gene is dispensable for circadian locomotor activity in mice [67]. Nevertheless, the activity of at least one of these two proteins is essential to circadian function [68, 69]. This activity appears to be that of a traditional transcriptional activator, directly or indirectly recruiting histone-modifying complexes, coactivators/adaptor complexes like p300/CBP, and thus RNA polymerase II itself [70–72].

In several respects, however, CLOCK does not behave as a "traditional" transcriptional activator. In addition to a PAS domain by which it probably interacts with its partner BMAL1, CLOCK possesses an intrinsic acetylase activity [73], which can act not only upon histones but upon its partner BMAL1, and is necessary to its activating function [74]. The same redox-sensitive SIRT1 protein that has been implicated in the deacetylation of PER2 protein has also been ascribed the function of deacteylating CLOCK [75]. Secondly, and in keeping with this connection to redox and cellular metabolism, the heterodimerisation of CLOCK and NPAS2 with BMAL1, and therefore its interaction with its target E-box DNA element, has been found to be redox-sensitive in vitro [76].

In mammals, the expression of the *Clock* gene is constant or very weakly circadian, but in *Drosophila* this gene shows a strong circadian amplitude. Its transcription is controlled by a pair of related transcription factors, PDP-1 (PAR-domaine protein 1) and VRILLE. Whereas the former protein activates transcription of *Clock* in flies, the latter represses it. In turn, the transcription of both of these factors is activated by dimers of CLOCK and its partner CYCLE (see below) [77, 78]. Both *Vrille* and *Pdp1* are essential for functional circadian oscillations in flies, and have a mammalian homolog, the E4BP4 protein, that probably plays a role in *Per2* expression [79, 80].

2.3.5 The Npas2 Gene

As mentioned in the immediately preceding section, this protein was initially identified as a homolog of the CLOCK protein, and appears to share or assume its functions in many tissues. Unlike CLOCK itself, however, the NPAS2 protein contains a heme-binding domain adjacent to its PAS domain responsible for interaction with the other circadian proteins. This heme-PAS combination is a common regulatory motif in a variety of enzymatic systems including histidine kinase and phosphodiesterase in mammals, as well as oxygen-sensing and nitrogen fixation proteins in plants and bacteria [81]. In the circadian oscillator, heme appears to modulate the activity of NPAS2 by preventing its DNA-binding in response to carbon monoxide [82, 83]. Thus, the NPAS2 protein might play a special role in circulatory or cardiac circadian clocks, but further research is required to clarify the nature of such a role [70].

Both CLOCK and NPAS2 are phosphorylated in vivo in circadian fashion. Although the identity of the responsible kinase is not known, this phosphorylation appears to facilitate DNA-binding and to be inhibited by the CRY proteins [84, 85]. Such a mechanism would therefore provide a mechanism for rhythmic transcriptional activation of circadian genes.

2.3.6 The Bmal1 Gene

This gene encodes the partner of CLOCK, and was initially identified in a yeast two-hybrid screen for proteins that interact with it [86]. Its fly homolog CYCLE possesses similar function [87]. As mentioned above, in mammals this protein is directly acetylated by its partner CLOCK, and these acetylated residues are critical to its ability to activate transcription [74]. Its interaction with its binding partner is also dictated in vitro by the redox potential of the incubation buffer [76]. In the cell, this state would be controlled principally by the concentrations of NAD+/NADH, NADP+/NADPH, and reduced and oxidized glutathione, opening a tempting link between the circadian clock and cellular metabolism. Although attempts to demonstrate a circadian oscillation of cellular redox state have so far proven unsuccessful, the SIRT1 "sirtuin" protein is a deacetylase activity that modulates circadian function by deactylating either BMAL1 or PER2, and its activity requires an NAD+ cofactor [30, 75]. Thus, two independent lines of evidence could tie the transcriptional activation of this dimer to cellular metabolism, and many more experiments underway in various laboratories will soon clarify this interesting subject.

The CLOCK-BMAL1 heterodimer also interacts physically with PER and CRY proteins [88], and this likely allows the repressive proteins described above to achieve their effects. Chromatin immunoprecipitation studies at clock gene promoters in vivo show rhythmic daily binding of CLOCK and BMAL1 to E-boxes, and their dissociation with these sites concomitant with the transient appearance of PER and CRY proteins [89]. Similarly, CLOCK, NPAS2, and BMAL1 undergo circadian phosphorylation concomitant with DNA-binding, and this phosphoryation appears inhibited by CRY proteins [84, 85]. The simplest model to explain these data would be that direct interaction of PER and CRY proteins with CLOCK/BMAL1 complex provokes their dephosphorylation, the dissociation of this complex from DNA, and the concomitant repression of target genes.

In addition to being phosphorylated and acetylated, the BMAL1 protein is also modified by sumoylation in circadian fashion. Although the effects of this modification for the function of the protein as a whole are not yet clear, overexpression in cells of a mutant BMAL1 protein that cannot be so modified shows altered circadian properties, implying that this post-translational modification also plays a functional role [90].

2.3.7 The Rev-Erbα and β Genes

The *Rev-Erbα* gene was originally identified via its binding activity upstream of the clock-gene *Bmal* [91, 92]. For the circadian mechanism itself, the important role of the REV-ERBα protein is its binding to *cis*-acting binding sites (the RREs, or Rev-Erbα-responsive elements) in the promoter of the *Bmal1* gene. This binding is essential to repression of *Bmal1*, and therefore to its rhythmic daily expression. Interestingly, such oscillation is not essential to circadian oscillation, and its disruption in mice results in only a small change in period length [91]. Thus, rhythmic expression of the positively-acting elements of the circadian clock is not essential to clock function. By contrast, overexpression of REV-ERBα has proven an effective genetic tool to silence circadian function, establishing the role of this gene, and of its targets, in the circadian clockwork [93].

The *Rev-Erbα* gene is a part of the nuclear orphan receptor superfamily. Although it lacks a traditional ligand-binding domain, like NPAS2 it is capable of interacting directly with a heme cofactor that is important for its repressive activity [94], and that can phase-shift the circadian oscillator [95]. Repression is likely carried out by the NCoR nuclear receptor corepressor complex [94]. This activity is also directly regulated by lithium ions commonly used to treat bipolar mania [96]. Hence, REV-ERBα may be important for conveying systemic signals from and/or to the circadian clock, and its close homolog REV-ERBβ likely plays a redundant role in these effects [97].

The *Rev-Erbα* gene itself contains multiple E-box regions necessary for its circadian transcription [98]. Therefore, it also represents a link in the mammalian circadian oscillator between the proteins controlling the Period and Cryptochrome negative elements and those controlling the positive elements Clock and Bmal1. For example, one likely way in which PER is an activator of *Bmal1* transcription is through its negative regulation of *Rev-Erbα* transcription.

2.3.8 The *Rorα*, *Rorβ*, and *Rorγ* Genes

The **R**etinoid-related **O**rphan **R**eceptor genes undoubtedly play a significant role in a large amount of nuclear hormone receptor-mediated physiology as well as in development and differentiation, both independently and by dimerising with other nuclear hormone receptor family members. In general, they function as transcriptional activators. Since they bind to the same elements as the REV-ERBα protein, they also affect circadian clock function by competing with REV-ERBα [99, 100]. Nevertheless, this activity appears nonessential to rhythmic *Bmal1* transcription [97]. What may be more important is the potential ability of ROR activators to introduce systemic influences upon the circadian oscillator. For example, PGC-1 is a coactivator of ROR proteins that also regulates energy metabolism, and mice lacking this gene not only show defects in *Bmal1* transcription patterns, but also abnormal diurnal activity patterns [101].

2.3.9 Clock-Associated Genes I: Kinases and Phosphatases

The previous paragraphs have discussed all known clock-dedicated proteins that play a transcriptional role within the feedback loop. Equally integral to clock function are an ever-growing number of kinases and phosphatases that modify clock proteins. These include casein kinase 1ε (known as Doubletime in flies) [25, 102], casein kinase 1δ [103], casein kinase 2 [22, 47], glycogen synthase kinase 3 (known as Shaggy in flies) [48], protein phosphatase 1 [104], protein phosphatase 2A [105], and protein phosphatase 5 [106, 107]. The casein kinase family likely phosphorylates Period and Cryptochrome proteins in multiple places leading to different effects, and the protein phosphatases mentioned above have been implicated in their dephosphorylation. Shaggy is likely the kinase responsible for phosphorylation of Timeless. The functions of most of these modifying proteins are as critical to clock function as the canonical clock-related transcription factors described above: their mutation severely attenuates or eliminates circadian function in metazoans from flies to human beings; and some like casein kinase 1ε appear to be stoichiometric members of clock protein transcription complexes [88, 108]. The first mammalian circadian clock mutation to be identified, the *Tau* mutation in the Syrian hamster, turned out to be in casein kinase 1ε! [25]. In short, the specific roles of each of these kinases and phosphatases are important enough that they are the subject of Chap. 3 in this book.

2.3.10 Clock-Associated Genes II: Chaperones

Even from theoretical grounds, it is easy to see that it would be impossible to have a functional circadian oscillator if its component proteins and RNAs were too long-lived. Hence, it is not surprising that many circadian proteins are targeted for proteasomic degradation, frequently after their phosphorylation by one of the kinases described above. Research by many labs has shown that clock proteins follow the traditional route to the proteasome: they are recognized by a particular class of chaperones containing an F-box motif, and that recruit a ubiquitin ligase complex. The clock protein is then ubiquitinated and later destroyed. For the most part, these chaperones have been discussed above in the context of their respective targets, and include SLIMB (targeting PER) and JETLAG (targeting TIM) in flies [17, 41], and FBXL3 [109–111], FBXL21 [112], and β-TrCP1 in mammals [16].

A second potentially emerging class of chaperone proteins important to the circadian clock are the heat shock proteins. It was recently discovered that Heat Shock Factor 1 (HSF1) binds to its target genes in circadian fashion and activates transcription at a wide number of chaperone loci at the onset of circadian night. Since mice carrying a mutant *HSF1* gene show an altered circadian period length, it is likely that this binding has functional consequences for the circadian clock [113], but further research is necessary to elucidate its target.

2.3.11 Clock-Associated Genes III: Chromatin-Modifying Proteins

One of the surprising recent discoveries within the circadian oscillator is that rhythmic circadian gene transcription is accompanied by corresponding rhythmic modification and demodification of surrounding chromatin in daily fashion. Thus, histone acetylation and histone methylation accompanies both the activation and the repression of clock genes and clock-controlled genes [70, 72, 89, 114]. It is likely that a large number of chromatin-modifying proteins that have been identified in other systems are also important to the circadian oscillator – histone methylases and demethylases, acetylases and deacetylases, and various classes of ATP-dependent chromatin reorganization machines. For the most part, however, these proteins have not yet been identified in the context of the circadian system. Three notable exceptions are WDR5, which is a histone methyltransferase adapter that interacts with PER proteins and is necessary for circadian histone methylation at multiple circadian loci [15]; the polycomb group protein EZH2, which probably facilitates the organization of a repressive chromatin structure during repressive phases of the circadian cycle [115]; and NCoR, the nuclear receptor corepressor complex that recruits histone deactylase HDAC3 to clock- and clock-controlled loci [116].

2.3.12 Clock-Associated Genes IV: Coactivators and Corepressors

A growing number of proteins have been isolated that are essential or important to the circadian clock mechanism, and whose actions are important for the transcriptional repression or activation of clock genes. Nevertheless, their exact functional roles have not yet been fully elucidated. For example, the mammalian CIPC gene appears to play a repressive role by antagonizing the CLOCK-BMAL-mediated activation independent of the cryptochromes. Its depletion results in a shortening of circadian period length [117]. Another repressor, the MYBBP1a protein, has been isolated through interaction with PER2 protein, and can be immunoprecipitated at the promoters of PER-regulated genes, where it appears to aid in transcriptional repression [18]. The NONO protein was also initially isolated via its interaction with PER proteins. Mutation of its homolog *NonA* in *Drosophila* or its depletion in mammalian cells results in arrhythmicity, confirming its importance to the circadian oscillator [15]. Nevertheless, the exact function of this protein remains unknown. Its two RNA-binding domains and previous implications in many different aspects of transcription and RNA processing, in both activating and repressing roles, leave many possibilities open.

In *Drosophila*, another important "mystery" repressor is encoded by the *Clockwork Orange (cwo)* gene. It was initially identified as a corepressor that acts together with PER to repress CLOCK-CYCLE-driven transcription of a large number of clock- and clock-controlled genes [118, 119]. Recent research suggests that at the same time that genes regulated by CWO show reduced peak expression levels, they show elevated trough levels, suggesting direct or indirect effects on both the activation and repression of clock genes [120]. Mammals possess two genes that are possible homologs of *Cwo*: *Dec1* and *Dec2,* which play a nonessential role in the repression of *Per1* and other clock-controlled genes [121].

2.3.13 Relating Clock Genes Together: Interlocking Feedback Loops

From the above description, exhausting but far from exhaustive, an idea of the various players of the circadian clock can be gleaned. In mammals, these proteins are organized into two major interlocking feedback loops, summarized in Fig. 2.2 In the first, *Cry*, *Per,* and *Rev-Erbα* transcription is activated by CLOCK or NPAS2 and BMAL1, and repressed by the CRY-PER complex. In the second, *Bmal1* transcription is repressed by REV-ERB proteins and activated by ROR proteins. *Clock* gene transcription is not rhythmic in the mammalian system. In *Drosophila*, a similar architecture exists, with CLOCK-BMAL1 substituted by CLOCK-CYCLE, and PER-CRY complexes probably substituted by PER-TIM complexes, with CRY playing an auxiliary role. Although the Bmal1-Rev Erbα interlocked loop does not exist in flies, a new feedback loop replaces it. The transcription of the *Clock* gene is strongly rhythmic, and is driven by an insect-specific second feedback loop in which *Clock* transcription is activated by PDP1 and repressed by VRILLE protein. In turn, the transcription of both *PDP1* and *Vrille* is activated by the CLOCK-CYCLE heterodimer [77]. Thus, the fundamental architecture of two interlocked loops is conserved across metazoans.

Given this complex structure, it is tempting to ask what within it is essential to circadian function. This question has assumed additional importance since the discovery of a circadian oscillator in cyanobacteria that is based entirely upon feedback loops of phosphorylation – i.e. in this organism the transcriptional feedback loops deemed essential to the metazoan oscillator are not necessary, since the entire clock can function in vitro in the absence of transcription. It has been speculated that a similar situation exists in mammals, and that transcriptional feedback is an "epiphenomenon" of an underlying ancient phosphorylation oscillator. Although post-translational modifications of clock proteins undoubtedly play a crucial role in all metazoans, absolutely no evidence exists to date to support a "post-translational-only" hypothesis, and a great deal against it.

Nevertheless, it is clear that several aspects of the metazoan oscillator are not required for its basic function. Since the *Rev-Erbα* gene can be deleted with only minor effects upon the core circadian oscillator and circadian behavior [91] – even though *Bmal1* transcription is almost constant as a result – rhythmic transcription of positive-limb components must be dispensable in mammals. On the other hand, the abundance of the positive-limb components CLOCK and BMAL1 is still critically important to circadian function, as well as to the overall period and amplitude of the circadian system. Inducible overexpression of wild-type CLOCK protein results in a shortening of period length in mice, and overexpression of a dominant negative mutant does the opposite [122]. The same is true for BMAL1, since reduction of its level in genetically engineered mice via REV-ERBα dampens or eliminates circadian rhythmicity [93]. Although the *Clock* gene displays rhythmic expression in flies, its protein level is constant [123]. Therefore, it is difficult to imagine that the cyclical nature of its transcription is a crucial feature of the circadian oscillator in flies, either. As in mammals, however, overall levels are important: elimination of either the repressor of this gene *Vrille* or its activator *PDP-1* results in behavioral arrythmicity [20, 77]. Since overexpression of *Clock* RNA per se does not affect circadian rhythms, some of this effect may be indirect [124].

Overall, for both mammals and flies, it is clear that the cyclical expression of positive elements within the circadian oscillator is dispensable, though their presence and abundance remains important. Negative elements pose a different question altogether. Mathematic modeling and experimental evidence all points to a crucial and necessary role of repressive components within the circadian oscillator. An excellent formal proof of this idea in mammals is provided by the fact that mutations in CLOCK and BMAL1 proteins that reduce their interaction with CRY proteins result in arrhythmicity at a cellular level [125]. Some studies have suggested in particular that levels or activities of these repressive components may be particularly important for setting the period length of the circadian oscillator [126]. Certainly, many *Per* mutations exist in flies and even in humans that alter period length, and overexpression of either CRY in mammals or either PER or TIM in flies disturbs the circadian period [127, 128]. Similarly, the expression of a CYCLE-VP16 fusion protein – which elevates the transcription of all CYCLE targets thanks to the strong VP16 transcriptional activation domain – severely shortens circadian period in flies [129]. Here as well, though, it is possible that cyclic transcription is dispensable. In mammals, expression of constant levels of CRY proteins does not visibly perturb rhythms [130]. In flies, constant transcription of both *Timeless* and *Period* also permits rhythmicity. It is possible, though, that these transcriptional perturbations are being compensated by post-transcriptional effects. In the latter example, PER and TIM protein levels continued to cycle in spite of their constant transcription! [127].

2.3.14 Summary: Redundancy is the Key Important Factor

Because of the interlocked nature of its various elements, it is perhaps not surprising that so many different aspects of circadian clock function can be ablated without abrogation of clock function. As mentioned above, circadian transcription of individual clock genes can be eliminated without serious effects, and these changes may be compensated by post-transcriptional effects. Many other examples of redundancy exist. For example, rhythmic histone methylation accompanies circadian oscillations of transcription in all clock- and clock-controlled genes examined so far. Nevertheless, the reduction of WDR5 protein levels in mammalian cells eliminates many of these oscillations, and has only a modest effect upon the circadian amplitude and none upon period length [15]. Similarly, disruption of the interaction between the NCoR repressor and the HDAC3 histone deacetylase changes the phase of some clock- and clock-controlled genes, but failure to recruit this histone deacetylase does not abrogate clock function [116].

Another example can be found in the redundancy of PER and CRY proteins in mammals. Given that two *Cry* and three *Per* genes exist in mammals, it is not surprising that the disruption of almost any one of these loci has only minor effects upon the clock. The only exception here is the *Per2* locus, which appears to play an essential and nonredundant role in the circadian oscillator. Nevertheless, the nefarious effects of a *Per2* gene disruption can be suppressed. . . .by a *Cry2* deletion! [59] Although *Cry1* gene disruption will not achieve this suppression normally, constant light conditions – which ordinarily degrade circadian rhythms in mice – will now allow such compensation to occur [131]. It is possible that the various PER and CRY proteins have similar roles in the cell – as transcriptional repressors, for example – but different potencies. Therefore, elimination of one member of the PER-CRY complex would change its potency, but elimination of another would change this balance again in a favorable direction. Nevertheless, existing mechanistic data do not argue in favor of functional equivalence of PER and CRY proteins. It is possible, however, that such compensation could also occur kinetically at completely different steps in the same pathway. In this case, a change in the potency of one step (for example, transcriptional repression) might be compensated by changing the effectiveness of a different step. (post-translational modification, nuclear export, etc.).

The overall implication of the redundancy, though, is increased robustness and precision. Perhaps, it is this redundancy that allows the circadian oscillator to continue to function indifferent of temperature and cell division. Most spectacularly, the circadian clock has even been shown to demonstrate transcriptional compensation: overall inhibition of RNA polymerase in a variety of ways does not alter the circadian period significantly! [132] How might such compensation work? Many models have been put forward, and their workings are the subject of Chap. 11 of this book. We shall close this section, though, by noting that temperature compensation and precision was also a problem for mechanical clocks. This inability to tell time accurately outdoors led early sailors repeatedly and tragically to misjudge their longitude. (Celestial indices were inadequate for this purpose due to the earth's rotation). The first reliable solutions were achieved by redundant mechanical gearing that allowed temperature-induced changes to act in opposite directions simultaneously. Perhaps a similar logic might govern the redundant and precise circadian biological clock.

2.4 Input and Phase Shifts

As we have seen in the two previous sections, the mammalian circadian oscillator suffices to generate rhythms with a free-running period of about 24 h. However, to be in resonance with the environment, an organism has to adjust its circadian clock, and consequently the circadian oscillators in the individual cells, every day to the external photoperiod. The flow of information to the circadian oscillator is termed the input. The synchronization of the organism to the environment is the main function of the SCN, which receives the relevant photic signals from the retina. The peripheral oscillators are subsequently synchronized by humoral and neuronal signals derived from the SCN. The readjustment of the circadian clock in response to an input signal is called *phase shift* and was originally investigated in animals (see also Chap. 4). This was useful to elaborate the *phase response curve* for a given Zeitgeber (german; "timing cue" which affects the phase of the circadian clock) but did not provide too much detail on the molecular mechanisms of the input pathways involved.

Solely the identification of clock genes and the recent advances of mammalian circadian in vitro systems allowed the investigation of signaling pathways that have an effect on the phase of the molecular oscillator. In principle, due to the organization of circadian oscillators as transcriptional and post-translational feedback loops, signaling pathways could directly influence the concentration or activity of certain oscillator components and consequently change the phase of the interconnected transcriptional network. Unfortunately, there were so many potential phase shifting agents identified that the overall picture at the moment is more confuse than concise. Therefore, the research nowadays attempts to combine data obtained from the animal and in vitro systems with appropriate computational models to identify the relevant input pathways to the circadian oscillator.

2.4.1 Induction of Genes by Light

A mammalian organism that is exposed to a light pulse at the beginning of its dark phase will adjust the phase of its circadian clock accordingly [133, 134]. Beginning from the next day, the phase of the circadian oscillator will be delayed (Fig. 2.3). In contrast, an animal receiving light information towards the end of its dark phase is forced to advance its circadian clock for the next day. Light will thus affect the phase of the circadian oscillator dependent on the exposure time during the dark phase. The entity of phase shifts of the oscillator in response to light (or any other Zeitgeber) is called a *phase response curve*. Typically, in animals a type-1 phase response curve is observed [134]. During the light phase or subjective light phase under constant dark conditions, it is not possible to provoke a phase shift in animals. This part of the phase response curve is sometimes referred to as the "dead zone". The light input to the SCN emanates from specialized cells in the retina and reaches the core region of the SCN as a glutamate or *pituitary adenylate cyclase activating peptide* (PACAP) signal (see Chap. 4). During the dead zone the SCN secretes the neuropeptides *Transforming Growth Factor* α (TGF α), *Cardiotropin-Like Cytokine* (CLC), and *Prokineticin 2* (PK2), which suppress the locomoter activity of mice and probably also prevent the inadequate phase shifts by light [135–137].

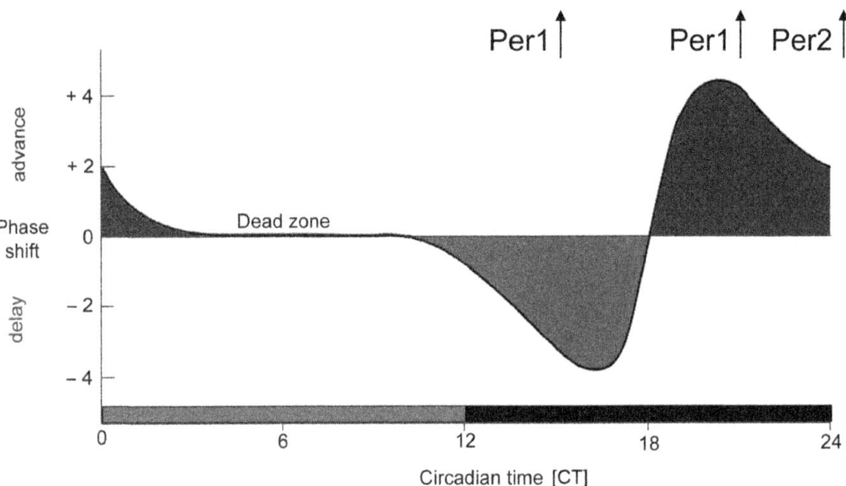

Fig. 2.3 Principles of phase shifting and phase response curves. A light signal (or another specific Zeitgeber) will effect the phase of the circadian oscillator. In a certain period, the oscillator is not responsive to a stimulus. This period is called "Dead zone". At the beginning of the subjective night phase, a light pulse causes a stable phase delay by up to 4 h. Thereafter, the phase of the oscillator will advance. Concomitant with the behavioral phenotype, a selective induction of the *Per* genes and of other genes like *c-Fos* is observed in the SCN. Courtesy of Isabelle Schmutz, University of Fribourg, Switzerland

This is a difference to the oscillators in the periphery, which can always respond to a resetting signal. The phase response curve for glucocorticoids on the circadian oscillator of the liver, for example, resembles the one shown in Fig. 2.3 but without a dead zone [138]. This is crucial because the periphery should respond to signals from the SCN at any time. Since the circadian oscillator is based on transcriptional feedback loops, the induction and consequently the accumulation of an oscillator component e.g. by light could directly influence the phase of the circadian oscillator.

On the molecular level, *c-fos* was the first gene identified to be induced by light in the SCN [139]. As a typical immediate-early gene, *c-fos* induction had a peak about 30 min after the light pulse and then its expression gradually declined. Most importantly, the induction of *c-fos* strongly correlated with the phase shifting behavior of hamsters by light. The upstream regulator of *c-fos* is the *cAMP response element binding* protein or CREB [140]. After phosphorylation of CREB at its serine residues 133 and 142 in response to light, this protein is capable of binding to CRE-sites within the *c-fos* gene and of activating its transcription [141, 142]. Later on, the binding of ICER, a negative regulator of CREB factors, abolishes the activity of CREB and the transcription of *c-fos* ceases [143–145].

Unexpectedly, mice deficient for the c-FOS protein display a completely normal phase shifting behavior [146]. Therefore, the function of *c-fos* and other immediate-early genes like *junB* and *egr-1*, which were identified in a screen for light-inducible transcripts in the SCN [147], are overall less important for the phase shift behavior of mice but they provide excellent markers to identify the neuronal activity and to reveal a light response in the SCN. Another consequence of a light signal is the drastic increase in serine 10 phosphorylation of histone H3 in the SCN [148]. This specific histone modification correlates with a facilitated accessibility of transcriptional regulatory sites within the chromatin, which may be the reason for the activation of many genes that are not directly involved in the phase shift response.

Shortly after the discovery of the *Period* genes (see Sect. 2.3.1), it was found that those genes were induced in response to a light pulse with a peak 1–2 h after the stimulus [7, 31, 32, 149–152]. The *Per1* gene was induced at the beginning and at the end of the dark phase, while the *Per2* gene was more restricted to the end of the light phase. In spite of this, some research groups also found induction of the *Per2* gene at the beginning of the dark phase. This discrepancy is explained by the different experimental setups employed [153] (genetic backgrounds, light intensities and light conditions used before the experiment, i.e. constant *versus* light-dark conditions). It appears that *Per2* needs more specialized conditions at the beginning of the dark phase for a successful induction by light. Although it appears that the induction of the *Per* genes occurs in different parts of the SCN and with different kinetics [154, 155], in this chapter,we will consider the SCN an entity to facilitate our argumentation.

Similarly, the phenotypes of *Per1* and *Per2* single deficient mice differed. Originally, *Per1* knockout mice were found unable to perform a phase advance in response to a light pulse at the end of the dark phase, while *Per2* knockout mice had a similar problem at the beginning of the dark phase [31]. They were incapable of performing the expected phase delays. This clear distinction between *Per1* and *Per2* was less evident in other mouse strains [156, 157]. Meanwhile, some researchers interpret the genetic experiments in a way that *Per2* has a more prominent function on the core oscillator, while *Per1* is more important for phase shifts. However, for a definite answer further experiments are necessary.

The induction of the *Per* genes by light appears to be a prerequisite for a phase shift. Interestingly, the *Per1* gene bears a functional CRE-site in its regulatory region and is consequently a target for the activated transcription factor CREB [141, 158]. The induction of *Per1* and *c-fos* occurs with different kinetics in the SCN. This is not completely understood at the moment but suggests that there are other factors that shape the expression of either gene as well. These could be coregulators of the ATF family known to bind together with CREB to CRE-sites or different repressors of the ICER family [159–161]. As a conclusion, the induction of *Per1* or *c-fos* in the SCN by light both rely on CREB binding but the reasons for the different kinetics and the modes of downregulation of both genes are currently unknown. In addition, the induction of the *Per1* gene is sensitive towards inhibitors of histone acetylation and deacetylation but these may be very general processes involved in transcriptional activation and repression, respectively [162]. The induction of the *Per2* gene by light is less well understood. Some experiments suggest a role of either the CREB protein [158] or the PER1 protein in the induction process [163]. Other experiments, mainly in vitro, favor an activation of the *Per* genes by a Ca^{2+} dependent protein kinase C pathway and the direct activation of the CLOCK transcription factor [164].

How would the induction of the *Per* genes cause different phase shifts at different times of the dark phase? This is clearly an unsolved issue. A condition for the different effects is the underlying circadian oscillator. At the beginning of the dark phase, the expression of the *Per* genes in the SCN declines, but there are still high levels of hyperphosphorylated PER proteins and CRY proteins present. In contrast, at the end of the dark phase, the transcription of the *Per* genes recommences but there are only low amounts of hypophosphorylated PER proteins detectable in the SCN. As a speculation, the induction of *Per* genes at the beginning of the dark phase extends the time of active PER proteins being present in the nuclei of the SCN neurons and lengthens the circadian cycle. Therefore, we obtain a stable phase delay for the following days. On the other side, the induction of the *Per* genes at the end of the dark phase mimics the concentrations of PER proteins found later on during the circadian cycle and consequently the following cycles advance. In addition to the *Per* genes, the *Dec1* gene is also light-inducible [121]. This factor was originally identified in a screen to find inhibitors or competitors of BMAL1 and CLOCK-mediated transcriptional activation. Since DEC1 can compete with BMAL1 and CLOCK for binding to regulatory E-box motifs, the induction of the *Dec1* gene by light could immediately modulate the phase of the circadian oscillator in concert with the PER proteins.

2.4.2 Input Signals for Peripheral Oscillators

For a long time, researchers considered the SCN the only real clock generating robust circadian rhythms. The circadian clocks in the periphery were regarded as "slave oscillators" that were incapable of maintaining rhythms without a permanent input from the SCN. This picture changed with the advances of organ cultures from transgenic rats and mice and with the upcoming mammalian in vitro models [33, 165–168]. The peripheral oscillators are as robust as the oscillator in the SCN [167, 168]. However, the input to both types of oscillators may be different. The major Zeitgeber for the SCN is the environmental light-dark phase but for the periphery, Zeitgebers like food uptake, body temperature, and neuronal and humoral signals have to be taken into consideration.

Explantation studies of different tissues from transgenic *Per2:luc* mice revealed two supplementary facts about peripheral oscillators [166]. First, the period of each tissue varied. This would indicate that there are tissue-specific variants of peripheral oscillators and the regulated transcriptional networks. Secondly, in mice, in which the SCN was ablated and consequently not functional, the organs continued to be rhythmic but they were no longer synchronized amongst each other. This would indicate that the main purpose of the SCN is to synchronize the peripheral oscillators but not to drive circadian rhythms overall. However, there is still evidence for signals that can drive rhythms in peripheral oscillators [93]. In transgenic mice without a functional oscillator in the liver, rhythmic transcripts including those of the *Per2* gene persisted. These rhythms rapidly declined after placing liver slices in culture demonstrating that those rhythms were solely driven by systemic cues.

A considerable progress of our understanding of the input pathways to the peripheral oscillators derived from mammalian circadian in vitro systems. In 1998, Aurelio Balsalobre in Geneva realized that the expression of the *Dbp* gene, an output transcription factor (see Sect. 2.5), transiently decreased after a serum shock in Rat-1 fibroblasts [33]. About 24 h after the shock, the expression levels were up again but continued to decrease thereafter. A careful analysis revealed that this rhythmic behavior proceeded for multiple days and that this was not specific for this gene but that many circadian markers followed the same pattern. The phase differences between all the circadian markers faithfully reflected what was known about the phase differences found in the SCN and peripheral oscillators. In addition, immediately after the serum shock, an induction of *Per1* and *Per2* occurred. Therefore, it was concluded that a serum shock induced free-running circadian rhythms with a period length of 22 h in Rat-1 fibroblasts, which have not been in contact with the SCN for at least 20 years.

Subsequent experiments demonstrated that free-running circadian rhythms could also be induced in *mouse embryonic fibroblasts* (MEF) derived from different genetic backgrounds [169]. Under these experimental conditions, the period of the MEFs in vitro resembled the period of the different mutant mouse strains. For that reason, the mammalian in vitro systems closely reflect the animal models. One major question remained. Are the circadian rhythms in the tissue culture cells newly induced, or are the circadian oscillations of each single cell synchronized? This question was answered by the inspection of individual cells in culture using rhythmically expressed, short-lived fluorescent protein [167]. Under normal culture conditions, the individual cells display circadian rhythms in different phases. After a serum shock, all the different cells become synchronized. This is possible because tissue culture cells show a typical type-0 phase response. Independent of the position of the oscillator within the circadian cycle a strong signal resets the oscillator always to the same point. Therefore, the oscillators in a culture start cycling from the same point after a serum shock. Using a similar culturing system expressing rhythmically luciferase protein and computer derived simulations, it was proven that the oscillators in cultured fibroblasts were capable of generating robust circadian rhythms similar to the SCN neurons [167, 168].

From early on, the mammalian in vitro systems were used to identify input pathways to the circadian oscillator. One of the first applications was to monitor the influence of dexamethasone, a glucocorticoid hormone analog, on the circadian oscillator. This drug is a potent means to synchronize the circadian oscillators in fibroblasts [34]. These data were compared to the influence of dexamethasone on the livers of animals [138]. As mentioned above, dexamethasone shifts the circadian oscillator of the liver without the presence of a dead zone. However, in tissue culture cells, the phase response to dexamethasone was a typical type-0 phase response. The discrepancy between the effects of dexamethasone on both experimental systems is not known. It is tempting to speculate that due to the absence of moderating hormonal inputs to the cells in the tissue culture, their circadian oscillators are more sensitive to a resetting stimulus. A further reduction of the concentration of dexamethasone to synchronize the tissue culture cells probably will provoke a type-1 phase response.

Interestingly, corticosterone, the natural compound of dexamethasone found in rats and mice, has a direct effect on the phase shift response of the liver circadian oscillator but not on the SCN [138]. The phases of the oscillators in the SCN and in the livers can be separated by up to 12 h using an inverted feeding regimen, a process during which the adaptation of the liver oscillator to the new feeding schedule takes about a week [170, 171]. In mice deficient for the glucocorticoid receptor in the liver or adrenalectomized mice without the capability to secrete corticosterones into the bloodstream, this readjustment occurs in about 2 days suggesting that the signals mediated by the glucocorticoid receptor normally prevent large phase shifts of the liver circadian oscillator [172]. In contrast, after the reconstitution of normal feeding conditions, the liver oscillator requires a couple of days to resynchronize to the phase dictated by the SCN, which is completely independent of the glucocorticoid hormone signaling.

The signaling pathways that were associated with the synchronization of circadian oscillators in vitro were manifold. In addition to a serum shock or glucocorticoids, researchers found an impact of activators of cAMP/CREB signaling (forskolin, dibutyryl cAMP), protein kinase A and C signaling (e.g. phorbol-12-myristate-13-acetate), Ca^{2+} signaling, IL-6 signaling, MAP kinase signaling, and of PPARα agonists (fenofibrate) on *Per1* induction and/ or the subsequent synchronization of the circadian oscillators in various tissue culture cell models [33, 34, 138, 173–177]. A further breakthrough was the coupling of the mammalian in vitro systems with real-time bioluminescence monitoring. In these systems, a luciferase reporter gene is driven by a circadian regulatory element. Different systems exploit the regulatory region of the *Per1*, *Per2*, *Bmal1*,

Dbp or *Rev-Erbα* gene. After the synchronization of the circadian oscillators, it is possible to measure the effect of a given treatment on the magnitude, amplitude or phase of a given reporter gene over the course of multiple circadian cycles. It is possible to exploit these techniques for the high-throughput screening of compounds [178, 179]. The experiments can easily be converted into cotransfection assays to reveal the function of a certain protein on the oscillator, or coupled to RNA interference to monitor the effect of the lack of certain protein on the oscillator (e.g. as described in Brown 2005 [15]). A recent variation of this technique is the transfer of circadian reporter genes by lentiviral-mediated infection. This allows the stable integration of circadian reporter genes even in cells that are normally not easy to transfect. In this manner, it was possible to measure the period of human fibroblasts derived from skin biopsies indicating that the human fibroblasts behave similarly as mouse and rat fibroblasts [180].

Are the input pathways used by light fundamentally different from the ones immerging into the peripheral oscillators? Surprisingly, the answer is no. In an elegant series of experiments, fibroblasts were stably transfected with an expression vector for the photoreceptor melanopsin [181]. These fibroblasts displayed a type-1 phase shift behavior in response to low light intensities and a type-0 phase shift behavior in response to higher light intensities. The phase shift behavior could be blocked by inhibitors of Ca^{2+} signaling or phospholipase C. This indicates that the signaling pathways in fibroblasts mediating the light or hormonal (e.g. a serum shock) input are very similar but specific receptors for a light response are normally missing. Nevertheless, it is tempting to speculate that the process of phase shifting by both types of phase shifting agents in general is essentially the same. Both kinds of phase shifting agents use the induction of specific components to affect the phase of the oscillator for the next circadian cycle.

2.4.3 Integration of the Input Signals

The major Zeitgeber for the SCN is light. The light signal activates the transcription factor CREB by phosphorylation. Upon binding of activated CREB to its relevant binding elements in the *Per1* and *Dec1* genes, these become transiently induced. In addition, under certain circumstances, the *Per2* gene is also induced. Depending on the phase of the underlying circadian oscillator in the SCN, a stable phase advance or phase delay results for the next circadian cycles. For sure, this is a very simplified summary of the processes that occur during the phase shift of a mammalian organism in response to light. Many more signaling molecules and pathways have been characterized to affect the circadian oscillator in the SCN including Vasoactive intestinal polypeptide, Neuropeptide Y, calcium/calmodulin-protein kinase, cGMP-dependent protein kinase II, GABA, glutamate, Gastrin-releasing peptide, and Pituitary adenylate cyclase-activating polypeptide [182–194]. However, this is a rapidly evolving field and it is too early to draw definite conclusions.

Some specific aspects will be further elaborated in Chap. 4. The phase shift is communicated to the various peripheral oscillators via signaling cues some of which remain to be verified in vivo.

One potent Zeitgeber for peripheral circadian oscillators is feeding. As mentioned above, it is possible to completely uncouple the liver circadian oscillator from the SCN by an inverted feeding regimen. Restricting the food access to the light phase (when rodents are normally inactive) is sufficient for the uncoupling of both types of circadian oscillators. It is currently unknown whether the feeding behavior of mammals under normal conditions is dictated by the SCN. If this is true, it would provide an elegant link between the SCN and the periphery allowing a tight coupling of the two different systems on one hand but a rapid uncoupling in the case of a limited food access on the other hand.

Another potent Zeitgeber for the periphery is temperature. The body temperature of mammals varies in a circadian fashion. When exactly these kinds of temperature variations were simulated in tissue culture, these temperature rhythms were sufficient to maintain circadian rhythms in Rat-1 fibroblasts [195]. Meanwhile, researchers chose even conditions to synchronize the circadian oscillators of primary human fibroblasts by temperature ramping indicating that rhythmic changes in the body temperature could be a general Zeitgeber for the periphery [180].

How is it possible to integrate the impact of all the different Zeitgebers on the circadian oscillators? To address this question the circadian *transcriptomes* of different tissues were compared [196]. The subsets of genes that were rhythmic in multiple tissues were analyzed for similarities in their regulation. Finally, the proteins expressed by these genes were arranged into regulatory cascades. The overall picture of these theoretical regulatory networks is shown in Fig. 2.4. A light signal to the SCN would activate the protein kinase A. This enzyme would phosphorylate CREB and some other regulatory components of the circadian oscillator. CREB in turn would induce the *Per1* gene, whose gene product (together with PER2 when feasible) would interfere with BMAL1 and CLOCK-mediated transcription to provoke a phase shift.

In response to food uptake, the adrenal gland would produce and secrete glucocorticoid, which would bind to and activate the glucocorticoid receptor. This activated protein can induce both *Per1* and *Per2* and therefore exert the same function as CREB. In the temperature response, the modeling suggests that the transcription factor HSF1 is activated and induces the transcription of many heat shock genes including Hsp90aa1. This protein and others form complexes to inactivate the glucocorticoid receptor, which in parallel could be activated by glucocorticoid due to a general stress response. The rapid inactivation of the glucocorticoid receptor would modulate the induction of the *Per1* and *Per2* genes. Similar to these examples, other signaling pathways could feed into the circadian oscillator by the induction or repression of the genes coding for oscillator components, or directly via the stabilization or degradation of some oscillator components. Only further work will tell us how the circadian oscillator can respond to so many different phase shifting cues at the same time.

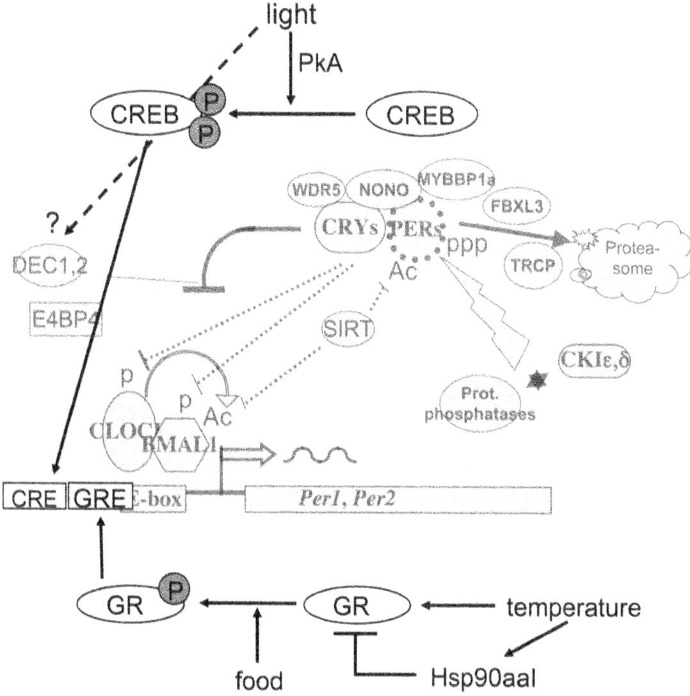

Fig. 2.4 Input to the mammalian circadian oscillator. The input pathways effect the phase of the circadian oscillator in different ways. A light pulse activates the transcription factor CREB by phosphorylation via PkA. This factor can subsequently activate the *Per1* gene. Light also induces under certain circumstances the *Per2* and *Dec* genes. A food-derived signal activates the glucocorticoid receptor (GR), which can activate both *Per* genes. Temperature uses a similar strategy but there is a modulating activity mediated by Hsp90aal, which appears to inhibit the GR and therefore modulates the response

2.5 Output and Clock Regulated Genes

In contrast to the input, the output is the effector part of the circadian oscillator. All circadian changes in the physiology, metabolism, and behavior are probably linked more or less directly to rhythmic gene expression. Many target genes are hardwired to the circadian oscillator and subsequently expressed in a rhythmic fashion. The organization of the molecular oscillator facilitates the direct coupling of circadian target genes to the transcriptional network. In principle, those rhythmic genes could be activated directly by the BMAL1 and CLOCK or BMAL1 and NPAS2 transcriptional activators, or repressed by the nuclear hormone receptor REV-ERBα (see Sect. 2.3). Indeed, many response elements for these kinds of transcriptional regulators are found in the circadian regulatory regions of rhythmic target genes.

Nevertheless, the situation is more complicated. Many target genes are regulated by rhythmically expressed transcription factors as intermediaries. These factors appear to be preferentially members of the PAR-bZip or nuclear hormone receptor families but examples are found in nearly all kinds of transcriptional regulator families. Using rhythmically expressed transcription factors as intermediaries allows an amplification of the output, the expression of genes in different phases, and tissue specific gene expression. Therefore, it is not surprising that up to 10% of a given transcriptome (>3,000 genes) is linked to the circadian oscillator but the overlap of rhythmically expressed genes in two different tissues may be less than 100. The next challenge will be the understanding of tissue specific circadian networks.

2.5.1 Regulation of Circadian Target Genes

There are now many genes known to be expressed in a circadian manner. The characterization of these genes unraveled many regulatory mechanisms responsible for the rhythmic transcription of these specific genes. Due to the vast number of circadian target genes, we will present here only a very limited number of examples. Interested readers may refer to the original work done by the different research groups. Here, we would like to focus on some basic principles of the regulation of circadian target genes.

The first example of a circadian target gene directly regulated by the circadian oscillator was the *arginine vasopressin* gene [197]. Originally, this hormone was characterized as a regulator of the salt and water balance in mammals. It is predominantly expressed in the vasopressinergic neurons of the paraventricular nuclei and the supraoptical nuclei, and the final hormone is stored in vesicles in the posterior pituitary. During hypertonic conditions, it is released into the bloodstream to increase water reabsorption in the kidneys. In the SCN, however, this hormone acts as a local neuropeptide. It is released from some SCN neurons to modulate the firing rate of other SCN neurons in the vicinity bearing the V1a receptor. In mice with a homozygous, dominant-negative mutation of the CLOCK protein, the expression of the *vasopressin arginine* gene in the SCN was abolished. Subsequent analysis revealed the existence of an E-box motif (see Sect. 2.3) in its promoter region. In cotransfection experiments, BMAL1 and CLOCK were capable to activate transcription via this E-box motif. Taken together, the genetic and biochemical experiments showed that this gene is hardwired to the circadian oscillator.

The question remains, why the expression of this gene is circadian in the SCN but regulated differently in the other regions of brain like the supraoptical nuclei, where its expression is constant over the day. The mutant CLOCK protein, for example, did not affect the *vasopressin arginine* expression in the supraoptical nuclei. This may be due to the fact that in this region there are only very low levels of BMAL1 detectable and consequently not enough heterodimers are formed to interfere with the expression of this gene. Nevertheless, we learn one important point about gene regulation: one gene can be expressed in a circadian manner, in a tissue specific manner, or in a combination of both. Specific regulatory elements in the promoters and enhancers of the genes have to govern this diversity.

Although there are now many examples of genes regulated by BMAL1 and CLOCK, the effect of a defect of the circadian oscillator on gene expression may be not as prominent as the effect observed for the *vasopressin arginine* gene. Though, there may be interesting phenotypes after all. In *Per2* mutant mice, there is an increase in the intracerebral dopamine levels in the ventral tegmental area and the nucleus accumbens area [198]. As a consequence, these mice display a more depression-resistant like phenotype. In this case, the increase in dopamine levels was associated with a slight downregulation of the *monoamine oxidase A* gene in the *Per2* mutant mice. The monoamine oxidase A is involved in the degradation of dopamine in the mitochondria. Downregulation of this gene indirectly augments the concentration of dopamine. In the regulatory region of this gene, there was again an E-box motif mediating the effects of BMAL1 and NPAS2 in vitro. In addition, in chromatin-immunoprecipitation assays was observed a rhythmic binding of BMAL1 to the promoter region of the *monoamine oxidase A* gene in the ventral tegmental area region. Again, the combination of genetic and biochemical experiments suggests that the *monoamine oxidase A* gene is hardwired to the circadian oscillator in a very defined brain region.

In the same mice, two more phenotypes have been discovered. The PER2 protein acts as a tumor suppressor gene [199]. Mice deficient for the PER2 protein, when irradiated with γ-rays, developed more tumors than their wild-type littermates. This particular phenotype was linked to a deregulation of genes involved in cell-cycle regulation and tumor suppression. In the brains of these mice, also a hyperglutamergic state was observed within the central nervous system [200]. This effect was probably due to a slight downregulation of an astrocyte-specific transporter for glutamate. The resulting phenotype was quite complex. The animals consumed more ethanol and were more resistant to the health-hazardous effects of ethanol. This phenotype could be reverted by the administration of acamprostate, a drug that regulates the intracerebral glutamate levels. Therefore, this neurotransmitter is involved in this phenotype. However, in contrast to the *monoamine oxidase A* gene, it is not known yet, whether this glutamate transporter is a direct target of the circadian oscillator. We have selected these examples to demonstrate that the phenotypes of mutant mice for specific oscillator components may be linked to the circadian oscillator itself or to specific functions of this component independent of the clock. In general, both options are difficult to distinguish.

The mouse *Dbp* gene represents a model system to understand the expression of target genes that are hardwired to the circadian oscillator. It was previously identified as a transcriptional regulator of the *albumin* gene in the liver [201]. Later on, its expression was found to occur with high circadian amplitude in multiple tissues including the liver, the brain, and the SCN [202]. Expression of this gene was abolished in mice with a homozygous, dominant-negative mutation of the CLOCK protein [203]. In addition, the gene contained multiple E-box motifs as potential targets of BMAL1 and CLOCK. A careful analysis of the regulatory region of *Dbp* revealed rhythmic binding of BMAL1 and CLOCK to three distinct regulatory regions [89]. Concomitant with the rhythmic binding of both transcriptional activators, the local chromatin structure changed accordingly. During the activity phase of this gene, the chromatin was in an open state, while during inactivity the chromatin resembled a heterochromatic, inaccessible state. The oscillator may direct the reversible acetylation of histone H3. The histone acetyl transferase activity of CLOCK [73], upon binding of this factor, may directly modify the local nucleosomes, while the NAD^+ dependent histone deacetylase SIRT1 may counteract the activity of CLOCK either on the level of histone acetylation or on the acetylation of the BMAL1 protein [75]. Recently, it was described that in the SCN of CLOCK deficient mice the *Dbp* gene was amongst a very limited number of genes, whose expression was abolished [67, 68]. Taken together, the genetic and biochemical experiments strongly suggest that *Dbp* is a genuine target gene for BMAL1 and CLOCK in essentially all the cells with functional oscillator.

Mice deficient for DBP and the related transcription factors TEF and HLF have only subtle effects on the circadian oscillator. However, similar to the *Per2* mutant mice, they exert an interesting phenotype again due to the deregulation of an enzymatic activity. In the brains of these mice, there occurs a slight downregulation of the expression of the *pyridoxal kinase* gene [204]. Its gene product is necessary for the phosphorylation of vitamin B6 derivates to generate pyridoxal phosphate. Pyridoxal phosphate is a cofactor required by many enzymes, some of which are involved in the metabolism of neurotransmitters, e.g. in the synthesis of dopamine from DOPA, or in the conversion of the excitatory neurotransmitter glutamate to the inhibitory neurotransmitter GABA. It appears that a diminution of the pyridoxal kinase-activity in the brain causes a concomitant reduction of serotonin and dopamine levels and consequently spontaneous epileptic seizures in these mice. In the context of this paper, also an interesting hypothesis was posed. In the brain, there occur only slight oscillations as compared to the other tissues. This may be associated with the fact that large variation of neurotransmitter concentrations would have a harmful impact on brain function.

At the same time, these mice also had a reduced life expectancy due to the deregulation of other enzymes in the liver [205]. Here, mainly enzymes involved in the detoxification pathways for xenobiotic compounds and in drug metabolism were affected. As a result, these mice were very sensitive to the toxic effects of xenobiotic compounds and this may play a part to their reduced life expectancy. Why would detoxification enzymes be expressed in a circadian manner? It is known that cytochrome P450-containing detoxification enzymes produce reactive oxygen species in the absence of their suitable substrates. This would cause severe damage to the enzyme itself and potentially also to the entire cell. The liver cells choose two ways to cope with this problem: first, there is a circadian basal expression of the detoxification enzymes to anticipate the beginning activity phase and the potential uptake of food and xenobiotics. Secondly, there are inducible mechanisms, which can drastically upregulate detoxification enzymes in the presence of higher quantities of xenobiotic compounds. Taken together, here we described an example of rhythmically expressed genes that are regulated by transcription factors, which themselves are hardwired to the circadian oscillator. However, on top of this circadian regulation, there may be inducible regulation to bolster up the expression of these genes, as well.

There exist many more classes of transcriptional regulators that can connect the circadian oscillator to the rhythmic output. A conclusion from DNA-microarray experiments, for example, was that REV-ERBα response elements were identified in the promoter regions of many rhythmic genes expressed in the SCN during the subjective night phase [92]. In the same kind of analysis, cyclic AMP response elements for CREB were found in many genes expressed in the SCN during the subjective light phase. Therefore, it was concluded that these particular kinds of response elements mediate rhythmic transcription of target genes in different phases. In later studies, this range of response elements was extended by E boxes as binding sites for BMAL1 and CLOCK, and D-elements as binding sites for the PAR-zip transcription factors and E4BP4 (Fig. 2.5) [128, 206–209]. However, circadian gene regulation may be even more complicated. The gene for the *Cholesterol 7α-hydroxylase*, for example, has in vitro binding sites for DEC2, E4BP4/DBP, PPARα and REV-ERBα and β [210]. All of these factors have to collaborate to fine-tune the expression of this particular gene in the liver.

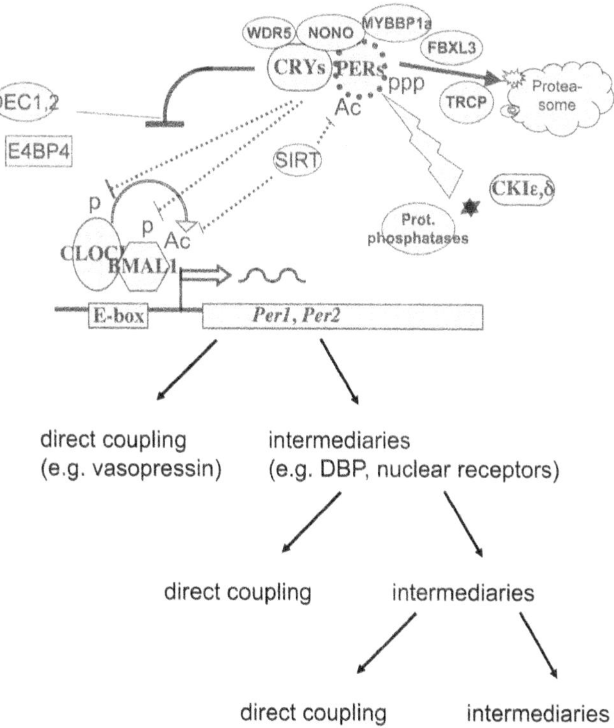

Fig. 2.5 Output from the mammalian circadian oscillator. Some target genes are directly hardwired to the circadian oscillator, either via BMAL1 and CLOCK, or REV-ERBα. Others rely on rhythmically expressed transcription factors like DBP, or nuclear receptors. Note that this simple make-up facilitates tissue-specific gene expression. Rhythmic signals are easily amplified by tissue-specific transcription factors expressed in a rhythmic fashion

If we combine all the possible ways to regulate the circadian target genes, we end up with very complicated networks of rhythmic gene expression (Fig. 2.5). At the center of these networks, we have the circadian oscillator. To this oscillator are connected a couple of direct target genes including transcriptional regulators. In the next layer, we have target genes that are indirectly regulated by the circadian oscillator with the help of these transcription factors as intermediaries. Since amongst those indirectly regulated genes there are other transcriptional regulators as well, the system creates more and more layers of rhythmically expressed genes. Interestingly, these networks establish also many additional feedback loops, which allow an even more precise regulation of gene activity and, last but not least, may feedback to the circadian oscillator. However, because these networks have so many dynamic layers of gene expression, it becomes increasingly complicated to distinguish between direct or real target genes of the circadian oscillator and a plethora of bystanders, which are rhythmically regulated but do not have any consequence for circadian changes in the physiology or metabolism. In the next section, we will have a closer look on some of these circadian networks and their interconnections. For the particularly well-characterized interaction between the circadian oscillator and the metabolism we have dedicated an entire chapter later on (see Chap. 5).

2.5.2 Analysis of Circadian Transcriptional Networks

How is it possible to have a glimpse on circadian transcriptional networks? The method of choice is the use of DNA microarrays. Briefly, RNA is extracted from a given tissue. This RNA is copied into complementary DNA using standard molecular biology methods. During this process the resulting DNA is normally amplified and marked with radioactivity or fluorescence and can be used as a probe. This pool of labeled DNA fragments, which still represents the proportions of the original pool of mRNA, is now hybridized to a DNA microarray. On the surface of a DNA microarray gene specific probes are fixed in well-defined patterns. In this fashion, it is possible to identify a specific transcript out of the pool, and since hybridization is a quantitative process, we also obtain information regarding the amount of a given transcript.

At the beginning, it was possible to analyze for the presence of a couple of thousand different transcripts in a single experiment. Nowadays, DNA microarrays cover the entire potential transcriptome of an organism, even with multiple probe sets. The data for the analysis of circadian networks are normally grouped into time frames, filtered and evaluated for rhythmic patterns. In the early days, this "data-mining" was not very standardized. Therefore, the overlap of rhythmic transcripts found by different research groups in the same tissue was limited. Again, there was a lot of progress achieved and the results from DNA microarrays are much more robust today. For instance, increasing the number of time points allowed a more detailed analysis of the circadian transcripts in the pituitary gland [211]. As another example, in the first experiments the expression of circadian genes in the liver showed a clear enrichment towards the transitions from the subjective light-to-dark and dark-to-light phases [212]. In subsequent experiments with higher-capacity microarrays, a very even distribution of the expression peaks of circadian mRNA accumulation was observed in the same tissue [213]. Even so, the percentage of rhythmically expressed genes in both studies was always about 10% of the expressed genes.

The first DNA microarray experiments were already conducted in 2002 [212, 214, 215]. However, most of the conclusions drawn at that time are still valid today. A major surprise was the discovery of tissue-specific gene networks. Depending on the experiment performed, between 5 and 10% of rhythmic transcripts in the transcriptome of a given tissue were identified. Tissue-to-tissue comparisons revealed only a restraint number of common genes. In the first experiments, there were a couple of dozen common transcripts, nowadays, there are close to 100. The implications of these findings are that there is a general circadian mechanism at the base of all the different kinds of circadian oscillators but depending on the cell type, these are linked to tissue-specific networks. The use of a general circadian mechanism does not incriminate that all of the circadian oscillators have to be in the same phase. In the brain, for example, circadian oscillations are observed in many regions albeit with different phases. This can probably be achieved either by modulating the input to the circadian oscillator by tissue-specific receptors or transcriptional regulators, or by tissue-specific regulatory feedback loops.

How are these tissue-specific regulatory networks organized? Again we will focus here only on one set of data. Interested readers may refer to the original work published by other groups, too. In addition, you will find a detailed description of the connections between the circadian clock and the metabolism in Chap. 5. The expression of rhythmic transcripts in the SCN was one of the first to be investigated [214, 215]. It was established then that the circadian oscillator could coordinate intracellular processes. One of the major functions of the SCN is the secretion of neuropeptides and neurotransmitters. Consequently the synthesis, processing, and degradation pathways for some neurotransmitters are temporarily optimized in the SCN neurons. This is also reflected in a coordinated expression of components of the ribosomes. Increasing the number of functional ribosomes can augment translation and consequently the production of peptide hormones. Interestingly, there was also observed a temporarily switch of the expression of components of the nascent polypeptide-associated complex and of the signal recognition particle. The first complex directs translation into the cytoplasm, the latter redirects and fixes the ribosomes to the membrane of the endoplasmatic reticulum, a prerequisite for protein secretion. Concomitant with the increase of translation through the membrane of the endoplasmatic reticulum, there was an upregulation of proteins involved in vesicle formation and trafficking. Taken together, the microarray data uncovered a step-by-step organization of neurotransmitter synthesis, starting from the transcription of these genes per se to the secretion of the final product. This may be the advantage of the circadian oscillator: to allow the temporal organization of synthesizing and degrading processes within a single cell.

To sum up, the circadian output is organized in tissue-specific transcriptional networks. It will be very interesting in the near future, to solve all the connections of the transcriptional networks in the different cell types. At the same token, it will become possible to analyze the interaction of the circadian oscillators between different tissues. However, data and models based solely on mRNA accumulations have to be granted with caution. There exists a second layer of complexity in the organization of circadian oscillators. This layer is made up of post-translational modifications of the gene products. The nature of these modifications and how these modifications affect the circadian oscillator will be in the focus of the next chapter.

References

1. Hardin PE, Hall JC, Rosbash M (1990) Feedback of the Drosophila period gene product on circadian cycling of its messenger RNA levels. Nature 343:536–540
2. Elowitz MB, Leibler S (2000) A synthetic oscillatory network of transcriptional regulators. Nature 403:335–338
3. Palmeirim I, Henrique D, Ish-Horowicz D, Pourquie O (1997) Avian hairy gene expression identifies a molecular clock linked to vertebrate segmentation and somitogenesis. Cell 91:639–648
4. Tigges M, Marquez-Lago TT, Stelling J, Fussenegger M (2009) A tunable synthetic mammalian oscillator. Nature 457:309–312
5. Stricker J, Cookson S, Bennett MR, Mather WH, Tsimring LS, Hasty J (2008) A fast, robust and tunable synthetic gene oscillator. Nature 456:516–519
6. Konopka RJ, Benzer S (1971) Clock mutants of Drosophila melanogaster. Proc Natl Acad Sci USA 68:2112–2116
7. Zylka MJ, Shearman LP, Weaver DR, Reppert SM (1998) Three period homologs in mammals: differential light responses in the suprachiasmatic circadian clock and oscillating transcripts outside of brain. Neuron 20:1103–1110
8. Shearman LP, Sriram S, Weaver DR, Maywood ES, Chaves I, Zheng B, Kume K, Lee CC, van der Horst GT, Hastings MH et al (2000) Interacting molecular loops in the mammalian circadian clock. Science 288:1013–1019
9. Toh KL, Jones CR, He Y, Eide EJ, Hinz WA, Virshup DM, Ptacek LJ, Fu YH (2001) An hPer2 phosphorylation site mutation in familial advanced sleep phase syndrome. Science 291:1040–1043
10. Baylies MK, Bargiello TA, Jackson FR, Young MW (1987) Changes in abundance or structure of the per gene product can alter periodicity of the Drosophila clock. Nature 326:390–392
11. Huang ZJ, Curtin KD, Rosbash M (1995) PER protein interactions and temperature compensation of a circadian clock in Drosophila. Science 267:1169–1172
12. Yildiz O, Doi M, Yujnovsky I, Cardone L, Berndt A, Hennig S, Schulze S, Urbanke C, Sassone-Corsi P, Wolf E (2005) Crystal structure and interactions of the PAS repeat region of the Drosophila clock protein PERIOD. Mol Cell 17:69–82
13. Taylor BL, Zhulin IB (1999) PAS domains: internal sensors of oxygen, redox potential, and light. Microbiol Mol Biol Rev 63:479–506
14. Vielhaber EL, Duricka D, Ullman KS, Virshup DM (2001) Nuclear export of mammalian PERIOD proteins. J Biol Chem 276:45921–45927
15. Brown SA, Ripperger J, Kadener S, Fleury-Olela F, Vilbois F, Rosbash M, Schibler U (2005) PERIOD1-associated proteins modulate the negative limb of the mammalian circadian oscillator. Science 308:693–696
16. Reischl S, Vanselow K, Westermark PO, Thierfelder N, Maier B, Herzel H, Kramer A (2007) Beta-TrCP1-mediated degradation of PERIOD2 is essential for circadian dynamics. J Biol Rhythms 22:375–386

17. Ko HW, Jiang J, Edery I (2002) Role for Slimb in the degradation of Drosophila period protein phosphorylated by Doubletime. Nature 420:673–678
18. Hara Y, Onishi Y, Oishi K, Miyazaki K, Fukamizu A, Ishida N (2009) Molecular characterization of Mybbp1a as a co-repressor on the Period2 promoter. Nucleic Acids Res 37:1115–1126
19. Ohno T, Onishi Y, Ishida N (2007) The negative transcription factor E4BP4 is associated with circadian clock protein PERIOD2. Biochem Biophys Res Commun 354:1010–1015
20. Blau J, Young MW (1999) Cycling vrille expression is required for a functional Drosophila clock. Cell 99:661–671
21. Kojima S, Matsumoto K, Hirose M, Shimada M, Nagano M, Shigeyoshi Y, Hoshino S, Ui-Tei K, Saigo K, Green CB et al (2007) LARK activates posttranscriptional expression of an essential mammalian clock protein, PERIOD1. Proc Natl Acad Sci USA 104:1859–1864
22. Lin JM, Kilman VL, Keegan K, Paddock B, Emery-Le M, Rosbash M, Allada R (2002) A role for casein kinase 2alpha in the Drosophila circadian clock. Nature 420:816–820
23. Lin JM, Schroeder A, Allada R (2005) In vivo circadian function of casein kinase 2 phosphorylation sites in Drosophila PERIOD. J Neurosci 25:11175–11183
24. Fan JY, Preuss F, Muskus MJ, Bjes ES, Price JL (2009) Drosophila and vertebrate casein kinase I{delta} exhibits evolutionary conservation of circadian function. Genetics 181:139–152
25. Lowrey PL, Shimomura K, Antoch MP, Yamazaki S, Zemenides PD, Ralph MR, Menaker M, Takahashi JS (2000) Positional syntenic cloning and functional characterization of the mammalian circadian mutation tau. Science 288:483–492
26. Price JL, Blau J, Rothenfluh A, Abodeely M, Kloss B, Young MW (1998) Double-time is a novel Drosophila clock gene that regulates PERIOD protein accumulation. Cell 94:83–95
27. Kim EY, Ko HW, Yu W, Hardin PE, Edery I (2007) A DOUBLETIME kinase binding domain on the Drosophila PERIOD protein is essential for its hyperphosphorylation, transcriptional repression, and circadian clock function. Mol Cell Biol 27:5014–5028
28. Nawathean P, Stoleru D, Rosbash M (2007) A small conserved domain of Drosophila PERIOD is important for circadian phosphorylation, nuclear localization, and transcriptional repressor activity. Mol Cell Biol 27:5002–5013
29. Vanselow K, Vanselow JT, Westermark PO, Reischl S, Maier B, Korte T, Herrmann A, Herzel H, Schlosser A, Kramer A (2006) Differential effects of PER2 phosphorylation: molecular basis for the human familial advanced sleep phase syndrome (FASPS). Genes Dev 20:2660–2672
30. Asher G, Gatfield D, Stratmann M, Reinke H, Dibner C, Kreppel F, Mostoslavsky R, Alt FW, Schibler U (2008) SIRT1 regulates circadian clock gene expression through PER2 deacetylation. Cell 134:317–328
31. Albrecht U, Sun ZS, Eichele G, Lee CC (1997) A differential response of two putative mammalian circadian regulators, mper1 and mper2, to light. Cell 91:1055–1064
32. Shearman LP, Zylka MJ, Weaver DR, Kolakowski LF Jr, Reppert SM (1997) Two period homologs: circadian expression and photic regulation in the suprachiasmatic nuclei. Neuron 19:1261–1269
33. Balsalobre A, Damiola F, Schibler U (1998) A serum shock induces circadian gene expression in mammalian tissue culture cells. Cell 93:929–937
34. Balsalobre A, Marcacci L, Schibler U (2000) Multiple signaling pathways elicit circadian gene expression in cultured Rat-1 fibroblasts. Curr Biol 10:1291–1294
35. Vallone D, Gondi SB, Whitmore D, Foulkes NS (2004) E-box function in a period gene repressed by light. Proc Natl Acad Sci USA 101:4106–4111
36. Ceriani MF, Darlington TK, Staknis D, Mas P, Petti AA, Weitz CJ, Kay SA (1999) Light-dependent sequestration of TIMELESS by CRYPTOCHROME. Science 285:553–556
37. Sehgal A, Price JL, Man B, Young MW (1994) Loss of circadian behavioral rhythms and per RNA oscillations in the Drosophila mutant timeless. Science 263:1603–1606
38. Vosshall LB, Price JL, Sehgal A, Saez L, Young MW (1994) Block in nuclear localization of period protein by a second clock mutation, timeless. Science 263:1606–1609

39. Meyer P, Saez L, Young MW (2006) PER-TIM interactions in living Drosophila cells: an interval timer for the circadian clock. Science 311:226–229
40. Rothenfluh A, Young MW, Saez L (2000) A TIMELESS-independent function for PERIOD proteins in the Drosophila clock. Neuron 26:505–514
41. Koh K, Zheng X, Sehgal A (2006) JETLAG resets the Drosophila circadian clock by promoting light-induced degradation of TIMELESS. Science 312:1809–1812
42. Sangoram AM, Saez L, Antoch MP, Gekakis N, Staknis D, Whiteley A, Fruechte EM, Vitaterna MH, Shimomura K, King DP et al (1998) Mammalian circadian autoregulatory loop: a timeless ortholog and mPer1 interact and negatively regulate CLOCK-BMAL1-induced transcription. Neuron 21:1101–1113
43. Zylka MJ, Shearman LP, Levine JD, Jin X, Weaver DR, Reppert SM (1998) Molecular analysis of mammalian timeless. Neuron 21:1115–1122
44. Barnes JW, Tischkau SA, Barnes JA, Mitchell JW, Burgoon PW, Hickok JR, Gillette MU (2003) Requirement of mammalian timeless for circadian rhythmicity. Science 302:439–442
45. Gotter AL (2006) A Timeless debate: resolving TIM's noncircadian roles with possible clock function. Neuroreport 17:1229–1233
46. Gotter AL, Manganaro T, Weaver DR, Kolakowski LF Jr, Possidente B, Sriram S, MacLaughlin DT, Reppert SM (2000) A time-less function for mouse timeless. Nat Neurosci 3:755–756
47. Meissner RA, Kilman VL, Lin JM, Allada R (2008) TIMELESS is an important mediator of CK2 effects on circadian clock function in vivo. J Neurosci 28:9732–9740
48. Martinek S, Inonog S, Manoukian AS, Young MW (2001) A role for the segment polarity gene shaggy/GSK-3 in the Drosophila circadian clock. Cell 105:769–779
49. Iitaka C, Miyazaki K, Akaike T, Ishida N (2005) A role for glycogen synthase kinase-3beta in the mammalian circadian clock. J Biol Chem 280:29397–29402
50. Thresher RJ, Vitaterna MH, Miyamoto Y, Kazantsev A, Hsu DS, Petit C, Selby CP, Dawut L, Smithies O, Takahashi JS et al (1998) Role of mouse cryptochrome blue-light photoreceptor in circadian photoresponses. Science 282:1490–1494
51. van der Horst GT, Muijtjens M, Kobayashi K, Takano R, Kanno S, Takao M, de Wit J, Verkerk A, Eker AP, van Leenen D et al (1999) Mammalian Cry1 and Cry2 are essential for maintenance of circadian rhythms. Nature 398:627–630
52. Kume K, Zylka MJ, Sriram S, Shearman LP, Weaver DR, Jin X, Maywood ES, Hastings MH, Reppert SM (1999) mCRY1 and mCRY2 are essential components of the negative limb of the circadian clock feedback loop. Cell 98:193–205
53. Hattar S, Lucas RJ, Mrosovsky N, Thompson S, Douglas RH, Hankins MW, Lem J, Biel M, Hofmann F, Foster RG et al (2003) Melanopsin and rod-cone photoreceptive systems account for all major accessory visual functions in mice. Nature 424:76–81
54. Stanewsky R, Kaneko M, Emery P, Beretta B, Wager-Smith K, Kay SA, Rosbash M, Hall JC (1998) The cryb mutation identifies cryptochrome as a circadian photoreceptor in Drosophila. Cell 95:681–692
55. Krishnan B, Levine JD, Lynch MK, Dowse HB, Funes P, Hall JC, Hardin PE, Dryer SE (2001) A new role for cryptochrome in a Drosophila circadian oscillator. Nature 411:313–317
56. Zhu H, Conte F, Green CB (2003) Nuclear localization and transcriptional repression are confined to separable domains in the circadian protein CRYPTOCHROME. Curr Biol 13:1653–1658
57. Busza A, Emery-Le M, Rosbash M, Emery P (2004) Roles of the two Drosophila CRYPTOCHROME structural domains in circadian photoreception. Science 304:1503–1506
58. Collins B, Mazzoni EO, Stanewsky R, Blau J (2006) Drosophila CRYPTOCHROME is a circadian transcriptional repressor. Curr Biol 16:441–449
59. Oster H, Yasui A, van der Horst GT, Albrecht U (2002) Disruption of mCry2 restores circadian rhythmicity in mPer2 mutant mice. Genes Dev 16:2633–2638
60. Gegear RJ, Casselman A, Waddell S, Reppert SM (2008) Cryptochrome mediates light-dependent magnetosensitivity in Drosophila. Nature 454:1014–1018
61. Zhu H, Sauman I, Yuan Q, Casselman A, Emery-Le M, Emery P, Reppert SM (2008) Cryptochromes define a novel circadian clock mechanism in monarch butterflies that may underlie sun compass navigation. PLoS Biol 6:e4

62. Vitaterna MH, King DP, Chang AM, Kornhauser JM, Lowrey PL, McDonald JD, Dove WF, Pinto LH, Turek FW, Takahashi JS (1994) Mutagenesis and mapping of a mouse gene, Clock, essential for circadian behavior. Science 264:719–725
63. King DP, Zhao Y, Sangoram AM, Wilsbacher LD, Tanaka M, Antoch MP, Steeves TD, Vitaterna MH, Kornhauser JM, Lowrey PL et al (1997) Positional cloning of the mouse circadian clock gene. Cell 89:641–653
64. Allada R, White NE, So WV, Hall JC, Rosbash M (1998) A mutant Drosophila homolog of mammalian Clock disrupts circadian rhythms and transcription of period and timeless. Cell 93:791–804
65. Darlington TK, Wager-Smith K, Ceriani MF, Staknis D, Gekakis N, Steeves TD, Weitz CJ, Takahashi JS, Kay SA (1998) Closing the circadian loop: CLOCK-induced transcription of its own inhibitors per and tim. Science 280:1599–1603
66. Reick M, Garcia JA, Dudley C, McKnight SL (2001) NPAS2: an analog of clock operative in the mammalian forebrain. Science 293:506–509
67. DeBruyne JP, Noton E, Lambert CM, Maywood ES, Weaver DR, Reppert SM (2006) A clock shock: mouse CLOCK is not required for circadian oscillator function. Neuron 50:465–477
68. DeBruyne JP, Weaver DR, Reppert SM (2007) Peripheral circadian oscillators require CLOCK. Curr Biol 17:R538–R539
69. DeBruyne JP, Weaver DR, Reppert SM (2007) CLOCK and NPAS2 have overlapping roles in the suprachiasmatic circadian clock. Nat Neurosci 10:543–545
70. Curtis AM, Seo SB, Westgate EJ, Rudic RD, Smyth EM, Chakravarti D, FitzGerald GA, McNamara P (2004) Histone acetyltransferase-dependent chromatin remodeling and the vascular clock. J Biol Chem 279:7091–7097
71. Takahata S, Ozaki T, Mimura J, Kikuchi Y, Sogawa K, Fujii-Kuriyama Y (2000) Transactivation mechanisms of mouse clock transcription factors, mClock and mArnt3. Genes Cells 5:739–747
72. Etchegaray JP, Lee C, Wade PA, Reppert SM (2003) Rhythmic histone acetylation underlies transcription in the mammalian circadian clock. Nature 421:177–182
73. Doi M, Hirayama J, Sassone-Corsi P (2006) Circadian regulator CLOCK is a histone acetyltransferase. Cell 125:497–508
74. Hirayama J, Sahar S, Grimaldi B, Tamaru T, Takamatsu K, Nakahata Y, Sassone-Corsi P (2007) CLOCK-mediated acetylation of BMAL1 controls circadian function. Nature 450:1086–1090
75. Nakahata Y, Kaluzova M, Grimaldi B, Sahar S, Hirayama J, Chen D, Guarente LP, Sassone-Corsi P (2008) The NAD+-dependent deacetylase SIRT1 modulates CLOCK-mediated chromatin remodeling and circadian control. Cell 134:329–340
76. Rutter J, Reick M, Wu LC, McKnight SL (2001) Regulation of clock and NPAS2 DNA binding by the redox state of NAD cofactors. Science 293:510–514
77. Cyran SA, Buchsbaum AM, Reddy KL, Lin MC, Glossop NR, Hardin PE, Young MW, Storti RV, Blau J (2003) vrille, Pdp1, and dClock form a second feedback loop in the Drosophila circadian clock. Cell 112:329–341
78. Glossop NR, Houl JH, Zheng H, Ng FS, Dudek SM, Hardin PE (2003) VRILLE feeds back to control circadian transcription of Clock in the Drosophila circadian oscillator. Neuron 37:249–261
79. Mitsui S, Yamaguchi S, Matsuo T, Ishida Y, Okamura H (2001) Antagonistic role of E4BP4 and PAR proteins in the circadian oscillatory mechanism. Genes Dev 15:995–1006
80. Ohno T, Onishi Y, Ishida N (2007) A novel E4BP4 element drives circadian expression of mPeriod2. Nucleic Acids Res 35:648–655
81. Gilles-Gonzalez MA, Gonzalez G (2004) Signal transduction by heme-containing PAS-domain proteins. J Appl Physiol 96:774–783
82. Kaasik K, Lee CC (2004) Reciprocal regulation of haem biosynthesis and the circadian clock in mammals. Nature 430:467–471
83. Dioum EM, Rutter J, Tuckerman JR, Gonzalez G, Gilles-Gonzalez MA, McKnight SL (2002) NPAS2: a gas-responsive transcription factor. Science 298:2385–2387
84. Dardente H, Fortier EE, Martineau V, Cermakian N (2007) Cryptochromes impair phosphorylation of transcriptional activators in the clock: a general mechanism for circadian repression. Biochem J 402:525–536

85. Kondratov RV, Kondratova AA, Lee C, Gorbacheva VY, Chernov MV, Antoch MP (2006) Post-translational regulation of circadian transcriptional CLOCK(NPAS2)/BMAL1 complex by CRYPTOCHROMES. Cell Cycle 5:890–895
86. Gekakis N, Staknis D, Nguyen HB, Davis FC, Wilsbacher LD, King DP, Takahashi JS, Weitz CJ (1998) Role of the CLOCK protein in the mammalian circadian mechanism. Science 280:1564–1569
87. Rutila JE, Suri V, Le M, So WV, Rosbash M, Hall JC (1998) CYCLE is a second bHLH-PAS clock protein essential for circadian rhythmicity and transcription of Drosophila period and timeless. Cell 93:805–814
88. Lee C, Etchegaray JP, Cagampang FR, Loudon AS, Reppert SM (2001) Posttranslational mechanisms regulate the mammalian circadian clock. Cell 107:855–867
89. Ripperger JA, Schibler U (2006) Rhythmic CLOCK-BMAL1 binding to multiple E-box motifs drives circadian Dbp transcription and chromatin transitions. Nat Genet 38:369–374
90. Cardone L, Hirayama J, Giordano F, Tamaru T, Palvimo JJ, Sassone-Corsi P (2005) Circadian clock control by SUMOylation of BMAL1. Science 309:1390–1394
91. Preitner N, Damiola F, Lopez-Molina L, Zakany J, Duboule D, Albrecht U, Schibler U (2002) The orphan nuclear receptor REV-ERBalpha controls circadian transcription within the positive limb of the mammalian circadian oscillator. Cell 110:251–260
92. Ueda HR, Chen W, Adachi A, Wakamatsu H, Hayashi S, Takasugi T, Nagano M, Nakahama K, Suzuki Y, Sugano S et al (2002) A transcription factor response element for gene expression during circadian night. Nature 418:534–539
93. Kornmann B, Schaad O, Bujard H, Takahashi JS, Schibler U (2007) System-driven and oscillator-dependent circadian transcription in mice with a conditionally active liver clock. PLoS Biol 5:e34
94. Raghuram S, Stayrook KR, Huang P, Rogers PM, Nosie AK, McClure DB, Burris LL, Khorasanizadeh S, Burris TP, Rastinejad F (2007) Identification of heme as the ligand for the orphan nuclear receptors REV-ERBalpha and REV-ERBbeta. Nat Struct Mol Biol 14:1207–1213
95. Meng QJ, McMaster A, Beesley S, Lu WQ, Gibbs J, Parks D, Collins J, Farrow S, Donn R, Ray D et al (2008) Ligand modulation of REV-ERBalpha function resets the peripheral circadian clock in a phasic manner. J Cell Sci 121:3629–3635
96. Yin L, Wang J, Klein PS, Lazar MA (2006) Nuclear receptor Rev-erbalpha is a critical lithium-sensitive component of the circadian clock. Science 311:1002–1005
97. Liu AC, Tran HG, Zhang EE, Priest AA, Welsh DK, Kay SA (2008) Redundant function of REV-ERBalpha and beta and non-essential role for Bmal1 cycling in transcriptional regulation of intracellular circadian rhythms. PLoS Genet 4:e1000023
98. Ripperger JA (2006) Mapping of binding regions for the circadian regulators BMAL1 and CLOCK within the mouse Rev-erbalpha gene. Chronobiol Int 23:135–142
99. Sato TK, Panda S, Miraglia LJ, Reyes TM, Rudic RD, McNamara P, Naik KA, FitzGerald GA, Kay SA, Hogenesch JB (2004) A functional genomics strategy reveals Rora as a component of the mammalian circadian clock. Neuron 43:527–537
100. Guillaumond F, Dardente H, Giguere V, Cermakian N (2005) Differential control of Bmal1 circadian transcription by REV-ERB and ROR nuclear receptors. J Biol Rhythms 20:391–403
101. Liu C, Li S, Liu T, Borjigin J, Lin JD (2007) Transcriptional coactivator PGC-1alpha integrates the mammalian clock and energy metabolism. Nature 447:477–481
102. Kloss B, Price JL, Saez L, Blau J, Rothenfluh A, Wesley CS, Young MW (1998) The Drosophila clock gene double-time encodes a protein closely related to human casein kinase Iepsilon. Cell 94:97–107
103. Xu Y, Padiath QS, Shapiro RE, Jones CR, Wu SC, Saigoh N, Saigoh K, Ptacek LJ, Fu YH (2005) Functional consequences of a CKIdelta mutation causing familial advanced sleep phase syndrome. Nature 434:640–644
104. Fang Y, Sathyanarayanan S, Sehgal A (2007) Post-translational regulation of the Drosophila circadian clock requires protein phosphatase 1 (PP1). Genes Dev 21:1506–1518

105. Sathyanarayanan S, Zheng X, Xiao R, Sehgal A (2004) Posttranslational regulation of Drosophila PERIOD protein by protein phosphatase 2A. Cell 116:603–615
106. Virshup DM, Eide EJ, Forger DB, Gallego M, Harnish EV (2007) Reversible protein phosphorylation regulates circadian rhythms. Cold Spring Harb Symp Quant Biol 72:413–420
107. Partch CL, Shields KF, Thompson CL, Selby CP, Sancar A (2006) Posttranslational regulation of the mammalian circadian clock by cryptochrome and protein phosphatase 5. Proc Natl Acad Sci USA 103:10467–10472
108. Kloss B, Rothenfluh A, Young MW, Saez L (2001) Phosphorylation of period is influenced by cycling physical associations of double-time, period, and timeless in the Drosophila clock. Neuron 30:699–706
109. Busino L, Bassermann F, Maiolica A, Lee C, Nolan PM, Godinho SI, Draetta GF, Pagano M (2007) SCFFbxl3 controls the oscillation of the circadian clock by directing the degradation of cryptochrome proteins. Science 316:900–904
110. Godinho SI, Maywood ES, Shaw L, Tucci V, Barnard AR, Busino L, Pagano M, Kendall R, Quwailid MM, Romero MR et al (2007) The after-hours mutant reveals a role for Fbxl3 in determining mammalian circadian period. Science 316:897–900
111. Siepka SM, Yoo SH, Park J, Song W, Kumar V, Hu Y, Lee C, Takahashi JS (2007) Circadian mutant Overtime reveals F-box protein FBXL3 regulation of cryptochrome and period gene expression. Cell 129:1011–1023
112. Dardente H, Mendoza J, Fustin JM, Challet E, Hazlerigg DG (2008) Implication of the F-Box Protein FBXL21 in circadian pacemaker function in mammals. PLoS ONE 3:e3530
113. Reinke H, Saini C, Fleury-Olela F, Dibner C, Benjamin IJ, Schibler U (2008) Differential display of DNA-binding proteins reveals heat-shock factor 1 as a circadian transcription factor. Genes Dev 22:331–345
114. Taylor P, Hardin PE (2008) Rhythmic E-box binding by CLK-CYC controls daily cycles in per and tim transcription and chromatin modifications. Mol Cell Biol 28:4642–4652
115. Etchegaray JP, Yang X, DeBruyne JP, Peters AH, Weaver DR, Jenuwein T, Reppert SM (2006) The polycomb group protein EZH2 is required for mammalian circadian clock function. J Biol Chem 281:21209–21215
116. Alenghat T, Meyers K, Mullican SE, Leitner K, Adeniji-Adele A, Avila J, Bucan M, Ahima RS, Kaestner KH, Lazar MA (2008) Nuclear receptor corepressor and histone deacetylase 3 govern circadian metabolic physiology. Nature 456:997–1000
117. Zhao WN, Malinin N, Yang FC, Staknis D, Gekakis N, Maier B, Reischl S, Kramer A, Weitz CJ (2007) CIPC is a mammalian circadian clock protein without invertebrate homologues. Nat Cell Biol 9:268–275
118. Kadener S, Stoleru D, McDonald M, Nawathean P, Rosbash M (2007) Clockwork Orange is a transcriptional repressor and a new Drosophila circadian pacemaker component. Genes Dev 21:1675–1686
119. Lim C, Chung BY, Pitman JL, McGill JJ, Pradhan S, Lee J, Keegan KP, Choe J, Allada R (2007) Clockwork orange encodes a transcriptional repressor important for circadian-clock amplitude in Drosophila. Curr Biol 17:1082–1089
120. Richier B, Michard-Vanhee C, Lamouroux A, Papin C, Rouyer F (2008) The clockwork orange Drosophila protein functions as both an activator and a repressor of clock gene expression. J Biol Rhythms 23:103–116
121. Honma S, Kawamoto T, Takagi Y, Fujimoto K, Sato F, Noshiro M, Kato Y, Honma K (2002) Dec1 and Dec2 are regulators of the mammalian molecular clock. Nature 419:841–844
122. Hong HK, Chong JL, Song W, Song EJ, Jyawook AA, Schook AC, Ko CH, Takahashi JS (2007) Inducible and reversible Clock gene expression in brain using the tTA system for the study of circadian behavior. PLoS Genet 3:e33
123. Houl JH, Yu W, Dudek SM, Hardin PE (2006) Drosophila CLOCK is constitutively expressed in circadian oscillator and non-oscillator cells. J Biol Rhythms 21:93–103
124. Kim EY, Bae K, Ng FS, Glossop NR, Hardin PE, Edery I (2002) Drosophila CLOCK protein is under posttranscriptional control and influences light-induced activity. Neuron 34:69–81

125. Sato TK, Yamada RG, Ukai H, Baggs JE, Miraglia LJ, Kobayashi TJ, Welsh DK, Kay SA, Ueda HR, Hogenesch JB (2006) Feedback repression is required for mammalian circadian clock function. Nat Genet 38:312–319
126. Wilkins AK, Barton PI, Tidor B (2007) The Per2 negative feedback loop sets the period in the mammalian circadian clock mechanism. PLoS Comput Biol 3:e242
127. Yang Z, Sehgal A (2001) Role of molecular oscillations in generating behavioral rhythms in Drosophila. Neuron 29:453–467
128. Ueda HR, Hayashi S, Chen W, Sano M, Machida M, Shigeyoshi Y, Iino M, Hashimoto S (2005) System-level identification of transcriptional circuits underlying mammalian circadian clocks. Nat Genet 37:187–192
129. Kadener S, Menet JS, Schoer R, Rosbash M (2008) Circadian transcription contributes to core period determination in Drosophila. PLoS Biol 6:e119
130. Fan Y, Hida A, Anderson DA, Izumo M, Johnson CH (2007) Cycling of CRYPTOCHROME proteins is not necessary for circadian-clock function in mammalian fibroblasts. Curr Biol 17:1091–1100
131. Abraham D, Dallmann R, Steinlechner S, Albrecht U, Eichele G, Oster H (2006) Restoration of circadian rhythmicity in circadian clock-deficient mice in constant light. J Biol Rhythms 21:169–176
132. Dibner C, Sage D, Unser M, Bauer C, d'Eysmond T, Naef F, Schibler U (2009) Circadian gene expression is resilient to large fluctuations in overall transcription rates. EMBO J 28:123–134
133. Aschoff J (1965) Response curves in circadian periodicity. In: Aschoff J (ed) Circadian clocks. Amsterdam, North-Holland Publishing Co, pp 95–111
134. Daan S, Pittendrigh CS (1976) A functional analysis of circadian pacemakers in rodents. II. The variability of phase response curves. J Comp Physiol 106:253–266
135. Cheng MY, Bullock CM, Li C, Lee AG, Bermak JC, Belluzzi J, Weaver DR, Leslie FM, Zhou QY (2002) Prokineticin 2 transmits the behavioural circadian rhythm of the suprachiasmatic nucleus. Nature 417:405–410
136. Kramer A, Yang FC, Snodgrass P, Li X, Scammell TE, Davis FC, Weitz CJ (2001) Regulation of daily locomotor activity and sleep by hypothalamic EGF receptor signaling. Science 294:2511–2515
137. Kraves S, Weitz CJ (2006) A role for cardiotropin-like cytokine in the circadian control of mammalian locomotor activity. Nat Neurosci 9:212–219
138. Balsalobre A, Brown SA, Marcacci L, Tronche F, Kellendonk C, Reichardt HM, Schutz G, Schibler U (2000) Resetting of circadian time in peripheral tissues by glucocorticoid signaling. Science 289:2344–2347
139. Aronin N, Sagar SM, Sharp FR, Schwartz WJ (1990) Light regulates expression of a Fos-related protein in rat suprachiasmatic nuclei. Proc Natl Acad Sci USA 87:5959–5962
140. Kornhauser JM, Nelson DE, Mayo KE, Takahashi JS (1990) Photic and circadian regulation of c-fos gene expression in the hamster suprachiasmatic nucleus. Neuron 5:127–134
141. Gau D, Lemberger T, von Gall C, Kretz O, Le Minh N, Gass P, Schmid W, Schibler U, Korf HW, Schutz G (2002) Phosphorylation of CREB Ser142 regulates light-induced phase shifts of the circadian clock. Neuron 34:245–253
142. Ginty DD, Kornhauser JM, Thompson MA, Bading H, Mayo KE, Takahashi JS, Greenberg ME (1993) Regulation of CREB phosphorylation in the suprachiasmatic nucleus by light and a circadian clock. Science 260:238–241
143. Foulkes NS, Borjigin J, Snyder SH, Sassone-Corsi P (1996) Transcriptional control of circadian hormone synthesis via the CREM feedback loop. Proc Natl Acad Sci USA 93:14140–14145
144. Schwartz WJ, Aronin N, Sassone-Corsi P (2005) Photoinducible and rhythmic ICER-CREM immunoreactivity in the rat suprachiasmatic nucleus. Neurosci Lett 385:87–91
145. Stehle JH, Pfeffer M, Kuhn R, Korf HW (1996) Light-induced expression of transcription factor ICER (inducible cAMP early repressor) in rat suprachiasmatic nucleus is phase-restricted. Neurosci Lett 217:169–172

146. Honrado GI, Johnson RS, Golombek DA, Spiegelman BM, Papaioannou VE, Ralph MR (199) The circadian system of c-fos deficient mice. J Comp Physiol [A] 178:563–570
147. Morris ME, Viswanathan N, Kuhlman S, Davis FC, Weitz CJ (1998) A screen for genes induced in the suprachiasmatic nucleus by light. Science 279:1544–1547
148. Crosio C, Cermakian N, Allis CD, Sassone-Corsi P (2000) Light induces chromatin modification in cells of the mammalian circadian clock. Nat Neurosci 3:1241–1247
149. Yan L, Takekida S, Shigeyoshi Y, Okamura H (1999) Per1 and Per2 gene expression in the rat suprachiasmatic nucleus: circadian profile and the compartment-specific response to light. Neuroscience 94:141–150
150. Field MD, Maywood ES, O'Brien JA, Weaver DR, Reppert SM, Hastings MH (2000) Analysis of clock proteins in mouse SCN demonstrates phylogenetic divergence of the circadian clockwork and resetting mechanisms. Neuron 25:437–447
151. Yan L, Okamura H (2002) Gradients in the circadian expression of Per1 and Per2 genes in the rat suprachiasmatic nucleus. Eur J Neurosci 15:1153–1162
152. Yan L, Silver R (2002) Differential induction and localization of mPer1 and mPer2 during advancing and delaying phase shifts. Eur J Neurosci 16:1531–1540
153. Spoelstra K, Albrecht U, van der Horst GT, Brauer V, Daan S (2004) Phase responses to light pulses in mice lacking functional per or cry genes. J Biol Rhythms 19:518–529
154. Yan L, Hochstetler KJ, Silver R, Bult-Ito A (2003) Phase shifts and Per gene expression in mouse suprachiasmatic nucleus. Neuroreport 14:1247–1251
155. Yan L, Silver R (2004) Resetting the brain clock: time course and localization of mPER1 and mPER2 protein expression in suprachiasmatic nuclei during phase shifts. Eur J Neurosci 19:1105–1109
156. Cermakian N, Monaco L, Pando MP, Dierich A, Sassone-Corsi P (2001) Altered behavioral rhythms and clock gene expression in mice with a targeted mutation in the Period1 gene. EMBO J 20:3967–3974
157. Bae K, Weaver DR (2003) Light-induced phase shifts in mice lacking mPER1 or mPER2. J Biol Rhythms 18:123–133
158. Travnickova-Bendova Z, Cermakian N, Reppert SM, Sassone-Corsi P (2002) Bimodal regulation of mPeriod promoters by CREB-dependent signaling and CLOCK/BMAL1 activity. Proc Natl Acad Sci USA 99:7728–7733
159. Kako K, Wakamatsu H, Ishida N (1996) c-fos CRE-binding activity of CREB/ATF family in the SCN is regulated by light but not a circadian clock. Neurosci Lett 216:159–162
160. Mioduszewska B, Jaworski J, Kaczmarek L (2003) Inducible cAMP early repressor (ICER) in the nervous system – a transcriptional regulator of neuronal plasticity and programmed cell death. J Neurochem 87:1313–1320
161. Shimizu F, Fukada Y (2007) Circadian phosphorylation of ATF-2, a potential activator of Period2 gene transcription in the chick pineal gland. J Neurochem 103:1834–1842
162. Naruse Y, Oh-hashi K, Iijima N, Naruse M, Yoshioka H, Tanaka M (2004) Circadian and light-induced transcription of clock gene Per1 depends on histone acetylation and deacetylation. Mol Cell Biol 24:6278–6287
163. Masubuchi S, Kataoka N, Sassone-Corsi P, Okamura H (2005) Mouse Period1 (mPER1) acts as a circadian adaptor to entrain the oscillator to environmental light/dark cycles by regulating mPER2 protein. J Neurosci 25:4719–4724
164. Shim HS, Kim H, Lee J, Son GH, Cho S, Oh TH, Kang SH, Seen DS, Lee KH, Kim K (2007) Rapid activation of CLOCK by Ca2+-dependent protein kinase C mediates resetting of the mammalian circadian clock. EMBO Rep 8:366–371
165. Wilsbacher LD, Yamazaki S, Herzog ED, Song EJ, Radcliffe LA, Abe M, Block G, Spitznagel E, Menaker M, Takahashi JS (2002) Photic and circadian expression of luciferase in mPeriod1-luc transgenic mice invivo. Proc Natl Acad Sci USA 99:489–494
166. Yoo SH, Yamazaki S, Lowrey PL, Shimomura K, Ko CH, Buhr ED, Siepka SM, Hong HK, Oh WJ, Yoo OJ et al (2004) PERIOD2::LUCIFERASE real-time reporting of circadian dynamics reveals persistent circadian oscillations in mouse peripheral tissues. Proc Natl Acad Sci USA 101:5339–5346

167. Nagoshi E, Saini C, Bauer C, Laroche T, Naef F, Schibler U (2004) Circadian gene expression in individual fibroblasts: cell-autonomous and self-sustained oscillators pass time to daughter cells. Cell 119:693–705
168. Welsh DK, Yoo SH, Liu AC, Takahashi JS, Kay SA (2004) Bioluminescence imaging of individual fibroblasts reveals persistent, independently phased circadian rhythms of clock gene expression. Curr Biol 14:2289–2295
169. Yagita K, Tamanini F, van Der Horst GT, Okamura H (2001) Molecular mechanisms of the biological clock in cultured fibroblasts. Science 292:278–281
170. Damiola F, Le Minh N, Preitner N, Kornmann B, Fleury-Olela F, Schibler U (2000) Restricted feeding uncouples circadian oscillators in peripheral tissues from the central pacemaker in the suprachiasmatic nucleus. Genes Dev 14:2950–2961
171. Stokkan KA, Yamazaki S, Tei H, Sakaki Y, Menaker M (2001) Entrainment of the circadian clock in the liver by feeding. Science 291:490–493
172. Le Minh N, Damiola F, Tronche F, Schutz G, Schibler U (2001) Glucocorticoid hormones inhibit food-induced phase-shifting of peripheral circadian oscillators. EMBO J 20:7128–7136
173. Akashi M, Nishida E (2000) Involvement of the MAP kinase cascade in resetting of the mammalian circadian clock. Genes Dev 14:645–649
174. Motzkus D, Albrecht U, Maronde E (2002) The human PER1 gene is inducible by interleukin-6. J Mol Neurosci 18:105–109
175. Motzkus D, Loumi S, Cadenas C, Vinson C, Forssmann WG, Maronde E (2007) Activation of human period-1 by PKA or CLOCK/BMAL1 is conferred by separate signal transduction pathways. Chronobiol Int 24:783–792
176. Motzkus D, Maronde E, Grunenberg U, Lee CC, Forssmann W, Albrecht U (2000) The human PER1 gene is transcriptionally regulated by multiple signaling pathways. FEBS Lett 486:315–319
177. Canaple L, Rambaud J, Dkhissi-Benyahya O, Rayet B, Tan NS, Michalik L, Delaunay F, Wahli W, Laudet V (2006) Reciprocal regulation of brain and muscle Arnt-like protein 1 and peroxisome proliferator-activated receptor alpha defines a novel positive feedback loop in the rodent liver circadian clock. Mol Endocrinol 20:1715–1727
178. Koinuma S, Yagita K, Fujioka A, Takashima N, Takumi T, Shigeyoshi Y (2009) The resetting of the circadian rhythm by Prostaglandin J2 is distinctly phase-dependent. FEBS Lett 583:413–418
179. Nakahata Y, Akashi M, Trcka D, Yasuda A, Takumi T (2006) The in vitro real-time oscillation monitoring system identifies potential entrainment factors for circadian clocks. BMC Mol Biol 7:5
180. Brown SA, Kunz D, Dumas A, Westermark PO, Vanselow K, Tilmann-Wahnschaffe A, Herzel H, Kramer A (2008) Molecular insights into human daily behavior. Proc Natl Acad Sci USA 105:1602–1607
181. Ukai H, Kobayashi TJ, Nagano M, Masumoto KH, Sujino M, Kondo T, Yagita K, Shigeyoshi Y, Ueda HR (2007) Melanopsin-dependent photo-perturbation reveals desynchronization underlying the singularity of mammalian circadian clocks. Nat Cell Biol 9:1327–1334
182. Akiyama M, Kouzu Y, Takahashi S, Wakamatsu H, Moriya T, Maetani M, Watanabe S, Tei H, Sakaki Y, Shibata S (1999) Inhibition of light- or glutamate-induced mPer1 expression represses the phase shifts into the mouse circadian locomotor and suprachiasmatic firing rhythms. J Neurosci 19:1115–1121
183. Brewer JM, Yannielli PC, Harrington ME (2002) Neuropeptide Y differentially suppresses per1 and per2 mRNA induced by light in the suprachiasmatic nuclei of the golden hamster. J Biol Rhythms 17:28–39
184. Dziema H, Oatis B, Butcher GQ, Yates R, Hoyt KR, Obrietan K (2003) The ERK/MAP kinase pathway couples light to immediate-early gene expression in the suprachiasmatic nucleus. Eur J Neurosci 17:1617–1627

185. Gamble KL, Allen GC, Zhou T, McMahon DG (2007) Gastrin-releasing peptide mediates light-like resetting of the suprachiasmatic nucleus circadian pacemaker through cAMP response element-binding protein and Per1 activation. J Neurosci 27:12078–12087
186. Gillespie CF, Van Der Beek EM, Mintz EM, Mickley NC, Jasnow AM, Huhman KL, Albers HE (1999) GABAergic regulation of light-induced c-Fos immunoreactivity within the suprachiasmatic nucleus. J Comp Neurol 411:683–692
187. Nakaya M, Sanada K, Fukada Y (2003) Spatial and temporal regulation of mitogen-activated protein kinase phosphorylation in the mouse suprachiasmatic nucleus. Biochem Biophys Res Commun 305:494–501
188. Nielsen HS, Hannibal J, Fahrenkrug J (2002) Vasoactive intestinal polypeptide induces per1 and per2 gene expression in the rat suprachiasmatic nucleus late at night. Eur J Neurosci 15:570–574
189. Nielsen HS, Hannibal J, Knudsen SM, Fahrenkrug J (2001) Pituitary adenylate cyclase-activating polypeptide induces period1 and period2 gene expression in the rat suprachiasmatic nucleus during late night. Neuroscience 103:433–441
190. Nomura K, Takeuchi Y, Yamaguchi S, Okamura H, Fukunaga K (2003) Involvement of calcium/calmodulin-dependent protein kinase II in the induction of mPer1. J Neurosci Res 72:384–392
191. Oster H, Werner C, Magnone MC, Mayser H, Feil R, Seeliger MW, Hofmann F, Albrecht U (2003) cGMP-dependent protein kinase II modulates mPer1 and mPer2 gene induction and influences phase shifts of the circadian clock. Curr Biol 13:725–733
192. Steenhard BM, Besharse JC (2000) Phase shifting the retinal circadian clock: xPer2 mRNA induction by light and dopamine. J Neurosci 20:8572–8577
193. Yokota S, Yamamoto M, Moriya T, Akiyama M, Fukunaga K, Miyamoto E, Shibata S (2001) Involvement of calcium-calmodulin protein kinase but not mitogen-activated protein kinase in light-induced phase delays and Per gene expression in the suprachiasmatic nucleus of the hamster. J Neurochem 77:618–627
194. Liu C, Reppert SM (2000) GABA synchronizes clock cells within the suprachiasmatic circadian clock. Neuron 25:123–128
195. Brown SA, Zumbrunn G, Fleury-Olela F, Preitner N, Schibler U (2002) Rhythms of mammalian body temperature can sustain peripheral circadian clocks. Curr Biol 12:1574–1583
196. Yan J, Wang H, Liu Y, Shao C (2008) Analysis of gene regulatory networks in the mammalian circadian rhythm. PLoS Comput Biol 4:e1000193
197. Jin X, Shearman LP, Weaver DR, Zylka MJ, de Vries GJ, Reppert SM (1999) A molecular mechanism regulating rhythmic output from the suprachiasmatic circadian clock. Cell 96:57–68
198. Hampp G, Ripperger JA, Houben T, Schmutz I, Blex C, Perreau-Lenz S, Brunk I, Spanagel R, Ahnert-Hilger G, Meijer JH et al (2008) Regulation of monoamine oxidase A by circadian-clock components implies clock influence on mood. Curr Biol 18:678–683
199. Fu L, Pelicano H, Liu J, Huang P, Lee C (2002) The circadian gene Period2 plays an important role in tumor suppression and DNA damage response in vivo. Cell 111:41–50
200. Spanagel R, Pendyala G, Abarca C, Zghoul T, Sanchis-Segura C, Magnone MC, Lascorz J, Depner M, Holzberg D, Soyka M et al (2005) The clock gene Per2 influences the glutamatergic system and modulates alcohol consumption. Nat Med 11:35–42
201. Wuarin J, Schibler U (1990) Expression of the liver-enriched transcriptional activator protein DBP follows a stringent circadian rhythm. Cell 63:1257–1266
202. Lopez-Molina L, Conquet F, Dubois-Dauphin M, Schibler U (1997) The DBP gene is expressed according to a circadian rhythm in the suprachiasmatic nucleus and influences circadian behavior. EMBO J 16:6762–6771
203. Ripperger JA, Shearman LP, Reppert SM, Schibler U (2000) CLOCK, an essential pacemaker component, controls expression of the circadian transcription factor DBP. Genes Dev 14:679–689

204. Gachon F, Fonjallaz P, Damiola F, Gos P, Kodama T, Zakany J, Duboule D, Petit B, Tafti M, Schibler U (2004) The loss of circadian PAR bZip transcription factors results in epilepsy. Genes Dev 18:1397–1412
205. Gachon F, Olela FF, Schaad O, Descombes P, Schibler U (2006) The circadian PAR-domain basic leucine zipper transcription factors DBP, TEF, and HLF modulate basal and inducible xenobiotic detoxification. Cell Metab 4:25–36
206. Bozek K, Kielbasa SM, Kramer A, Herzel H (2007) Promoter analysis of Mammalian clock controlled genes. Genome Inform 18:65–74
207. Nakahata Y, Yoshida M, Takano A, Soma H, Yamamoto T, Yasuda A, Nakatsu T, Takumi T (2008) A direct repeat of E-box-like elements is required for cell-autonomous circadian rhythm of clock genes. BMC Mol Biol 9:1
208. Paquet ER, Rey G, Naef F (2008) Modeling an evolutionary conserved circadian cis-element. PLoS Comput Biol 4:e38
209. Kumaki Y, Ukai-Tadenuma M, Uno KD, Nishio J, Masumoto KH, Nagano M, Komori T, Shigeyoshi Y, Hogenesch JB, Ueda HR (2008) Analysis and synthesis of high-amplitude Cis-elements in the mammalian circadian clock. Proc Natl Acad Sci USA 105:14946–14951
210. Noshiro M, Usui E, Kawamoto T, Kubo H, Fujimoto K, Furukawa M, Honma S, Makishima M, Honma K, Kato Y (2007) Multiple mechanisms regulate circadian expression of the gene for cholesterol 7alpha-hydroxylase (Cyp7a), a key enzyme in hepatic bile acid biosynthesis. J Biol Rhythms 22:299–311
211. Hughes M, Deharo L, Pulivarthy SR, Gu J, Hayes K, Panda S, Hogenesch JB (2007) High-resolution time course analysis of gene expression from pituitary. Cold Spring Harb Symp Quant Biol 72:381–386
212. Storch KF, Lipan O, Leykin I, Viswanathan N, Davis FC, Wong WH, Weitz CJ (2002) Extensive and divergent circadian gene expression in liver and heart. Nature 417:78–83
213. Miller BH, McDearmon EL, Panda S, Hayes KR, Zhang J, Andrews JL, Antoch MP, Walker JR, Esser KA, Hogenesch JB et al (2007) Circadian and CLOCK-controlled regulation of the mouse transcriptome and cell proliferation. Proc Natl Acad Sci USA 104:3342–3347
214. Panda S, Antoch MP, Miller BH, Su AI, Schook AB, Straume M, Schultz PG, Kay SA, Takahashi JS, Hogenesch JB (2002) Coordinated transcription of key pathways in the mouse by the circadian clock. Cell 109:307–320
215. Akhtar RA, Reddy AB, Maywood ES, Clayton JD, King VM, Smith AG, Gant TW, Hastings MH, Kyriacou CP (2002) Circadian cycling of the mouse liver transcriptome, as revealed by cDNA microarray, is driven by the suprachiasmatic nucleus. Curr Biol 12:540–550

Chapter 3
Posttranslational Regulation of Circadian Clocks

Jens T. Vanselow and Achim Kramer

3.1 Introduction

Circadian clocks are biological timing systems with a period of about 24 h. In many organisms they represent internally the external time of our 24 h world. They regulate a variety of processes ranging from gene expression to behavior and thereby enable organisms to anticipate daily recurring changes in their environment. Although circadian clocks are widespread among organisms as diverse as cyanobacteria, plants, *Neurospora*, *Drosophila*, and mammals, fundamental mechanisms essential for a functional clock are highly conserved (see Chap. 7). All circadian clocks investigated so far are based on cell-autonomous molecular oscillators composed of interlocked transcriptional-translational feedback loops [1].

The central mechanism of the mammalian circadian clock is a delayed negative transcriptional-translational feedback loop (see Chap. 2). The transcription factors CLOCK (in the SCN alternatively NPAS2 [2]) and BMAL1, containing bHLH (basic helix-loop-helix) and PAS (Per-Arnt-Sim) domains, heterodimerize and transactivate expression of clock genes, e.g. *Period* (*Per1*, *Per2* and *Per3*) and *Cryptochromes* (*Cry1* and *Cry2*), by binding to E-box enhancer elements in their promoter regions [3]. PER and CRY proteins form large multiprotein complexes, including Casein Kinase 1 isoforms ε and δ (CK1ε/δ) [4, 5] and inhibit their own gene expression after a delay of several hours by entering the nucleus and directly interfering with the CLOCK-BMAL1 heterodimer. The exact composition of this "repressor complex" is not yet known, but multiple additional proteins have been suggested to contribute to the periodic inhibition of CLOCK-BMAL1 activity. These include mammalian Timeless [6, 7], the bHLH transcription factors DEC1/2 [8], the PER1-associated proteins NONO and WDR5 [9] as well as the CLOCK-interacting protein CIPC [10].

J.T. Vanselow
Center for Experimental BioInformatics (CEBI), University of Southern Denmark, Odense, Denmark

A. Kramer (✉)
Laboratory of Chronobiology, Charité Universitätsmedizin, Berlin, Germany
e-mail: achim.kramer@charite.de

U. Albrecht (ed.), *The Circadian Clock*, Protein Reviews 12,
DOI 10.1007/978-1-4419-1262-6_3, © Springer Science+Business Media, LLC 2010

Proteasomal degradation of PER proteins upon phosphorylation by CK1ε/δ probably releases the inhibition of CLOCK-BMAL, thereby closing the negative feedback loop, and a new cycle with expression of *Per* and *Cry* begins. It has been recognized, that CRY proteins inhibit the CLOCK-BMAL1 mediated transactivation to a much stronger extent [11, 12] than PER proteins, which rather seem to have a regulatory function. PER levels are presumably limiting the formation of PER-CRY complexes and their translocation to the nucleus, respectively. In mice which express no functional PER1 and PER2, no nuclear CRY could be detected [5]. The total amount of CRYs in a cell oscillates only weakly on a relatively high abundance level, whereas PER protein levels show a strong circadian rhythm. In the nucleus, however, both CRY and PER proteins are highly abundant between CT15 and CT18 [5], suggesting that their abundance as well as their subcellular localization are tightly regulated by posttranslational mechanisms.

Posttranslational modifications (PTMs) of several core clock proteins have been recognized as essential for establishing and maintaining circadian rhythms [13, 14]. Those modifications include but are not limited to protein phosphorylation, acetylation, ubiquitination, and sumoylation (see Fig. 3.1). While posttranslational regulation is also important for signal transduction processes in input and output pathways to and from the molecular clockwork, this review focuses on posttranslational regulations in the core circadian clockwork.

3.2 Protein Phosphorylation in the Negative Feedback Loop

Protein phosphorylation is the most widely studied PTM. It is involved in presumably all signaling pathways and essential processes within eukaryotic cells [15] including proteasomal degradation, subcellular localization, complex formation, and protein activity.

There are more than 500 protein kinases encoded in the human and mouse genome, catalyzing the reversible phosphorylation of proteins [16, 17]. Most kinases differ in their substrate specificity, regarding both the phosphoacceptor

Fig. 3.1 (continued) phorylation sites at PER1/2 are present and control subcellular localization and stability of the proteins (3.4). CK1ε/δ itself possesses inhibitory autophosphorylation activity (3.3), which is countered by PP5 (3.7.1). Interestingly, CRY1/2 binding to PP5 inhibits the dephosphorylation of CK1ε/δ, thereby CK1ε/δ activity is reduced. GSK3β phosphorylates CRY1/2 proteins (3.6.2), which leads to their ubiquitination and proteasomal degradation. The F-box protein FBXL3 has been recognized as important for the ubiquitination of CRY1/2 (3.5). Phosphorylation of CLOCK and BMAL1 is coincident with their dimerization and their highest transcriptional activity (3.8.1). BMAL1 is sumoylated (S) and ubiquitinated, which facilitates its proteasomal degradation (3.8.2). Furthermore, BMAL1 and PER2 are acetylated, which increases their stability. In addition, BMAL1 acetylation (probably mediated by histone acetyltransferase activity of CLOCK) seems to be important for the repressor activity of CRY1 (3.8.3). The histone deacetylase SIRT1 binds to CLOCK, BMAL1 and PER2 in a circadian manner and deacetylates PER2 and BMAL1

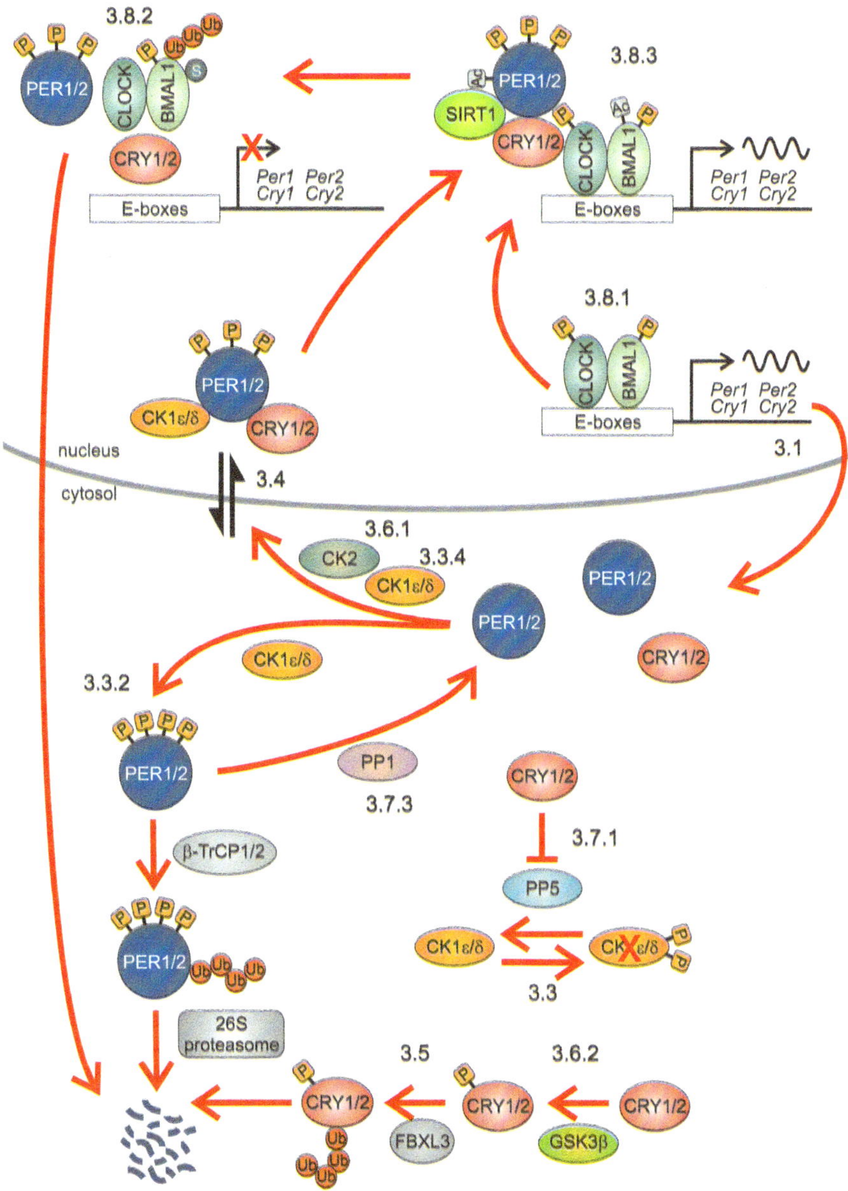

Fig. 3.1 Simplified schematic representation of posttranslational mechanisms in the mammalian circadian clock, particularly in the negative feedback loop. After CLOCK/BMAL1 mediated transcription (3.1 – numbers in the figure refer to the according section in the main text), PER1/2 proteins are accumulating in the cytoplasm. PER1/2 are phosphorylated (P) by CK1ε/δ, which triggers its subcellular localization (3.3.4) as well as its ubiquitination (Ub) and proteasomal degradation (3.3.2). PER1/2 ubiquitination is facilitated by β-TrCP1/2, an F-box protein of the E3 ubiquitin ligase complex. PP1 dephosphorylates and stabilizes PER2 (3.7.3). CK2 is also involved in controlling the nuclear localization of PER1/2 (3.6.1). Importantly, functional different phos

amino acid (mainly serine/threonine kinases or tyrosine kinases) and the context of the phosphoacceptor (consensus sequence).

Protein phosphorylation is made reversible by phosphatases. They can be classified by their substrate specificity in serine/threonine, tyrosine, or dual specific phosphatases. Whereas about 100 tyrosine phosphatases are known [18], the number of catalytic subunits of serine/threonine phosphatases is much lower [19]. Many of them are multimeric protein complexes and the combination with a variety of regulatory subunits mediates their substrate specificity and probably results in a similar number of serine/threonine phosphatase holoenzymes compared to kinases [20].

Whether the fundamental principle of circadian oscillations might be based mainly on rhythmic posttranslational mechanisms rather than on *transcriptional*-translational feedback loops [21] is currently under discussion. Recent results from cyanobacteria showed that a purely posttranslational mechanism based on rhythmic phosphorylation and dephosphorylation of the hexameric KaiC protein is sufficient to generate and sustain circadian rhythms at least in vitro [22, 23]. In addition, in flies and mammals there is a huge body of evidence demonstrating the particular importance of posttranslational mechanisms. Most core clock proteins are phosphorylated in vivo and show a circadian phosphorylation pattern – for the first time reported by Edery et al. for *Drosophila* PER in 1994 [24]. It was not until 2001, that Lee et al. showed that the mammalian clock proteins PER1, PER2, PER3, CLOCK, BMAL1 and possibly also CRY1 and CRY2 are phosphoproteins in vivo with a circadian regulation of their phosphorylation pattern [5].

Protein phosphorylation plays an important role for the regulation of proteasomal degradation of clock proteins. This is probably of crucial importance for the negative feedback loop. Although slowing the (premature) accumulation of repressors like PERs may contribute to the delay in the feedback loop, degradation of the repressors at the end of the cycle allows the circadian cycle to progress in the first place. Posttranslational mechanisms might also explain why constitutive expression of *dPer* in *per*^{0} flies rescues arrhythmic behavior and leads to rhythmic PER protein levels [25]. Similarly, mammalian PER protein abundance is rhythmic, even when mammalian *Per* genes are constitutively expressed in synchronized NIH3T3 or rat-1 fibroblasts, respectively [26, 27].

In conclusion, the observed rhythmic mPER protein abundance in vivo might be generated by a combination of rhythmic transcription of *mPer* genes and posttranslational events. At least in the case of flies, the posttranslational mechanisms seem to be sufficient for behavioral circadian rhythmicity.

3.3 Casein Kinase 1ε/δ: An Essential Regulator of the Negative Feedback Loop

In 1998, the *Drosophila* kinase DOUBLETIME (DBT) – an orthologue of mammalian CK1ε/δ [28–30] – was the first enzyme shown to be essential for the circadian clock mechanism. The CK1 family belongs to the serine/threonine kinases. So far, seven CK1 isoforms (α, β, χ, δ, ε, φ, and γ) have been found in mammals. They are involved in the regulation of many cellular processes, such as tumor suppression, Wnt-signaling, cell cycle progression, and apoptosis [31, 32]. Furthermore, splice variants of CK1ε have been found, which differ in their tissue specific expression and kinase activity in vitro [33]. All CK1 family members possess a highly conserved kinase domain, but differ strongly in their short N- and variable C-termini, both of which have regulatory functions [32].

Both CK1ε and CK1δ play a major role in the mammalian circadian oscillator [34], as will be described in the following sections. They are largely identical in their N-termini and their catalytic domains and both contain inhibitory autophosphorylation sites in their C-termini. Therefore it is widely assumed, that CK1ε and CK1δ have redundant or complementary functions in the molecular clockwork.

One role of CK1ε/δ is the promotion of proteasomal degradation of PER proteins presumably by activating recognition sites for β-TRCP1/2, a part of the E3 ubiquitin ligase complex; another role seems to be the regulation of nuclear localization of the PER/CRY complex. Thus, CK1ε/δ is involved in the temporal and spatial coordination of PER proteins, and this directly influences the negative feedback loop. When the sophisticated balance of components within the negative feedback loop is perturbed, e.g. by altered phosphorylation, changes in circadian dynamics are likely to occur.

3.3.1 Impact of DBT Mutations on Circadian Behavior of Flies

In *Drosophila*, mutations in the active centre of DBT lead to short (about 18 h, dbt^S) and long (about 27 h, dbt^L) behavioral phenotypes, respectively [28, 29]. DBT phosphorylates dPER and influences its stability and subcellular distribution. Phosphorylation of dPER by DBT induces its proteasomal degradation [35]. Furthermore, DBT is also found in complexes together with dPER and dTIM [36].

Interestingly, although both DBT^S and DBT^L have reduced kinase activities, they have opposite effects on the circadian period [37]. These mutations do not affect the binding capability towards dPER. Rather in dbt^S mutant flies, the nuclear accumulation dynamics of dPER is altered [38], whereas in dbt^L phosphorylation-dependent dPER turnover is decreased [29, 39].

In addition, without a functional DBT (dbt^P-larvae), dPER is hypophosphorylated, dPER levels are constitutively high and dPER and dTIM are not oscillating any more [28, 29], presumably due to the necessity of DBT for initiating the proteasomal degradation of dPER [35, 40]. The reduced function dbt^{AR} allele leads to arrhythmic behavior of flies and also to constitutively high dPER levels [39]. Moreover, additional DBT variants have been described, resulting in long behavioral phenotypes and altered *dPer* and *dTim* transcript and protein profiles [41].

3.3.2 CK1ε/δ and F-box Proteins Mediate Proteasomal PER Degradation

In *Drosophila*, hyperphosphorylated dPER is recognized by the F-box/WD40-repeat protein Slimb and subsequently ubiquitinated and degraded by the proteasome [35, 40]. F-box proteins are components of SCF-type E3 ubiquitin ligase complexes and mediate the substrate specificity of the ubiquitin ligase complex, eventually tagging proteins for degradation via the ubiquitin-proteasomal pathway. Slimb mutant flies are behaviorally arrhythmic, underlining the importance of time-specific degradation of dPER proteins for the generation and maintenance of circadian rhythms.

Interestingly, it is not the overall phosphorylation state of dPER but rather the phosphorylation on specific sites that seems to determine its stability. Recognition by Slimb is triggered by DBT mediated phosphorylation of Ser47 on dPER [42]. In flies expressing dPER S47A – a variant deficient in phosphorylation at this site – the stability of dPER is increased, which results in locomotor activity rhythms with lengthened period (~31 h in the mutant compared to ~24 h in the wild-type). The opposite effect is observed when constitutive phosphorylation is mimicked in dPER (by mutating Ser47 to aspartic acid): the protein is less stable and the period of behavioral rhythms gets shorter (~22 h vs. ~24 h wt).

In the mammalian circadian system corresponding investigations showed, that also mammalian PER proteins are degraded by a similar mechanism, which involves even homologous proteins. Phosphorylation of mammalian PER proteins by CK1δ and CK1ε marks them for proteasomal degradation and thereby reduces their stability [43–45]. The Slimb homologues β-TrCP1 and β-TrCP2 are F-box proteins and components of E3 ubiquitin ligase complexes. They recognize specific sequences within substrate proteins and for both mPER2 and mPER1, a β-TrCP1/2 binding motif (a phosphodegron) has been proposed, which has to be activated by phosphorylation [46, 47]. When endogenous β-TrCP1 and β-TrCP2 are downregulated by RNAi in synchronized tissue culture cells, the period of circadian rhythms is substantially lengthened [48, 49]. Furthermore, the expression of β-TrCP1/2 interaction-deficient mPER2 in synchronized fibroblasts leads to a dramatic stabilization of the PER2 protein and also to a disruption of circadian rhythmicity [48, 50].

3.3.3 CK1ε tau Mutation in Hamster

The first heritable circadian phenotype in mammals was discovered in 1988 [51]. Syrian hamsters bearing this mutation show a shortened period in wheel running: compared to ~24 h in wild-type animals, heterozygous mutants have a ~22 h period and homozygotes a ~20 h period. Therefore this mutation was called *tau* (the Greek symbol τ is used for period length). It took more than a decade to identify a mutation in the *CK1ε* gene that is responsible for the *tau* phenotype [52]. A single point mutation in the coding region of the *CK1ε* gene results in an exchange from alanine to cysteine at position 178, which is located in the phosphate recognition site of the kinase.

In vitro, CK1ε-*tau* has a reduced activity towards PER proteins as well as towards canonical CK1ε substrates, while the binding affinity to mPER1 and mPER2 is not altered [52]. Because of the decreased in vitro kinase activity of CK1ε-*tau*, it was initially proposed that major substrates of CK1ε, i.e. the PER proteins, are hypophosphorylated and therefore less affected by proteasomal degradation (see above). It has been assumed that this leads to an earlier repression of the CLOCK-BMAL1 transactivation at E-boxes in the *tau* mutants relative to wild-type animals. This explanation was concordant with the finding that *Per1* mRNA in the SCN of *tau* homozygotes was both reduced and phase-advanced [52]. However, the duration of the negative feedback was not prolonged as would be expected from a slower PER degradation. Without analyzing the PER protein itself in vivo it was concluded, that the *tau* mutation mainly reduces the delay of PER protein accumulation in the early part of the circadian cycle resulting in an advanced negative feedback. However, the *tau* mutant hamsters exhibit in vivo robust circadian rhythms in both PER1 and PER2 protein abundance and phosphorylation. Furthermore, no substantial difference in the overall phosphorylation state was observed, although the total amount of PER1 and PER2 protein was reduced [5].

The molecular mechanism of *tau*, which leads to the strong period shortening in vivo, is still under discussion. One possible explanation is based on a simplified mathematical model of the negative feedback loop, which claimed an *increased* kinase activity (gain-of-function) in vivo for CK1ε-*tau* specifically on PER proteins as the only possible explanation for the short period phenotype [53]. Gallego et al. indeed found a higher destabilizing effect of CK1ε-*tau* on PER1 and PER2 in cell culture. To further elucidate the mechanism behind the *tau* phenotype, Meng et al. recently generated CK1ε-null and CK1ε-*tau* mice [54]. The locomotor activity rhythms of wild-type ($CK1\varepsilon^{+/+}$) hetero- and homozygous knock-out ($CK1\varepsilon^{+/-}$ and $CK1\varepsilon^{-/-}$) and hetero-, homo- and hemizygous *tau* mutated ($CK1\varepsilon^{+/tau}$, $CK1\varepsilon^{tau/tau}$ and $CK1\varepsilon^{tau/-}$) mice were studied. As in *tau* hamsters, the CK1ε–*tau* mutation in mice has a dominant effect on period length resulting dose dependently in a period shortening to about 22 h in heterozygous $CK1\varepsilon^{+/tau}$ and about 20 h in homozygous $CK1\varepsilon^{tau/tau}$ animals. Interestingly, there was no difference in wheel running period between heterozygous and hemizygous $CK1\varepsilon^{tau/-}$ animals. Furthermore, reducing gene dosage of CK1ε-wt in $CK1\varepsilon^{+/-}$ and $CK1\varepsilon^{-/-}$ mice had no effect on period length. Thus, the lack of CK1ε may be compensated by another kinase, probably by CK1δ.

Although these results show that CK1ε-*tau* is a dominant mutation with a dose dependent mode of action, these results do not unambiguously prove that CK1ε-*tau* is a gain-of-function mutation. A gain-of-function mutation in the case of a kinase would mean, that the mutant form phosphorylates more substrate sequences or with an increased reaction rate. However, rather the opposite has been shown in vitro for several substrates of CK1ε including PER2. Furthermore the binding affinity of CK1ε-*tau* towards PER proteins has been shown to be not altered [52].

In conclusion, it is still unclear, whether the observed increased proteasomal degradation of PERs in *tau* animal is based on an increased (i.e. gain-of-function) kinase activity of CK1ε-*tau*. In analogy to a model describing the molecular background of FASPS (see below), we assume that different phosphorylation sites on PER proteins either stabilize or destabilize the protein when phosphorylated and that the *tau* mutated CK1ε/δ might have a reduced ability to phosphorylate stabilizing phosphorylation sites on PER proteins [55].

How could an altered (possibly decreased) overall activity of CK1ε lead to increased proteasomal degradation? Dey et al. reported an accelerated nuclear clearance of PER proteins in the SCN of *tau* hamsters [56]. This resulted in an advanced release of transcriptional inhibition and an earlier rise of e.g. *Per1/2* mRNA in the *next* circadian cycle, which in the end could explain the shortened free-running period of the mutant animals. An accelerated nuclear export of PER proteins might also be the reason for increased proteasomal degradation, given that PER proteins are degraded mainly in the cytoplasm. This mechanism is supported by the finding, that CK1ε/δ activity regulates other functions of mammalian PER proteins in addition to marking them for degradation, such as control of their subcellular localization (see below).

In cell culture, the stability of overexpressed PER2 is strongly decreased by coexpression of wild-type CK1ε and to an even higher extent by CK1ε-*tau* [55]. Interestingly, Meng et al. reported that overexpressed PER2 trapped in the nucleus of tissue culture cells (e.g. by treating the cells with the nuclear export inhibitor leptomycine B (LMB)), partly rescues PER2 from CK1ε–mediated degradation, but not from the increased destabilization triggered by CK1ε-*tau* [54]. On the basis of this, they concluded, that CK1ε-*tau* destabilizes PER2 regardless of its subcellular localization. However, in experimental conditions that are closer to the in vivo situation, this conclusion does not completely hold true. For example, the authors find that in primary lung fibroblasts LMB treatment leads to PER2 stabilization in both wild-type as well as CK1ε-*tau* mice, whereas in the SCN, LMB has not even a stabilizing effect on PER2 in wild-type animals. In conclusion, the exact mode of CK1ε-*tau* action seems still to be elusive.

3.3.4 Control of Subcellular Localization of PER Proteins

The timing of nuclear localization of the PER/CRY complex is important for determining the circadian period length. The subcellular localization of clock proteins in mammals was studied mostly by overexpressing the respective proteins, which often leads to inconclusive and arguable results, may be because of masking by the strength of overexpression. Nevertheless, data from many groups hint towards the importance of subcellular localization for oscillator properties. This is exemplified in a mouse model where overexpression of rPER2 with a deleted "nuclear localizing domain" (NLD) leads to delayed nuclear entry of rPER2(-NLD) in the SCN and to a longer behavioral period compared to overexpression of wild-type rPER2, which results in shorter period phenotype [57]. The nuclear localization of PER proteins is – at least in part – controlled by CK1ε/δ as we have discussed above for the observed accelerated nuclear clearance of PER proteins in the SCN of *tau* hamsters [56].

Nuclear localization signals found in mPER1 and mPER3 are masked by CK1δ/ε-mediated phosphorylation, which results in arrested nuclear translocation [45, 58]. Treatment of cells with the specific CK1δ/ε inhibitor CKI-7 stabilizes hPER1 and prolongs its nuclear retention [59]. The opposite was observed in other studies, where mPER1 and rCK1ε were localized in the cytoplasm when expressed individually and predominantly in the nucleus when coexpressed together [60]. A phosphorylation cluster in mPER1 which is phosphorylated by rCK1ε (and which is not conserved in mPER2) is responsible for its nuclear import [61]. The confusing differences in the subcellular localization of PER proteins reported in these studies might be explained by different expression levels and/or different cell types used. Nevertheless, in cell extracts from murine liver, mCKIε/δ as well as mPER1 are located both in the nucleus and in the cytoplasm, whereas mPER2 was found to be almost exclusively located in the nucleus [5].

3.4 FASPS: Differential Effects of PER2 Phosphorylation

In the *tau* hamster, a protein kinase with a potentially very high number of substrates, is mutated. For a deeper understanding of the molecular oscillator, a severe human circadian phenotype called "familial advanced sleep-phase syndrome" (FASPS) was of particular interest, because at least one type of FASPS is associated with the mutation of a single phosphoacceptor site in hPER2, i.e. only a single kinase substrate is affected.

FASPS is an autosomal inherited human sleep disorder. Affected individuals are – unlike their unaffected siblings – extreme early birds ("larks") with a phase advance of about 4–5 h regarding sleep onset and offset, core body temperature, and hormone rhythms (e.g. melatonin). Furthermore, the endogenous period has been measured for one affected individual and indicates a period shortening of about 1 h compared to the average population [62, 63]. Two molecular causes of FASPS have been described so far, linking a human behavioral phenotype to disturbed PTMs of clock proteins. FASPS is associated with a mutation either in human CK1δ [64] or with a mutation of a phosphoacceptor amino acid in hPER2 [63, 65]. The mutation in the *hCK1δ* gene results in an amino acid exchange (T44A) and – similar to the above described CK1ε-*tau*- and DBT-mutants – in a reduced enzymatic activity of the kinase [64]. The opposite situation has been described for a mutation in the *hCK1ε* gene leading to the *delayed* sleep phase syndrome (DSPS). Here, the mutation S408N resides in the autoregulatory C-terminus of the kinase and due to the missing phosphorylation site, the kinase is presumably less (auto-)inhibited and therefore more active [66].

The mutation in hPER2 resulting in FASPS is mechanistically very interesting because here a single putative phosphorylation site of a specific kinase substrate is mutated. In hPER2-FASPS, serine 662 located in the CK1δ/ε-binding region of hPER2 is mutated to glycine and in vitro, hPER2-FASPS is less phosphorylated by CK1ε [65]. Serine 662 resides directly N-terminal to a cluster of five serine and threonine residues (659-KAESVASLTSQCSYSSTIVH-678), which form CK1ε/δ substrate consensus motifs (pS/pT-X_{1-2}-S/T), only when the C-terminal serine or threonine is phosphorylated [32, 67, 68]. Indeed, it has been shown that peptides from the FASPS region are less phosphorylated by CK1ε in vitro when the FASPS site has not been phosphorylated beforehand. This suggests that a priming phosphorylation at serine 662 (possibly by a yet unknown kinase) is necessary [69]. Therefore it was initially assumed that the lack of this priming phosphorylation site in hPER2-S662G might prevent subsequent phosphorylations by CK1ε/δ at downstream sites [45]. On the basis of this hypothesis, it has been proposed, that inadequate phosphorylation of FASPS-hPER2 might attenuate its CK1ε/δ-mediated degradation and thus accelerate its autorepression, leading to phase-advanced molecular rhythms [65]. An alternative hypothesis was suggested later by Vanselow et al. [55]. An altered phosphorylation profile of hPER2 could affect the timing of nuclear localization of hPER2 similar to what has been described in *dbt* mutants [38] or the *tau* mutant hamster [56].

3.4.1 Identification of PER2 Phosphorylation Sites

Until recently, protein phosphorylation was investigated mainly by looking at the overall phosphorylation state of a protein. Thereby, the positional information, i.e. at which particular amino acid a protein is modified was neglected. Advances in peptide mass spectrometry in the last years made it possible to identify PTMs of proteins in a high resolution, e.g. identifying phosphorylated residues of proteins. To map PER2 phosphorylation sites, we studied mPER2 phosphorylated in vitro by recombinant candidate kinases or expressed in HEK293 cells relying on phosphorylation by endogenous or cotransfected kinases [55, 70, 71]. For comprehensive identification of mPER2 phosphorylation sites, the epitope-tagged protein was immunoprecipitated and digested in-gel by four different proteases. This resulted in a large variety of peptides covering most of the protein sequence. To reduce complexity, phosphopeptides were enriched using metal-affinity purification on TiO_2-beads before peptides were separated and analyzed by nanoliter flow liquid chromatography coupled to tandem mass spectrometry (nanoLC-MS/MS). In these experiments, mPER2 was found to be phosphorylated by endogenous kinases in cell culture on 21 residues (out of 275 potential phosphoacceptor amino acids). Many of the endogenously phosphorylated sites were also phosphorylated in vitro by CK1δ, CK2 and/or GSK3β [49].

Among the residues phosphorylated by endogenous kinases we also found the serine residue, which is mutated in hPER2-FASPS [55]. In these experiments also the first serine in the following phosphoacceptor site cluster was detected to be phosphorylated. The FASPS phenotype therefore relies on the loss of a functional phosphorylation site.

3.4.2 *Elucidating the Molecular Basis of FASPS*

Several fibroblast cell lines, e.g. rat-1 or NIH3T3, show self-sustained circadian oscillations over several days resembling those observed in the SCN or in peripheral tissues [72–74]. To study the effect of PER2 phosphorylation site mutants – and in particular of the FASPS mutation – for circadian dynamics in a comparable manner, we established a reductionist model system [55]. It is based on the expression of mPER2 phosphorylation site mutants in synchronized fibroblasts from the same chromosomal location in the fibroblast genome.

Cells expressing mPER2-FASPS have a phenotype that resembles that of FASPS patients, i.e. they show an early phase of entrainment and a short free running period compared to wild-type mPER2 expressing cells. How does the phosphorylation at the FASPS position and most likely also at the following phosphoacceptor sites affect the dynamics of the circadian oscillator? PER2 stability and/or subcellular localization could be altered, which would in turn affect the timing and duration of feedback inhibition. Another possibility, a direct effect of the FASPS mutation on the ability of PER2 to inhibit CLOCK-BMAL1 transactivation, has not been found [55].

3.4.3 *Phosphorylation at the FASPS Position Antagonizes CK1ε/δ Mediated Destabilization of PER2*

While phosphorylation of PER2 was mostly associated with its proteasomal degradation, we unexpectedly found that the stability of PER2-FASPS is not increased but – in contrast – is substantially reduced. In our cell culture model PER2-wt half-life is about 3 h, whereas the FASPS mutation reduced it to about 1.5 h. When all phosphoacceptor sites in the cluster are mutated to alanine, the destabilization is even more pronounced.

Coexpression of PER2 with CK1ε leads to its rapid proteasomal degradation [43, 44, 46] and is even more increased for the FASPS mutant PER2 [55]. Notably, similar observations were made when analyzing the corresponding sites in PER1 [61, 75]. Thus, phosphorylation of PER proteins at the FASPS position and the following Ser/Thr-residues leads to a stabilization of the protein, whereas phosphorylation at other residues promotes proteasomal degradation. Thus phosphorylation of PER proteins can have different, even opposing functions [13].

3.4.4 Phosphorylation at the FASPS Position Controls Subcellular Localization of PER2

How does phosphorylation in the FASPS region stabilize PER2? One possibility is that the observed stabilization is an indirect consequence of an altered subcellular localization of PER2. While the nuclear import rate of PER2-FASPS is not changed, the nuclear clearance of the PER2-FASPS protein is accelerated [55] similar to what is reported for PER1 and PER2 in the SCN of *tau* mutant hamsters [56]. Thus, phosphorylation in the FASPS region increases the nuclear retention of PER2 which in turn leads to a stabilization of the protein – presumably by protecting it from proteasomal degradation in the cytoplasm (see above). Conversely, FASPS mutant PER2 protein would be exported earlier from the nucleus and thus prematurely degraded resulting in an earlier release of CLOCK-BMAL1 inhibition and a phase advance of the next circadian cycle.

The findings of our cell culture model were in part supported by a study of Xu et al. [69]. They generated transgenic mice, which express hPER2-FASPS in wild-type or *mPer2* knockout animals. These animals show phenotypes very similar to FASPS patients. The protein levels of PER2-FASPS (especially in the nuclear fraction) in the liver of these transgenic mice as well as in fibroblasts from skin biopsies of FASPS patients were found to be reduced. In agreement with our study, this indicates a decreased stability of the PER2-FASPS protein. However, Xu et al. did not find a stabilization of hPER2-FASPS upon treatment of the fibroblasts with proteasomal inhibitor. Therefore they excluded the possibility of an altered stability of PER2-FASPS as the primary consequence of the mutation. Instead, they found a reduction of the *Per2* mRNA in FASPS and concluded that phosphorylation at the FASPS position has an impact on *Per2* transcription, a result we did not observe in our cell culture model.

Furthermore, modulating *CK1δ* gene dosage in wild-type animals had no effect on the free-running period τ, suggesting that the overall activity of CK1ε/δ is compensated, possibly by the combined action or controlled total activity of both CK1δ and CK1ε. Heterozygous CK1ε knockout-mice do not show altered wheel-running period and even homozygous CK1ε knockouts have only a slightly longer period than their wild-type littermates [54].

However, in PER2 mutant background, Xu et al. found that altering the *CK1δ* gene dosage does have an effect on behavioral rhythms [69], further supporting the model of functionally different phosphorylation sites on PER proteins and the involvement of CK1δ in phosphorylating positions other than FASPS position on the PER2 protein. Decreasing the *CK1δ* gene dosage of mice expressing either the hPER2 FASPS protein (phosphoacceptor loss S659G) or a hPER2-S659D variant mimicking phosphorylation at the FASPS position lengthened τ in both cases. An increase of *CK1δ* gene dosage further shortened the already short τ of hPER2-FASPS animals. Hence, while in the wild-type situation, overexpression of CK1δ does not seem to affect period length because of a fine balance and each compensating the shortening and lengthening processes of the other, in FASPS a period lengthening component of the circadian clockwork is seemingly lost, therefore CK1δ overexpression leads to an even shorter τ. This indicates a fine tuned balance between kinase and substrate to be necessary for proper circadian dynamics.

3.5 Posttranslational Regulation of Cryptochromes

Although there are numerous studies investigating the mechanisms of PER1/2 regulation, much less is known about the posttranslational regulation of Cryptochromes. CRY1/2 levels are much higher compared to PERs and do not show as high an amplitude oscillation as PER1/2 [5]. Therefore PER proteins are thought to act as scaffold proteins for the control of the inhibitory action of CRYs within the negative feedback loop.

Nevertheless, like the PER proteins, some properties of CRY1/2 are also at least in part regulated by protein phosphorylation. The proteasomal degradation of mCRY2 is regulated by the action of GSK3β [76]. Recently, the F-box protein FBXL3 has been reported to be involved in the proteasomal degradation of CRY proteins. Hypomorphic mouse mutations of *Fbxl3* as well as downregulating *Fbxl3* in cell culture lengthened the circadian period [49, 77–79]. Whether CRY requires PTM to bind FBXL3 is currently unknown. The activity of CRY1 and CRY2 as transcriptional repressors is regulated by MAPK and single phosphorylation sites have been identified which modulate this activity [80]. mCRY1, mCRY2, CK1ε, and PP5 can bind to each other and at least mCRY2 interaction with PP5 counteracts CK1ε activation by PP5 [81].

3.6 Other Kinases in the Circadian Clock

Although CK1ε/δ is an essential component of the molecular clockworks in many organisms, other kinases also seem to be important for the mammalian circadian oscillator.

3.6.1 Casein Kinase 2

CK2 is a tetrameric kinase, composed of two regulatory (α and α') and two catalytic subunits (β). In *Drosophila*, CK2 phosphorylates dPER and regulates its nuclear localization and also possibly the repression of dCLK-dCYK by dPER [82]. Flies with mutations in CK2α or CK2β subunits have a lengthened period and a delayed nuclear entry of dPER and dTIM [83, 84]. Although CK2 is conserved in many species and phosphorylate clock components in *Neurospora crassa*, *Arabidopsis thaliana*, and *Drosophila melanogaster*, we recently elucidated a role for CK2 in the mammalian circadian oscillator [49]. Disruption of circadian rhythms or period lengthening is observed, when CK2 subunits are downregulated by RNAi or when CK2 is blocked pharmacologically, whereas overexpression of CK2 results in period shortening. One molecular target of CK2 is PER2, which is phosphorylated by CK2 in vitro. Mutation of N-terminal phosphorylation sites of PER2, which are phosphorylated in living cells and in vitro only by CK2 but not by CK1δ or GSK3β leads to destabilization of PER2 and also to period lengthening. CK2 seems to stabilize PER2 in a phase specific manner, which is probably required for nuclear accumulation and thus, normal circadian rhythms [49].

3.6.2 GSK3β in the Mammalian Clock and Depression

Glycogen synthase kinase 3β (GSK3β) is involved in several signal transduction pathways (e.g. together with CK1ε in Wnt/β-catenin signaling) and many other cellular processes [85]. GSK3β is a target of lithium mimetica, substances, which since several decades have been used for treatment of depressions and bipolar disorders (see Chap. 10). Symptoms of specific forms of depression such as "seasonal affective disorder" (SAD) but also bipolar disorders are alleviated by light therapy or by sleep deprivation [86]. The activity of GSK3β is determined by phosphorylation of serine 9, which shows a circadian pattern in the SCN and liver of mice [76, 87]. Lithium and other inhibitors of GSK3β lengthen circadian period and delay the phase of *mPer2* transcription, respectively [88–90]. In contrast, it has recently been shown that inhibition of GSK3β by small molecules or its downregulation by siRNA results in a period shortening of circadian rhythms in cell culture [91].

In *Drosophila*, overexpression of GSK3β orthologue SHAGGY results in a shorter period, while reduction of kinase activity lengthens it [92]. Molecular targets of GSK3β in the mammalian clockwork are PER2 (which also interacts with the kinase), CRY2 as well as REV-ERBα and in *Drosophila* dTIM. GSK3β is involved in the nuclear localization of rPER2 [87] and the dPER/dTIM complex [92] and also in controlling proteasomal degradation of mCRY2 [76]. Furthermore, GSK3β phosphorylates and stabilizes the nuclear receptor REV-ERBα, a negative regulator of *Bmal1* transcription. Lithium induces the proteasomal degradation of REV-ERBα probably by inhibiting the stabilizing effect of GSK3β phosphorylation, which eventually activates the transcription of *Bmal1* [93].

3.6.3 MAPK

The ERK/MAP kinase cascade is found to be involved in photoentrainment of the circadian clock in the SCN. MAPK signaling leads to phosphorylation and activation of CREB (CRE-element binding protein), which in turn activates immediate early genes like *c-Jun* and *c-Fos* but also *Per1* and *Per2* [94–96]. CRY1 and CRY2 are also direct targets of MAPK and exchanging single phosphoacceptor sites on

CRY proteins reduces their inhibitory activity on CLOCK-BMAL1-mediated transactivation [80]. BMAL1 associates with MAPK and is phosphorylated by the kinase in vitro [97]. Pharmacological inhibition of MAPK in SCN slice cultures results in amplitude dampening of circadian oscillations [98]. Although these results point towards an involvement of MAPK also in the negative feedback loop, its exact role here is far from being understood.

3.7 An Emerging Role for Phosphatases in the Circadian Clockwork

Serine/threonine phosphatases belonging to the PPP-family (e.g. PP1, PP2A, PP5) are shown to be involved in the regulation of circadian oscillations [20]. Opposing actions of kinases and phosphatases might result in a sophisticated temporal regulation of the phosphorylation status of clock proteins, which is likely to be essential for generation and maintenance of robust circadian rhythms. Furthermore, phosphatases may also regulate kinase activity directly, which are often regulated by phosphorylation, resulting in (auto)-inhibition or – activation, respectively.

In the circadian oscillator of the cyanobacterium *Synechococcus elongatus* alternating phosphorylation and dephosphorylation activity of KaiC is sufficient to generate a circadian phosphorylation pattern of KaiC in vitro [22]. In addition, PP1, PP4 [99] and PP2A are crucial for adjusting the circadian period and phase of the filamentous fungus *Neurospora crassa* [100, 101]. In mammals, unspecific inhibition of protein phosphatases of the PPP family in cell culture by calyculin A leads to a destabilization of PER2 [46]. One function of phosphatases in the mammalian circadian clock therefore seems to be counteracting the CK1ε/δ mediated destabilization of PER proteins possibly by dephosphorylation of PER-destabilizing phosphorylation sites. Particular roles of PP5, PP2A and PP1 in the molecular clockwork are exemplified in the following sections.

3.7.1 PP5

Protein phosphatase 5 (PP5) is a unique member of the PPP family of serine/threonine phosphatases, which is based on the tetratricopeptide repeat (TPR) domains present within its structure [102]. Since its discovery, PP5 has been implicated in wide ranging cellular processes. Partch et al. found that PP5 is involved in an interesting regulatory mechanism of the mammalian circadian clockwork [81]. PP5 binds to CK1ε and – with a higher affinity – also to mCRY1 and mCRY2. While CRYs and CK1ε interact with different regions of PP5, PP5 activates CK1ε by dephosphorylating its autoinhibitory C-terminal tail. This results in increased mPER2 phosphorylation by CK1ε. mCRY2 interaction with PP5 counteracts the CK1ε activation. Moreover, downregulation of PP5 by RNAi results in hypophosphorylation of PER1 and PER2 and in decreased amplitudes of PER1 and PER2 protein oscillations in cell culture.

3.7.2 PP2A

Protein phosphatase 2A (PP2A) is a serine/threonine phosphatase with a heterotrimeric architecture. It is composed of a scaffold subunit, a catalytic subunit as well as a variable regulatory subunit. It is implicated in many cellular processes, which are regulated predominantly by the nature of the regulatory subunit conferring substrate specificity and subcellular localization as well as activity [103–105]. In the *Drosophila* circadian system, PP2A stabilizes dPER [106], dCLK [107] and possibly also dTIM by dephosphorylation, thereby preventing their proteasomal degradation [108]. Furthermore, PP2A activity is presumably involved in an increase of nuclear expression of dPER-dTIM complexes [106]. Also in the circadian system of *Neurospora*, a role for PP2A has been described; it dephosphorylates and thereby activates the transcriptional activators of the core circadian oscillator [101].

Up to now, a role for PP2A in the mammalian oscillator has not been described.

3.7.3 PP1

Protein phosphatase 1 (PP1) is a ubiquitous protein serine/threonine phosphatase regulating a huge variety of cellular functions by the interaction of its catalytic subunit with over fifty different established or putative regulatory subunits [109]. In the *Neurospora* circadian system, PP1 has been shown to dephosphorylate and stabilize the negative regulator FRQ [100]. In mammals, PP1 coimmunoprecipitates with mPER2 and specific inhibition of PP1 destabilizes mPER2. Furthermore, expression of dominant-negative catalytic subunits of PP1 lead to faster proteasomal degradation of mPER2 [110]. Since PP1 plays a role in the regulation of mPER2 stability, inhibition or genetic lesion of PP1 is predicted to influence the dynamics of mammalian circadian oscillations – an effect not reported so far.

3.8 Posttranslational Modifications of Transactivators CLOCK and BMAL1

CLOCK (alternatively NPAS2 in the SCN) and BMAL1 are the essential transcription factors in the circadian clockwork. They heterodimerize, bind to E-box enhancer elements in the promoter regions of several clock genes and activate their transcription. The repressor activity of PER and CRY proteins inhibits CLOCK-BMAL1 transactivation. During a circadian cycle, CLOCK and BMAL1 undergo PTMs including phosphorylation, acetylation, ubiquitination, and sumoylation, which modulate heterodimerization, nuclear localization, transcriptional activity, and degradation.

3.8.1 Phosphorylation and Nuclear Localization of CLOCK and BMAL1

CLOCK and BMAL1 are phosphoproteins in vivo exhibiting a circadian phosphorylation pattern. Although the total protein abundance of BMAL1 shows a circadian rhythm, CLOCK levels do not change significantly [5]. BMAL1 accumulates in the cytoplasm after its transcription, enters the nucleus and is exported again, mainly by nuclear export before the increase of *Bmal1* transcript levels [111]. Interestingly, a lower BMAL1 protein level, particularly in the nucleus, is observed at circadian times, when transcriptional activity is highest. During the circadian cycle the predominating CLOCK species in the nucleus is hyperphosphorylated. Levels of the nuclear CLOCK also seem to be reduced when BMAL1 protein abundance is low and transactivational activity is high.

CLOCK/BMAL1 heterodimers have been reported to stay bound to E-box elements of *Per1* and *Cry1* promoters during the whole circadian cycle [5, 112]. In contrast rhythmic DNA binding of CLOCK/BMAL1 has been observed on the *Dbp* promoter, which is a clock controlled gene [113]. In both cases, PTMs of CLOCK and BMAL1 might control their transcriptional activity, in particular because it has been shown that even constitutive expression of *Bmal1* is sufficient to maintain normal circadian rhythms [114].

Indeed, phosphorylation of CLOCK and BMAL1 is coincident with their dimerization, their highest transcriptional activity and the disappearance of nuclear BMAL1. The ectopic expression of increasing levels of *Clock* leads to accumulation and hyperphosphorylation of BMAL1, whereas ectopic expression of *Bmal1* reduces CLOCK levels. Furthermore, in *Bmal1*-deficient mice, no nuclear CLOCK can be detected [115], therefore the nuclear localization of CLOCK is dependent on BMAL1 and/or a functional circadian clock. However, because CLOCK is present in the nucleus of *Cry*-deficient mice, which also do not have a functioning circadian clock [111], the presence of BMAL1 rather than a functional clock seems to be required for the nuclear translocation of CLOCK and transcriptional activity of the heterodimer.

CRY1 and CRY2 inhibit the transcriptional activation of CLOCK-BMAL1 by binding to the transactivator complex [116]. Thereby, CRY1 and CRY2 stabilize unphosphorylated (transcriptionally inactive) forms of BMAL1 and may also reduce the hyperphosphorylation of CLOCK [117].

Neither bHLH nor PAS domains of CLOCK and BMAL1 are necessary for their heterodimerization, but are required for their transcriptional activity [117]. Interestingly, deletion of CLOCK and BMAL1 PAS domains abolishes the hyperphosphorylation which is normally observed upon their heterodimer formation.

Little is known about the kinases which phosphorylate CLOCK and BMAL1. Protein kinase C (PKC) phosphorylates CLOCK after serum shock and is important for immediate early gene expression and phase resetting, although its role for normal clock function under constant conditions has not been elucidated [118]. It has been further reported, that CK1ε phosphorylates BMAL1 in vitro and that downregulation of CK1ε or expression of a kinase-dead version of CK1ε reduces CLOCK-BMAL1 transcriptional activity [119]. Also MAPK binds to BMAL1 and phosphorylates it, which leads to negative regulation of CLOCK-BMAL1 activity [97].

3.8.2 BMAL1 Sumoylation and Ubiquitination Alters Transcriptional Activity

The small ubiquitin-like modifiers (SUMO) are structurally related to ubiquitin but do not directly target proteins for proteasomal degradation [120, 121]. Sumoylation regulates protein interaction and subcellular localization foremost. Three SUMO paralogues exist in mammals, SUMO-1 and the two nearly identical SUMO-2 and SUMO-3, which are conjugated by an E2-ligase to target proteins. Cardone et al. found that BMAL1 is sumoylated at lysine 259 when coexpressed with SUMO-1 or SUMO-2 [122]. The presence of a SUMO protease inhibitor in cell lysates stabilizes a low mobility sumoylated BMAL1, which is not observed when lysine 259 is mutated. This mutant stabilizes BMAL1 and – when expressed in *Bmal1*$^{-/-}$ mouse embryonic fibroblasts – does not rescue circadian BMAL1 levels. Furthermore, CLOCK is necessary to induce BMAL1 sumoylation, although binding to CLOCK alone is not sufficient. This is based on the finding that BMAL1 is not sumoylated, when it forms dimers with a CLOCK mutant deficient in transactivational activity. This suggests that BMAL1 is sumoylated after the peak of CLOCK-BMAL1 transcriptional activity [122]. BMAL1 modification by SUMO-2/3 at lysine 259 facilitates BMAL1 ubiquitination and proteasomal degradation. Furthermore, sumoylation of BMAL1 enhances the transcriptional activity of CLOCK-BMAL1 [123]. Interestingly, the highest CLOCK-BMAL1 activity has been observed at circadian times, when they have the lowest abundance.

3.8.3 (De-)Acetylation of BMAL1 and PER2

Histone modifications, i.e. acetylation, methylation, and phosphorylation, fundamentally control chromatin remodeling and thereby transcription [124]. It has been recognized that histone H3 bound to E-boxes of *mPer1* and *mPer2* promoters is rhythmically acetylated in phase with transcription of the *mPer* genes [112]. On the clock controlled *Dbp* gene, rhythmic histone acetylation is accompanied with CLOCK-BMAL1 binding and *Dbp* transcription [113]. Interestingly, it has been found that CLOCK itself has histone acetyl transferase (HAT) activity and is involved in enabling an activated chromatin state [125]. Furthermore, CLOCK is not only able to acetylate H3 and H4 histone, it also acetylates BMAL1 on lysine 237 [126], which seems to be important for the repressor activity of CRY1 and thus, for normal circadian rhythms.

This story was recently expanded by the findings of two independent groups, that the histone deacetylase SIRT1 is involved in regulating the amplitude of clock controlled gene transcription [127, 128]. SIRT1 is a NAD^+ dependent enzyme; therefore energy state and cell metabolism influence its activity. BMAL1 as well as PER2 have been identified as targets of SIRT1 deacetylase activity (in addition to histone H3). In vivo, SIRT1 is found in complex with CLOCK and BMAL1. It seems to be necessary for deacetylating BMAL1 at lysine 237 thereby destabilizing it [127]. The effect of SIRT1 levels on amplitude and/or magnitude of circadian oscillations of several clock genes seem to depend on model systems or experimental setup, and both publications also disagree on the fact, whether SIRT1 protein is circadian in its abundance.

Amazingly, PER2 is also regulated by acetylation, yet the acetyl transferase is unknown [128]. SIRT1 binds to CLOCK, BMAL1, and also PER2 in a circadian manner. PER2 is stabilized in the absence of SIRT1, maybe because acetylation inhibits PER2 ubiquitination and interferes with its proteasomal degradation.

3.9 Conclusions

Posttranslational mechanisms are fundamentally involved in circadian clock function (see Fig. 3.1). From the current point of view, the transcriptional activity of CLOCK/NPAS2-BMAL1 delivers the repressors (mainly PER1/2, CRY1/2) of this activity, which then form negative (and positive) feedback loops. But to fulfill this task and to enable circadian oscillations, their action has to be tightly controlled and also delayed with respect to their transcripts. It seems that this control is largely based on posttranslational regulation. The repressors are regulated by phosphorylation (and for PER2 also by acetylation), which induces their ubiquitination and proteasomal degradation thereby modulating their stability, and also their subcellular localization, complex formation, and activity. On the other hand, the transactivators CLOCK and BMAL1 are themselves controlled by phosphorylation, ubiquitination, sumoylation, and acetylation.

Several kinases, phosphatases and recently also (de-)acetylases, have been recognized as important cogs in the circadian clockwork. Moreover, in this highly dynamical system, the temporal and spatial availability of these compounds likely determines the oscillator properties.

The identification of core clock components provides a first layer of understanding the molecular composition of the circadian clock. Although many genes show a circadian expression pattern in the SCN and other tissues, only a limited number of them comprise the core clock. Importantly, the circadian clock is a dynamic system, which makes it necessary to closely investigate abundance, interactions as well as patterns of PTMs of clock proteins in a temporal and spatial resolution. Furthermore, since protein abundance itself may have a strong impact on the dynamics of circadian oscillations, it is important to carefully select model systems for investigating the biochemical properties of clock components, without undesirably overexpressing key components and thereby disrupting the fine balanced oscillator.

Several emerging techniques are beginning to provide deeper insight into the circadian clockwork. First of all, downregulation of endogenous clock components, e.g. by RNA interference or small molecules, and analyzing the impact on circadian dynamics allows to screen for new clock components. Furthermore, state-of-the-art mass spectrometry will allow the investigation of the quantitative changes in the proteome of oscillating cells, including temporal and spatial distribution and the dynamics of post-translational modifications of clock proteins, respectively. Similarly, dynamics in the protein–protein interactions have to be investigated in a temporal and spatial resolution.

Acknowledgments The authors would like to thank Katja Vanselow and Sabrina Wendt for critical reading and helpful comments on the manuscript. Work in A.K.'s laboratory is supported by the Deutsche Forschungsgemeinschaft as well as by the 6th EU framework program EUCLOCK.

References

1. Lowrey PL, Takahashi JS (2004) Mammalian circadian biology: elucidating genome-wide levels of temporal organization. Annu Rev Genomics Hum Genet 5:407–441
2. DeBruyne JP, Weaver DR, Reppert SM (2007) CLOCK and NPAS2 have overlapping roles in the suprachiasmatic circadian clock. Nat Neurosci 10:543–545
3. Reppert SM, Weaver DR (2002) Coordination of circadian timing in mammals. Nature 418:935–941
4. Lee C, Weaver DR, Reppert SM (2004) Direct association between mouse PERIOD and CKIepsilon is critical for a functioning circadian clock. Mol Cell Biol 24:584–594
5. Lee C, Etchegaray JP, Cagampang FR, Loudon AS, Reppert SM (2001) Posttranslational mechanisms regulate the mammalian circadian clock. Cell 107:855–867
6. Barnes JW, Tischkau SA, Barnes JA, Mitchell JW, Burgoon PW, Hickok JR, Gillette MU (2003) Requirement of mammalian timeless for circadian rhythmicity. Science 302:439–442
7. Sangoram AM, Saez L, Antoch MP, Gekakis N, Staknis D, Whiteley A, Fruechte EM, Vitaterna MH, Shimomura K, King DP et al (1998) Mammalian circadian autoregulatory loop: a timeless ortholog and mPer1 interact and negatively regulate CLOCK-BMAL1-induced transcription. Neuron 21:1101–1113
8. Honma S, Kawamoto T, Takagi Y, Fujimoto K, Sato F, Noshiro M, Kato Y, Honma K (2002) Dec1 and Dec2 are regulators of the mammalian molecular clock. Nature 419:841–844
9. Brown SA, Ripperger J, Kadener S, Fleury-Olela F, Vilbois F, Rosbash M, Schibler U (2005) PERIOD1-associated proteins modulate the negative limb of the mammalian circadian oscillator. Science 308:693–696
10. Zhao WN, Malinin N, Yang FC, Staknis D, Gekakis N, Maier B, Reischl S, Kramer A, Weitz CJ (2007) CIPC is a mammalian circadian clock protein without invertebrate homologues. Nat Cell Biol 9:268–275
11. Kume K, Zylka MJ, Sriram S, Shearman LP, Weaver DR, Jin X, Maywood ES, Hastings MH, Reppert SM (1999) mCRY1 and mCRY2 are essential components of the negative limb of the circadian clock feedback loop. Cell 98:193–205
12. Griffin EA Jr, Staknis D, Weitz CJ (1999) Light-independent role of CRY1 and CRY2 in the mammalian circadian clock. Science 286:768–771

13. Vanselow K, Kramer A (2007) Role of phosphorylation in the mammalian circadian clock. Cold Spring Harb Symp Quant Biol 72:167–176
14. Gallego M, Virshup DM (2007) Post-translational modifications regulate the ticking of the circadian clock. Nat Rev Mol Cell Biol 8:139–148
15. Hunter T (2000) Signaling – 2000 and beyond. Cell 100:113–127
16. Caenepeel S, Charydczak G, Sudarsanam S, Hunter T, Manning G (2004) The mouse kinome: discovery and comparative genomics of all mouse protein kinases. Proc Natl Acad Sci USA 101:11707–11712
17. Manning G, Whyte DB, Martinez R, Hunter T, Sudarsanam S (2002) The protein kinase complement of the human genome. Science 298:1912–1934
18. Alonso A, Sasin J, Bottini N, Friedberg I, Friedberg I, Osterman A, Godzik A, Hunter T, Dixon J, Mustelin T (2004) Protein tyrosine phosphatases in the human genome. Cell 117: 699–711
19. Honkanen RE, Golden T (2002) Regulators of serine/threonine protein phosphatases at the dawn of a clinical era? Curr Med Chem 9:2055–2075
20. Gallego M, Virshup DM (2005) Protein serine/threonine phosphatases: life, death, and sleeping. Curr Opin Cell Biol 17:197–202
21. Lakin-Thomas PL (2006) Transcriptional feedback oscillators: maybe, maybe not. J Biol Rhythms 21:83–92
22. Nakajima M, Imai K, Ito H, Nishiwaki T, Murayama Y, Iwasaki H, Oyama T, Kondo T (2005) Reconstitution of circadian oscillation of cyanobacterial KaiC phosphorylation in vitro. Science 308:414–415
23. Tomita J, Nakajima M, Kondo T, Iwasaki H (2005) No transcription-translation feedback in circadian rhythm of KaiC phosphorylation. Science 307:251–254
24. Edery I, Zwiebel LJ, Dembinska ME, Rosbash M (1994) Temporal phosphorylation of the Drosophila period protein. Proc Natl Acad Sci USA 91:2260–2264
25. Yang Z, Sehgal A (2001) Role of molecular oscillations in generating behavioral rhythms in Drosophila. Neuron 29:453–467
26. Fujimoto Y, Yagita K, Okamura H (2006) Does mPER2 protein oscillate without its coding mRNA cycling?: post-transcriptional regulation by cell clock. Genes Cells 11:525–530
27. Nishii K, Yamanaka I, Yasuda M, Kiyohara YB, Kitayama Y, Kondo T, Yagita K (2006) Rhythmic post-transcriptional regulation of the circadian clock protein mPER2 in mammalian cells: a real-time analysis. Neurosci Lett 401:44–48
28. Kloss B, Price JL, Saez L, Blau J, Rothenfluh A, Wesley CS, Young MW (1998) The Drosophila clock gene double-time encodes a protein closely related to human casein kinase Iepsilon. Cell 94:97–107
29. Price JL, Blau J, Rothenfluh A, Abodeely M, Kloss B, Young MW (1998) double-time is a novel Drosophila clock gene that regulates PERIOD protein accumulation. Cell 94: 83–95
30. Fan JY, Preuss F, Muskus MJ, Bjes ES, Price JL (2009) Drosophila and vertebrate casein kinase Idelta exhibits evolutionary conservation of circadian function. Genetics 181:139–152
31. Gross SD, Anderson RA (1998) Casein kinase I: spatial organization and positioning of a multifunctional protein kinase family. Cell Signal 10:699–711
32. Knippschild U, Gocht A, Wolff S, Huber N, Lohler J, Stoter M (2005) The casein kinase 1 family: participation in multiple cellular processes in eukaryotes. Cell Signal 17:675–689
33. Takano A, Hoe HS, Isojima Y, Nagai K (2004) Analysis of the expression, localization and activity of rat casein kinase 1epsilon-3. Neuroreport 15:1461–1464
34. Eide EJ, Kang H, Crapo S, Gallego M, Virshup DM (2005) Casein kinase I in the mammalian circadian clock. Methods Enzymol 393:408–418
35. Ko HW, Jiang J, Edery I (2002) Role for Slimb in the degradation of Drosophila Period protein phosphorylated by Doubletime. Nature 420:673–678
36. Kloss B, Rothenfluh A, Young MW, Saez L (2001) Phosphorylation of period is influenced by cycling physical associations of double-time, period, and timeless in the Drosophila clock. Neuron 30:699–706

37. Preuss F, Fan JY, Kalive M, Bao S, Schuenemann E, Bjes ES, Price JL (2004) Drosophila doubletime mutations which either shorten or lengthen the period of circadian rhythms decrease the protein kinase activity of casein kinase I. Mol Cell Biol 24:886–898
38. Bao S, Rihel J, Bjes E, Fan JY, Price JL (2001) The Drosophila double-timeS mutation delays the nuclear accumulation of period protein and affects the feedback regulation of period mRNA. J Neurosci 21:7117–7126
39. Rothenfluh A, Abodeely M, Young MW (2000) Short-period mutations of per affect a double-time-dependent step in the Drosophila circadian clock. Curr Biol 10:1399–1402
40. Grima B, Lamouroux A, Chelot E, Papin C, Limbourg-Bouchon B, Rouyer F (2002) The F-box protein slimb controls the levels of clock proteins period and timeless. Nature 420:178–182
41. Suri V, Hall JC, Rosbash M (2000) Two novel doubletime mutants alter circadian properties and eliminate the delay between RNA and protein in Drosophila. J Neurosci 20:7547–7555
42. Chiu JC, Vanselow JT, Kramer A, Edery I (2008) The phospho-occupancy of an atypical SLIMB-binding site on PERIOD that is phosphorylated by DOUBLETIME controls the pace of the clock. Genes Dev 22:1758–1772
43. Camacho F, Cilio M, Guo Y, Virshup DM, Patel K, Khorkova O, Styren S, Morse B, Yao Z, Keesler GA (2001) Human casein kinase Idelta phosphorylation of human circadian clock proteins period 1 and 2. FEBS Lett 489:159–165
44. Keesler GA, Camacho F, Guo Y, Virshup D, Mondadori C, Yao Z (2000) Phosphorylation and destabilization of human period I clock protein by human casein kinase I epsilon. Neuroreport 11:951–955
45. Akashi M, Tsuchiya Y, Yoshino T, Nishida E (2002) Control of intracellular dynamics of mammalian period proteins by casein kinase I epsilon (CKIepsilon) and CKIdelta in cultured cells. Mol Cell Biol 22:1693–1703
46. Eide EJ, Woolf MF, Kang H, Woolf P, Hurst W, Camacho F, Vielhaber EL, Giovanni A, Virshup DM (2005) Control of mammalian circadian rhythm by CKIepsilon-regulated proteasome-mediated PER2 degradation. Mol Cell Biol 25:2795–2807
47. Shirogane T, Jin J, Ang XL, Harper JW (2005) SCFbeta-TRCP controls clock-dependent transcription via casein kinase 1-dependent degradation of the mammalian period-1 (Per1) protein. J Biol Chem 280:26863–26872
48. Reischl S, Vanselow K, Westermark PO, Thierfelder N, Maier B, Herzel H, Kramer A (2007) Beta-TrCP1-mediated degradation of PERIOD2 is essential for circadian dynamics. J Biol Rhythms 22:375–386
49. Maier B, Wendt S, Vanselow JT, Wallach T, Reischl S, Oehmke S, Schlosser A, Kramer A (2009) A large-scale functional RNAi screen reveals a role for CK2 in the mammalian circadian clock. Genes Dev 23:708–718
50. Ohsaki K, Oishi K, Kozono Y, Nakayama K, Nakayama KI, Ishida N (2008) The role of {beta}-TrCP1 and {beta}-TrCP2 in circadian rhythm generation by mediating degradation of clock protein PER2. J Biochem 144:609–618
51. Ralph MR, Menaker M (1988) A mutation of the circadian system in golden hamsters. Science 241:1225–1227
52. Lowrey PL, Shimomura K, Antoch MP, Yamazaki S, Zemenides PD, Ralph MR, Menaker M, Takahashi JS (2000) Positional syntenic cloning and functional characterization of the mammalian circadian mutation tau. Science 288:483–492
53. Gallego M, Eide EJ, Woolf MF, Virshup DM, Forger DB (2006) An opposite role for tau in circadian rhythms revealed by mathematical modeling. Proc Natl Acad Sci USA 103:10618–10623
54. Meng QJ, Logunova L, Maywood ES, Gallego M, Lebiecki J, Brown TM, Sladek M, Semikhodskii AS, Glossop NR, Piggins HD et al (2008) Setting clock speed in mammals: the CK1 epsilon tau mutation in mice accelerates circadian pacemakers by selectively destabilizing PERIOD proteins. Neuron 58:78–88
55. Vanselow K, Vanselow JT, Westermark PO, Reischl S, Maier B, Korte T, Herrmann A, Herzel H, Schlosser A, Kramer A (2006) Differential effects of PER2 phosphorylation: molecular basis for the human familial advanced sleep phase syndrome (FASPS). Genes Dev 20: 2660–2672

56. Dey J, Carr AJ, Cagampang FR, Semikhodskii AS, Loudon AS, Hastings MH, Maywood ES (2005) The tau mutation in the Syrian hamster differentially reprograms the circadian clock in the SCN and peripheral tissues. J Biol Rhythms 20:99–110
57. Miyazaki K, Wakabayashi M, Chikahisa S, Sei H, Ishida N (2007) PER2 controls circadian periods through nuclear localization in the suprachiasmatic nucleus. Genes Cells 12:1225–1234
58. Vielhaber E, Eide E, Rivers A, Gao ZH, Virshup DM (2000) Nuclear entry of the circadian regulator mPER1 is controlled by mammalian casein kinase I epsilon. Mol Cell Biol 20:4888–4899
59. Miyazaki K, Nagase T, Mesaki M, Narukawa J, Ohara O, Ishida N (2004) Phosphorylation of clock protein PER1 regulates its circadian degradation in normal human fibroblasts. Biochem J 380:95–103
60. Takano A, Shimizu K, Kani S, Buijs RM, Okada M, Nagai K (2000) Cloning and characterization of rat casein kinase 1epsilon. FEBS Lett 477:106–112
61. Takano A, Isojima Y, Nagai K (2004) Identification of mPer1 phosphorylation sites responsible for the nuclear entry. J Biol Chem 279:32578–32585
62. Czeisler CA, Duffy JF, Shanahan TL, Brown EN, Mitchell JF, Rimmer DW, Ronda JM, Silva EJ, Allan JS, Emens JS et al (1999) Stability, precision, and near-24-hour period of the human circadian pacemaker. Science 284:2177–2181
63. Jones CR, Campbell SS, Zone SE, Cooper F, DeSano A, Murphy PJ, Jones B, Czajkowski L, Ptacek LJ (1999) Familial advanced sleep-phase syndrome: a short-period circadian rhythm variant in humans. Nat Med 5:1062–1065
64. Xu Y, Padiath QS, Shapiro RE, Jones CR, Wu SC, Saigoh N, Saigoh K, Ptacek LJ, Fu YH (2005) Functional consequences of a CKIdelta mutation causing familial advanced sleep phase syndrome. Nature 434:640–644
65. Toh KL, Jones CR, He Y, Eide EJ, Hinz WA, Virshup DM, Ptacek LJ, Fu YH (2001) An hPer2 phosphorylation site mutation in familial advanced sleep phase syndrome. Science 291:1040–1043
66. Takano A, Uchiyama M, Kajimura N, Mishima K, Inoue Y, Kamei Y, Kitajima T, Shibui K, Katoh M, Watanabe T et al (2004) A missense variation in human casein kinase I epsilon gene that induces functional alteration and shows an inverse association with circadian rhythm sleep disorders. Neuropsychopharmacology 29:1901–1909
67. Flotow H, Graves PR, Wang AQ, Fiol CJ, Roeske RW, Roach PJ (1990) Phosphate groups as substrate determinants for casein kinase I action. J Biol Chem 265:14264–14269
68. Flotow H, Roach PJ (1991) Role of acidic residues as substrate determinants for casein kinase I. J Biol Chem 266:3724–3727
69. Xu Y, Toh KL, Jones CR, Shin JY, Fu YH, Ptacek LJ (2007) Modeling of a human circadian mutation yields insights into clock regulation by PER2. Cell 128:59–70
70. Schlosser A, Vanselow JT, Kramer A (2005) Mapping of phosphorylation sites by a multi-protease approach with specific phosphopeptide enrichment and NanoLC-MS/MS analysis. Anal Chem 77:5243–5250
71. Schlosser A, Vanselow JT, Kramer A (2007) Comprehensive phosphorylation site analysis of individual phosphoproteins applying scoring schemes for MS/MS data. Anal Chem 79:7439–7449
72. Balsalobre A, Damiola F, Schibler U (1998) A serum shock induces circadian gene expression in mammalian tissue culture cells. Cell 93:929–937
73. Yagita K, Tamanini F, van Der Horst GT, Okamura H (2001) Molecular mechanisms of the biological clock in cultured fibroblasts. Science 292:278–281
74. Brown SA, Kunz D, Dumas A, Westermark PO, Vanselow K, Tilmann-Wahnschaffe A, Herzel H, Kramer A (2008) Molecular insights into human daily behavior. Proc Natl Acad Sci USA 105:1602–1607
75. Takano A, Nagai K (2006) Serine 714 might be implicated in the regulation of the phosphorylation in other areas of mPer1 protein. Biochem Biophys Res Commun 346:95–101
76. Harada Y, Sakai M, Kurabayashi N, Hirota T, Fukada Y (2005) Ser-557-phosphorylated mCRY2 is degraded upon synergistic phosphorylation by glycogen synthase kinase-3 beta. J Biol Chem 280:31714–31721

77. Busino L, Bassermann F, Maiolica A, Lee C, Nolan PM, Godinho SI, Draetta GF, Pagano M (2007) SCFFbxl3 controls the oscillation of the circadian clock by directing the degradation of cryptochrome proteins. Science 316:900–904
78. Siepka SM, Yoo SH, Park J, Song W, Kumar V, Hu Y, Lee C, Takahashi JS (2007) Circadian mutant Overtime reveals F-box protein FBXL3 regulation of cryptochrome and period gene expression. Cell 129:1011–1023
79. Godinho SI, Maywood ES, Shaw L, Tucci V, Barnard AR, Busino L, Pagano M, Kendall R, Quwailid MM, Romero MR et al (2007) The after-hours mutant reveals a role for Fbxl3 in determining mammalian circadian period. Science 316:897–900
80. Sanada K, Harada Y, Sakai M, Todo T, Fukada Y (2004) Serine phosphorylation of mCRY1 and mCRY2 by mitogen-activated protein kinase. Genes Cells 9:697–708
81. Partch CL, Shields KF, Thompson CL, Selby CP, Sancar A (2006) Posttranslational regulation of the mammalian circadian clock by cryptochrome and protein phosphatase 5. Proc Natl Acad Sci USA 103:10467–10472
82. Allada R, Meissner RA (2005) Casein kinase 2, circadian clocks, and the flight from mutagenic light. Mol Cell Biochem 274:141–149
83. Akten B, Jauch E, Genova GK, Kim EY, Edery I, Raabe T, Jackson FR (2003) A role for CK2 in the Drosophila circadian oscillator. Nat Neurosci 6:251–257
84. Lin JM, Kilman VL, Keegan K, Paddock B, Emery-Le M, Rosbash M, Allada R (2002) A role for casein kinase 2alpha in the Drosophila circadian clock. Nature 420:816–820
85. Doble BW, Woodgett JR (2003) GSK-3: tricks of the trade for a multi-tasking kinase. J Cell Sci 116:1175–1186
86. Gould TD, Manji HK (2005) Glycogen synthase kinase-3: a putative molecular target for lithium mimetic drugs. Neuropsychopharmacology 30:1223–1237
87. Iitaka C, Miyazaki K, Akaike T, Ishida N (2005) A role for glycogen synthase kinase-3beta in the mammalian circadian clock. J Biol Chem 280:29397–29402
88. Abe M, Herzog ED, Block GD (2000) Lithium lengthens the circadian period of individual suprachiasmatic nucleus neurons. Neuroreport 11:3261–3264
89. Iwahana E, Akiyama M, Miyakawa K, Uchida A, Kasahara J, Fukunaga K, Hamada T, Shibata S (2004) Effect of lithium on the circadian rhythms of locomotor activity and glycogen synthase kinase-3 protein expression in the mouse suprachiasmatic nuclei. Eur J Neurosci 19:2281–2287
90. Kaladchibachi SA, Doble B, Anthopoulos N, Woodgett JR, Manoukian AS (2007) Glycogen synthase kinase 3, circadian rhythms, and bipolar disorder: a molecular link in the therapeutic action of lithium. J Circadian Rhythms 5:3
91. Hirota T, Lewis WG, Liu AC, Lee JW, Schultz PG, Kay SA (2008) A chemical biology approach reveals period shortening of the mammalian circadian clock by specific inhibition of GSK-3beta. Proc Natl Acad Sci USA 105:20746–20751
92. Martinek S, Inonog S, Manoukian AS, Young MW (2001) A role for the segment polarity gene shaggy/GSK-3 in the Drosophila circadian clock. Cell 105:769–779
93. Yin L, Wang J, Klein PS, Lazar MA (2006) Nuclear receptor Rev-erbalpha is a critical lithium-sensitive component of the circadian clock. Science 311:1002–1005
94. Akashi M, Nishida E (2000) Involvement of the MAP kinase cascade in resetting of the mammalian circadian clock. Genes Dev 14:645–649
95. Butcher GQ, Doner J, Dziema H, Collamore M, Burgoon PW, Obrietan K (2002) The p42/44 mitogen-activated protein kinase pathway couples photic input to circadian clock entrainment. J Biol Chem 277:29519–29525
96. Obrietan K, Impey S, Storm DR (1998) Light and circadian rhythmicity regulate MAP kinase activation in the suprachiasmatic nuclei. Nat Neurosci 1:693–700
97. Sanada K, Okano T, Fukada Y (2002) Mitogen-activated protein kinase phosphorylates and negatively regulates basic helix-loop-helix-PAS transcription factor BMAL1. J Biol Chem 277:267–271
98. Akashi M, Hayasaka N, Yamazaki S, Node K (2008) Mitogen-activated protein kinase is a functional component of the autonomous circadian system in the suprachiasmatic nucleus. J Neurosci 28:4619–4623

99. Cha J, Chang SS, Huang G, Cheng P, Liu Y (2008) Control of WHITE COLLAR localization by phosphorylation is a critical step in the circadian negative feedback process. EMBO J 27:3246–3255
100. Yang Y, He Q, Cheng P, Wrage P, Yarden O, Liu Y (2004) Distinct roles for PP1 and PP2A in the Neurospora circadian clock. Genes Dev 18:255–260
101. Schafmeier T, Haase A, Kaldi K, Scholz J, Fuchs M, Brunner M (2005) Transcriptional feedback of Neurospora circadian clock gene by phosphorylation-dependent inactivation of its transcription factor. Cell 122:235–246
102. Swingle MR, Honkanen RE, Ciszak EM (2004) Structural basis for the catalytic activity of human serine/threonine protein phosphatase-5. J Biol Chem 279:33992–33999
103. Janssens V, Goris J, Van Hoof C (2005) PP2A: the expected tumor suppressor. Curr Opin Genet Dev 15:34–41
104. Eichhorn PJ, Creyghton MP, Bernards R (2009) Protein phosphatase 2A regulatory subunits and cancer. Biochim Biophys Acta 1795:1–15
105. Janssens V, Longin S, Goris J (2008) PP2A holoenzyme assembly: in cauda venenum (the sting is in the tail). Trends Biochem Sci 33:113–121
106. Sathyanarayanan S, Zheng X, Xiao R, Sehgal A (2004) Posttranslational regulation of Drosophila PERIOD protein by protein phosphatase 2A. Cell 116:603–615
107. Kim EY, Edery I (2006) Balance between DBT/CKIepsilon kinase and protein phosphatase activities regulate phosphorylation and stability of Drosophila CLOCK protein. Proc Natl Acad Sci USA 103:6178–6183
108. Fang Y, Sathyanarayanan S, Sehgal A (2007) Post-translational regulation of the Drosophila circadian clock requires protein phosphatase 1 (PP1). Genes Dev 21:1506–1518
109. Ceulemans H, Bollen M (2004) Functional diversity of protein phosphatase-1, a cellular economizer and reset button. Physiol Rev 84:1–39
110. Gallego M, Kang H, Virshup DM (2006) Protein phosphatase 1 regulates the stability of the circadian protein PER2. Biochem J 399:169–175
111. Tamaru T, Isojima Y, van der Horst GT, Takei K, Nagai K, Takamatsu K (2003) Nucleocytoplasmic shuttling and phosphorylation of BMAL1 are regulated by circadian clock in cultured fibroblasts. Genes Cells 8:973–983
112. Etchegaray JP, Lee C, Wade PA, Reppert SM (2003) Rhythmic histone acetylation underlies transcription in the mammalian circadian clock. Nature 421:177–182
113. Ripperger JA, Schibler U (2006) Rhythmic CLOCK-BMAL1 binding to multiple E-box motifs drives circadian Dbp transcription and chromatin transitions. Nat Genet 38:369–374
114. McDearmon EL, Patel KN, Ko CH, Walisser JA, Schook AC, Chong JL, Wilsbacher LD, Song EJ, Hong HK, Bradfield CA et al (2006) Dissecting the functions of the mammalian clock protein BMAL1 by tissue-specific rescue in mice. Science 314:1304–1308
115. Kondratov RV, Chernov MV, Kondratova AA, Gorbacheva VY, Gudkov AV, Antoch MP (2003) BMAL1-dependent circadian oscillation of nuclear CLOCK: posttranslational events induced by dimerization of transcriptional activators of the mammalian clock system. Genes Dev 17:1921–1932
116. Langmesser S, Tallone T, Bordon A, Rusconi S, Albrecht U (2008) Interaction of circadian clock proteins PER2 and CRY with BMAL1 and CLOCK. BMC Mol Biol 9:41
117. Dardente H, Fortier EE, Martineau V, Cermakian N (2007) Cryptochromes impair phosphorylation of transcriptional activators in the clock: a general mechanism for circadian repression. Biochem J 402:525–536
118. Shim HS, Kim H, Lee J, Son GH, Cho S, Oh TH, Kang SH, Seen DS, Lee KH, Kim K (2007) Rapid activation of CLOCK by Ca2+-dependent protein kinase C mediates resetting of the mammalian circadian clock. EMBO Rep 8:366–371
119. Eide EJ, Vielhaber EL, Hinz WA, Virshup DM (2002) The circadian regulatory proteins BMAL1 and cryptochromes are substrates of casein kinase Iepsilon. J Biol Chem 277:17248–17254
120. Hay RT (2005) SUMO: a history of modification. Mol Cell 18:1–12
121. Johnson ES (2004) Protein modification by SUMO. Annu Rev Biochem 73:355–382
122. Cardone L, Hirayama J, Giordano F, Tamaru T, Palvimo JJ, Sassone-Corsi P (2005) Circadian clock control by SUMOylation of BMAL1. Science 309:1390–1394

123. Lee J, Lee Y, Lee MJ, Park E, Kang SH, Chung CH, Lee KH, Kim K (2008) Dual modification of BMAL1 by SUMO2/3 and ubiquitin promotes circadian activation of the CLOCK/BMAL1 complex. Mol Cell Biol 28:6056–6065
124. Jenuwein T, Allis CD (2001) Translating the histone code. Science 293:1074–1080
125. Doi M, Hirayama J, Sassone-Corsi P (2006) Circadian regulator CLOCK is a histone acetyltransferase. Cell 125:497–508
126. Hirayama J, Sahar S, Grimaldi B, Tamaru T, Takamatsu K, Nakahata Y, Sassone-Corsi P (2007) CLOCK-mediated acetylation of BMAL1 controls circadian function. Nature 450:1086–1090
127. Nakahata Y, Kaluzova M, Grimaldi B, Sahar S, Hirayama J, Chen D, Guarente LP, Sassone-Corsi P (2008) The NAD+-dependent deacetylase SIRT1 modulates CLOCK-mediated chromatin remodeling and circadian control. Cell 134:329–340
128. Asher G, Gatfield D, Stratmann M, Reinke H, Dibner C, Kreppel F, Mostoslavsky R, Alt FW, Schibler U (2008) SIRT1 regulates circadian clock gene expression through PER2 deacetylation. Cell 134:317–328

Chapter 4
Non-image-Forming Photoreceptors

Stuart N. Peirson and Russell G. Foster

4.1 Introduction

An almost universal feature of living things is the possession of a biological representation of the 24 h day. This body clock, or more accurately circadian system, acts to "fine-tune" physiology and behaviour in advance of the varying demands of the day/night cycle. The ability to predict changes in the environment enables an organism to anticipate the changing conditions rather than passively responding to them, thus avoiding suboptimal transition states [1, 2]. Such an endogenous temporal program is useful only if the biological time remains synchronised to the solar time. In theory, multiple environmental cues could be used to set (entrain) circadian rhythms to the solar day, but in practice, the systematic daily change in the gross amount of light (irradiance) at dawn or dusk provides the most reliable indicator (zeitgeber) of the time of day. As a result, most organisms have evolved to use the twilight transition as their main zeitgeber, a process termed photoentrainment [3].

The site of the primary circadian clock in mammals is the suprachiasmatic nuclei (SCN) of the hypothalamus, which receives input from the retina via the retinohypothalamic tract (RHT) [4, 5]. The photoreceptors providing the light input to the mammalian circadian system are ocular. In this respect, mammals are quite unlike the other vertebrate classes which possess multiple photoreceptors located in structures such as the pineal complex and deep-brain [6]. Extra-retinal photoreceptors have, from time-to-time, been suggested to exist in mammals, including the idea that photoentrainment could occur from light exposure to the skin behind the knee of humans [7]. Although widely reported, this finding could not be replicated by other laboratories, even when the methodology was duplicated precisely [8]. The demonstration that the eyes provide the exclusive light-input to the circadian system has been clearly shown by enucleation [9].

R.G. Foster (✉)
Nuffield Laboratory of Ophthalmology, University of Oxford, Level 5 and 6 West Wing, The John Radcliffe Hospital, Headley Way, Headington, Oxford, OX3 9DU, UK
e-mail: russell.foster@eye.ox.ac.uk

U. Albrecht (ed.), *The Circadian Clock*, Protein Reviews 12,
DOI 10.1007/978-1-4419-1262-6_4,

4.2 Nonimage Forming Photoreception

Although clearly located within the eye, establishing which ocular cells bring about photoentrainment proved surprisingly difficult. Historically, the only components of the mammalian central nervous system thought to be directly light sensitive were the rods and cones. As a result, all light responses have been attributed to one or both of these photoreceptor classes. Rods mediate scotopic (dim light) vision, providing low-resolution but high sensitivity, whereas cones are involved in photopic (bright light) vision, and are found at high density in the human fovea, enabling high-resolution colour vision [10]. Photoreception by rods and cones is rapid, responding to bright, short duration stimuli and providing a topographically mapped output via the optic nerve. In addition, signals from the outer retina are processed by the inner retinal neurons, providing the initial stages of image construction, before being relayed via the axons of the retinal ganglion cells (RGCs) to the visual centres of the brain (Fig. 4.1). By contrast, the RHT projection to the SCN is comprised of a small number of morphologically distinct RGCs. These cells (~1% of the total) are evenly distributed across the entire retina and are not topographically organised [11]. Unlike the visual system, the circadian system needs relatively bright light of a long duration to achieve photoentrainment. Furthermore, the circadian system is capable of temporally integrating signals over some extended periods of time [12, 13].

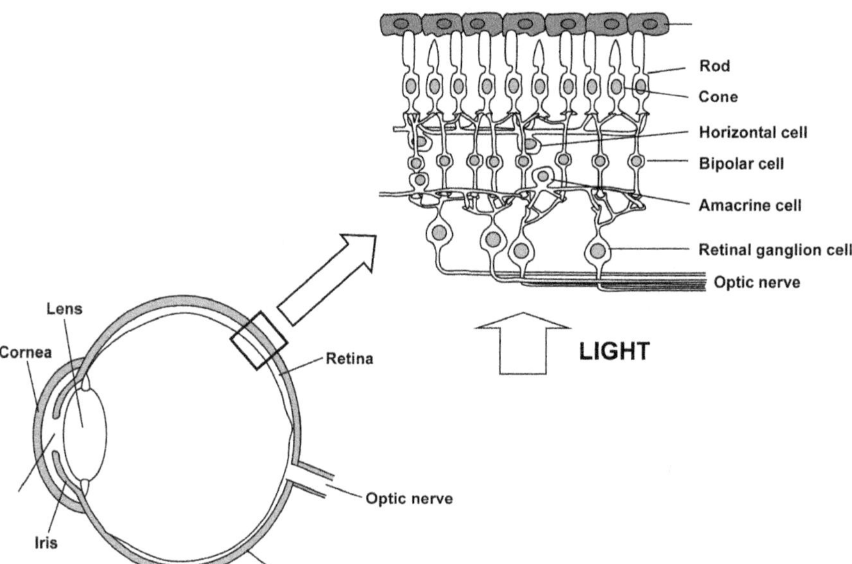

Fig. 4.1 The mammalian eye and retina. Light entering the eye is detected by photoreceptors in the outer retina, the rods and cones. These classical photoreceptors project to bipolar cells, which in turn synapse with retinal ganglion cells (RGCs). Horizontal cells and amacrine cells are involved in retinal processing. The axons of RGCs form the optic nerve projecting to the visual centres of the brain

Rather than detecting the patterns of incoming light from a specific source of light (radiance), the circadian system has been fine-tuned to detect gross changes in the environmental light (irradiance). These marked differences between image forming and circadian light detection were not easy to explain and led circadian biologists to investigate how rods and cones might be capable of mediating these two markedly different forms of light detection.

4.3 A Novel Retinal Photoreceptor

The initial studies addressing which retinal photoreceptors mediate circadian responses to light took advantage of a naturally occurring mutation in mice, termed retinal degeneration (*rd*/*rd*). These animals lack rods, and show a greatly reduced number of cones [14]. As might be expected, *rd*/*rd* mice fail to show any classical visual responses to light [15]. By contrast, the circadian system of *rd*/*rd* mice seemed unaffected by rod and cone photoreceptor loss. If a wild-type mouse is housed under a light dark (LD) cycle, its circadian wheel running behaviour is entrained, showing activity during the dark and relative inactivity during the light phase. Under conditions of constant darkness, circadian behaviour freeruns with a period around 23.5 h. If the mouse is exposed to a pulse of light shortly after activity onset, it will delay the onset of activity the following day (Fig. 4.2a). The magnitude

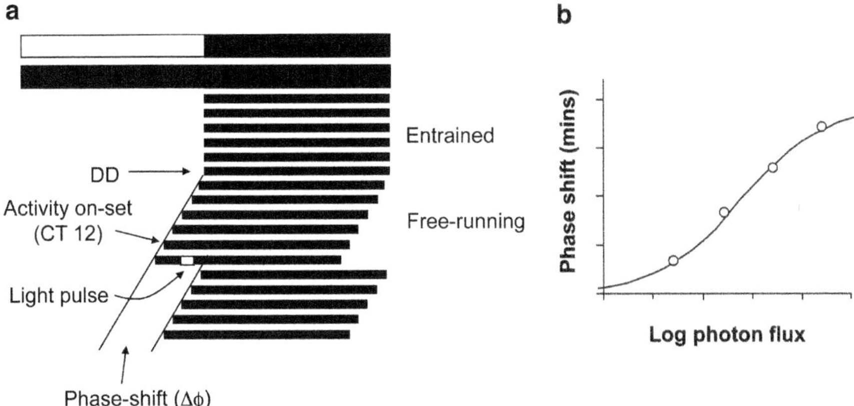

Fig. 4.2 Schematic actogram illustrating mouse locomotor activity. (**a**) Under a light-dark cycle (LD) activity is confined to the dark phase of the 24 h cycle. However, in constant darkness mice freerun with a period of slightly less than 24 h. If exposed to a brief light pulse in the early subjective evening, the animal will delay its activity onset the following day. Activity on-set under freerunning conditions is designated Circadian Time 12 (CT 12). A light pulse delivered 2 h after activity on-set is said to have been given at CT 14; A light pulse 4 h after activity on-set will be CT 16 and so on. (**b**) The phase-shift produced in this manner can be used to assess the sensitivity of the circadian system to light of different wavelengths and intensities. The magnitude of the phase-shift is proportional to the intensity of the irradiance provided by the light pulse, and can be plotted as an irradiance response curve

of this phase-delay ($\Delta\varphi$) is intensity dependent and can be described by an irradiance response curves comparable to a pharmacological dose-response curve. Both the wavelength and intensity of the light pulse can be systematically varied and the effects of these treatments on delaying wheel running behaviour can be assessed in a dose-dependent manner to determine the spectral sensitivity of the circadian system [16] (Fig. 4.2b). Remarkably, the massive loss of classical photoreceptors in *rd/rd* mice had little or no effect on the ability of mice to phase-shift their circadian rhythms in wheel running behaviour. Irradiance response curves to white light were indistinguishable from congenic wildtype controls [17–19]. These studies demonstrated that the processing of light for circadian and visual responses must be different and hinted that there may be another class of ocular photoreceptor. However, as ~5% of the cones survive in the peripheral retina of *rd/rd* mice [14], it was possible that these cells might be capable of mediating photoentrainment [20].

The development of a mouse model lacking all rods and cones, the *rd/rd cl* mouse resolved this issue. The *rd/rd cl* mouse was produced by crossing *rd/rd* mice with transgenic animals lacking cones. Remarkably, the functional loss of rods and cones had little or no impact on the ability of mice to entrain or phase-shift their circadian rhythms to light at all irradiances examined [21]. This finding provided the conceptual framework for researchers to investigate what other retinal neurons are photosensitive and which photopigments mediate these nonimage forming responses to light.

4.4 Photosensitive Retinal Ganglion Cells

The cellular identity of this novel retinal photoreceptor came from parallel studies in the rat and mouse which identified a subset of photosensitive retinal ganglion cells (pRGCs). In the rat, the approach involved injecting fluorescent microspheres into the SCN which then travelled down the axons of the RHT to retrogradely label retinal ganglion cells (RGCs). These RGCs showed a light-evoked depolarisation that persisted in the presence of a cocktail of drugs that blocked all retinal intercellular communication [22]. In mice, another approach was utilised involving the isolated *rd/rd cl* retina loaded with the Ca^{2+} sensitive FURA-2 AM dye. Fluorescent imaging identified light-induced Ca^{2+} changes in ~3.0 % of neurons within the RGC layer. Significantly, the gap junction blocker carbenoxolone reduced the number of RGCs responding to light to ~1.0 %. This suggested that pRGCs are coupled via gap junctions and form a syncitium of photosensitive and nonphotosensitive neurons [23]. In addition, three discrete classes of light-induced Ca^{2+} change were identified in the pRGCs: (a) Sustained; (b) Transient and (c) Repetitive. More recent studies using electrode arrays have confirmed these three response types [24]. It remains unclear whether these distinct pRGC subtypes project to specific retinorecipient regions of the brain or whether those brain regions that receive irradiance information receive a mixed set of irradiance signals from the pRGCs [23].

The *rd/rd cl* mouse also proved valuable in providing the initial characterization of the photopigment of the pRGCs. The known photopigments of animals consist of an opsin protein linked to a chromophore which is a specific form of vitamin A called 11-*cis*-retinal. All opsin/vitamin A-based photopigments have a characteristic absorption profile that allows these photopigments to be identified on the basis of their spectral responses to light (action spectra). The first completed action spectrum in *rd/rd cl* mice demonstrated the involvement of an opsin/vitamin A-based photopigment with maximum sensitivity (λ_{max}) close to 480 nm. This pigment was given the preliminary designation of OP^{479} (opsin photopigment λ_{max} 479 nm) [25]. The known photopigments of mice peak at ~360 nm (UV cone) [26], ~498 nm (rod) [27], and ~508 nm (green cone) [28], and show no significant fit to the pupil constriction action spectrum in *rd/rd cl* mice. Subsequent action spectra have all demonstrated the involvement of an opsin/vitamin A-based photopigment with a λ_{max} of ~480 nm e.g. [29]. Although the biochemistry of the photopigment had been deduced, the molecular identity of OP^{480} remained a mystery until the characterization of melanopsin (now officially designated *Opn4*).

The first member of the melanopsin photopigment family was isolated from the photosensitive dermal melanophores of the African clawed toad *(Xenopus)*, and was then subsequently identified in the mammalian retina in a subset of retinal ganglion cells [30, 31]. The observation that melanopsin is expressed in pRGCs was critically important, but on its own, insufficient to show that melanopsin is the "circadian photopigment". Support for this role came from melanopsin knockout mice (*Opn4*$^{-/-}$). Such mice show attenuated phase-shifting responses [32, 33], and in mice lacking functional rods and cones and melanopsin, all responses to light are lost [29, 34]. These studies demonstrate that rods, cones and pRGCs can fully account for the circadian light detection within the eye, and strongly implicated melanopsin as the photopigment molecule of the pRGCs.

Gene ablation alone can only indicate that a gene is important, whereas studies on the biochemistry of its protein product are necessary to define its function. For example, the phenotype of melanopsin knockout mice could also occur if melanopsin served an essential accessory function such as in chromophore regeneration for an unidentified pigment [35]. This concern was finally addressed by three different groups using heterologous expression of either human or murine melanopsin in Neuro2A cells [36], HEK293 cells [37] and *Xenopus* oocytes [38]. In each expression system, melanopsin expression was sufficient to drive a retinal dependent light response. For example, in neuroblastoma Neuro2A cells, the expression of melanopsin, in the presence of retinal chromophore (11-*cis*-retinal), transformed a nonphotosensitive neurone into a photoreceptor. Although all the three expression studies broadly showed the same result, inconsistencies relating to the spectral sensitivity of expressed human and mouse melanopsins emerged. The spectral maxima of the pigments varied between 440 nm (human) and 480 nm (mouse). The reason for this discrepancy still remains unclear but presumably relates to varied experimental procedures, including the host cell environment and chromophore availability.

Several recent studies have also provided information regarding the melanopsin signalling pathway [39–41]. In brief, the working model emerging proposes that light activated melanopsin resembles an invertebrate-like photopigment and interacts with a G_q/G_{11} G-protein that in-turn activates a phospholipase C (PLC-β).

PLC-β generates IP_3 and DAG, which ultimately modulate a TRPC channel, possibly via a protein kinase C (PKC). Most recently, a number of candidate genes/proteins have been identified that may form critical components of the melanopsin signalling pathway, including the atypical protein kinase C zeta (*Prkcz*). Remarkably, the genetic ablation of *Prkcz* mimics precisely the melanopsin knock-out phenotype in a battery of behavioural tests [40].

4.5 Photoreceptor Contributions to Nonimage Forming Responses to Light

Whilst the pRGCs provide a measure of environmental irradiance in mammals, this does not mean that the rods and cones play no role in these light-detecting tasks. Indeed, multiple lines of evidence suggest that both classical and novel photoreceptors interact in different nonimage forming response to light (Fig. 4.3, Table 4.1). This is clear in the case of the pupillary light response, where the loss of the

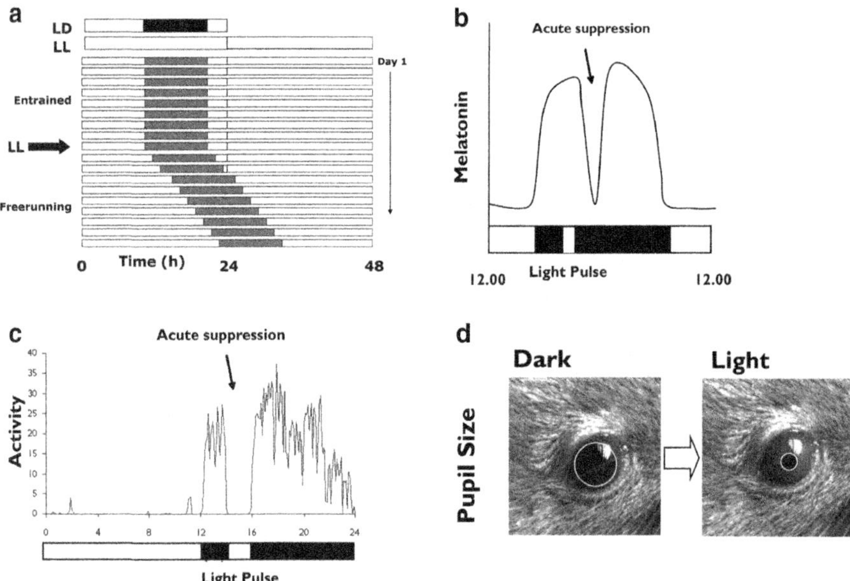

Fig. 4.3 Nonimage forming responses to light in the mouse. (**a**) When exposed to constant light (LL) a mouse freeruns, but with a period of longer than 24 h. The activity onset (CT 12) is thus delayed each day. The extent of this period lengthening is intensity-dependent, and can even exceed 26 h. (**b**) Nocturnal exposure to light acutely suppresses pineal melatonin synthesis. This response is retained in mice lacking rods and cones (*rd/rd cl*). (**c**) Nocturnal light exposure acutely suppresses activity – a process termed negative masking. This response is unaffected in *rd/rd cl* mice, but is attenuated in melanopsin-deficient animals ($Opn4^{-/-}$). (**d**) Pupillary constriction in response to light is mediated by both classical photoreceptors and pRGCs. In *rd/rd cl* mice, the pupillary light response (PLR) is less sensitive, but full constriction can still occur. In melanopsin knockout animals, pupillary responses appear normal under dim light, but under bright light conditions full constriction cannot be achieved

Table 4.1 Summary of the contribution of rods/cones and melanopsin-based pRGCs to a variety of nonimage forming responses to light. LL refers to constant light

Nonimage forming response to light	Rods/cone contribution	Melanopsin contribution	References
Circadian entrainment	Yes	Yes	[21, 29]
Suppression of pineal melatonin	?	Yes	[47]
Pupillary light response	Yes	Yes	[48, 49]
Period lengthening under LL	Yes	Yes	[33]
Negative masking	Yes	Yes	[43, 50]
Increase in heart-rate	Yes	No	[45]
Acute sleep induction	No	Yes	[46]

rods and cones leads to a marked decrease in sensitivity. In this case, it has been suggested that the rods and cones allow a rapid assessment of light transitions, and cause an immediate pupillary constriction or dilation. By contrast, the pRGCs are likely to provide sustained responses to the prolonged periods of environmental brightness [42]. In the case of the circadian system, the role of the rods and cones is less clear. The loss of these receptors causes only subtle effects on the circadian entrainment [21] and negative masking [43, 44]. By contrast, loss of melanopsin merely attenuates circadian responses to light [32, 33]. It is only when all the rods, cones, and melanopsin are genetically ablated that circadian responses to light are completely abolished [29]. Recent studies on other nonimage forming responses to light show that acute induction of heart-rate following an exposure to light during the animal's subjective day is dependent upon the classical photoreceptors, and is abolished in *rd/rd cl* animals [45]. By contrast, the acute induction of sleep following nocturnal light exposure is dependent upon melanopsin, and is abolished in *Opn4*$^{-/-}$ mice [46].

4.6 Future Perspectives

Our understanding of the photoreceptors involved in the regulation of circadian physiology has undergone a remarkable transformation in the last decade. The discovery of a previously unrecognised photoreceptor system in the mammalian eye has provided a greater understanding of how light modulates not only the circadian system, but a whole range of nonimage forming responses to light. As with any new discovery, numerous additional questions remain unanswered. Future studies are required to address the relative contribution of rods, cones and pRGCs to different irradiance detection tasks. We also need to develop a much greater understanding of the phototransduction mechanisms used by the pRGCs and how the signals from the retina are integrated by the SCN to bring about photoentrainment. Answering these questions will provide a greater understanding of how numerous aspects of our physiology are regulated by the light environment, and may provide future therapeutic strategies for dealing with the circadian disturbances associated with jet lag, shift work, sleep disorders and a range of mental health problems.

References

1. Aschoff J (1984) Circadian timing. Ann N Y Acad Sci 423:442–468
2. Pittendrigh CS (1993) Temporal organisation: reflections of a Darwinian Clock-Watcher. Annu Rev Physiol 55:17–54
3. Roenneberg T, Foster RG (1997) Twilight times: light and the circadian system. Photochem Photobiol 66:549–561
4. Moore R, Lenn N (1972) A retinohypothalamic projection in the rat. J Comp Neurol 146:1–14
5. Moore RY, Eichler VB (1972) Loss of a circadian adrenal corticosterone rhythm following suprachiasmatic lesions in the rat. Brain Res 42:201–206
6. Shand J, Foster RG (1999) The extraretinal photoreceptors of non-mammalian vertebrates. In: Archer SN, Djamgoz MBA, Loew ER, Partridge JC, Vallerga S (eds) Adaptive mechanisms in the ecology of vision. Dordrecht, Netherlands, Kluwer Academic Publishers, pp 197–222
7. Campbell S, Murphy P (1998) Extraocular phototransduction in humans. Science 179:396–398
8. Wright KP Jr, Czeisler CA (2002) Absence of circadian phase resetting in response to bright light behind the knees. Science 297:571
9. Nelson RJ, Zucker I (1981) Absence of extra-ocular photoreception in diurnal and nocturnal rodents exposed to direct sunlight. Comp Biochem Physiol 69A:145–148
10. Rodieck RW (1998) The first steps in seeing. Sinauer Associates, INC., Sunderland, MA
11. Provencio I, Cooper HM, Foster RG (1998) Retinal projections in mice with inherited retinal degeneration: implications for circadian photoentrainment. J Comp Neurol 395:417–439
12. Nelson DE, Takahashi JS (1991) Sensitivity and integration in a visual pathway for circadian entrainment in the hamster (*Mesocricetus auratus*). J Physiol 439:115–145
13. Nelson DE, Takahashi JS (1991) Comparison of visual sensitivity for suppression of pineal melatonin and circadian phase-shifting in the golden hamster. Brain Res 554:272–277
14. Carter-Dawson LD, LaVail MM, Sidman RL (1978) Differential effect of the rd mutation on rods and cones in the mouse retina. Invest Ophthalmol Vis Sci 17:489–498
15. Provencio I, Wong S, Lederman AB, Argamaso SM, Foster RG (1994) Visual and circadian responses to light in aged retinally degenerate mice. Vision Res 34:1799–1806
16. Peirson SN, Thompson S, Hankins MW, Foster RG (2005) Mammalian photoentrainment: results, methods, and approaches. Methods Enzymol 393:697–726
17. Foster RG, Provencio I, Hudson D, Fiske S, De Grip W, Menaker M (1991) Circadian photoreception in the retinally degenerate mouse (*rd*/*rd*). J Comp Physiol A 169:39–50
18. Provencio I, Foster RG (1995) Circadian rhythms in mice can be regulated by photoreceptors with cone-like characteristics. Brain Res 694:183–190
19. Yoshimura T, Ebihara S (1996) Spectral sensitivity of photoreceptors mediating phase-shifts of circadian rhythms in retinally degenerate CBA/J (*rd*/*rd*) and normal CBA/N (+/+)mice. J Comp Physiol A 178:797–802
20. Foster RG, Argamaso S, Coleman S, Colwell CS, Lederman A, Provencio I (1993) Photoreceptors regulating circadian behavior: a mouse model. J Biol Rhythms 8(Suppl):17–23
21. Freedman MS, Lucas RJ, Soni B, von Schantz M, Munoz M, David-Gray Z, Foster R (1999) Regulation of mammalian circadian behavior by non-rod, non-cone, ocular photoreceptors. Science 284:502–504
22. Berson DM, Dunn FA, Takao M (2002) Phototransduction by retinal ganglion cells that set the circadian clock. Science 295:1070–1073
23. Sekaran S, Foster RG, Lucas RJ, Hankins MW (2003) Calcium imaging reveals a network of intrinsically light-sensitive inner-retinal neurons. Curr Biol 13:1290–1298
24. Tu DC, Zhang D, Demas J, Slutsky EB, Provencio I, Holy TE, Van Gelder RN (2005) Physiologic diversity and development of intrinsically photosensitive retinal ganglion cells. Neuron 48:987–999
25. Lucas R, Douglas R, Foster R (2001) Characterization of an ocular photopigment capable of driving pupillary constriction in mice. Nat Neurosci 4:621–626
26. Jacobs GH, Neitz J, Deegan JFd (1991) Retinal receptors in rodents maximally sensitive to ultraviolet light. Nature 353:655–657

27. Bridges CDB (1959) The visual pigments of some common laboratory animals. Nature 184:1727–1728
28. Sun H, Macke JP, Nathans J (1997) Mechanisms of spectral tuning in the mouse green cone pigment. Proc Natal Acad Sci U S A 94:8860–8865
29. Hattar S, Lucas RJ, Mrosovsky N, Thompson S, Douglas RH, Hankins MW, Lem J, Biel M, Hofmann F, Foster RG et al (2003) Melanopsin and rod-cone photoreceptive systems account for all major accessory visual functions in mice. Nature 424:75–81
30. Provencio I, Jiang G, De Grip WJ, Hayes WP, Rollag MD (1998) Melanopsin: an opsin in melanophores, brain, and eye. Proc Natl Acad Sci U S A 95:340–345
31. Provencio I, Rodriguez IR, Jiang G, Hayes WP, Moreira EF, Rollag MD (2000) A novel human opsin in the inner retina. J Neurosci 20:600–605
32. Ruby NF, Brennan TJ, Xie X, Cao V, Franken P, Heller HC, O'Hara BF (2002) Role of melanopsin in circadian responses to light. Science 298:2211–2213
33. Panda S, Sato TK, Castrucci AM, Rollag MD, DeGrip WJ, Hogenesch JB, Provencio I, Kay SA (2002) Melanopsin (Opn4) requirement for normal light-induced circadian phase shifting. Science 298:2213–2216
34. Panda S, Provencio I, Tu DC, Pires SS, Rollag MD, Castrucci AM, Pletcher MT, Sato TK, Wiltshire T, Andahazy M et al (2003) Melanopsin is required for non-image-forming photic responses in blind mice. Science 301:525–527
35. Foster R, Bellingham J (2002) Opsins and melanopsins. Curr Biol 12:R543–544
36. Melyan Z, Tarttelin EE, Bellingham J, Lucas RJ, Hankins MW (2005) Addition of human melanopsin renders mammalian cells photoresponsive. Nature 433:741–745
37. Qiu X, Kumbalasiri T, Carlson SM, Wong KY, Krishna V, Provencio I, Berson DM (2005) Induction of photosensitivity by heterologous expression of melanopsin. Nature 433:745–749
38. Panda S, Nayak SK, Campo B, Walker JR, Hogenesch JB, Jegla T (2005) Illumination of the melanopsin signaling pathway. Science 307:600–604
39. Sekaran S, Lall GS, Ralphs KL, Wolstenholme AJ, Lucas RJ, Foster RG, Hankins MW (2007) 2-Aminoethoxydiphenylborane is an acute inhibitor of directly photosensitive retinal ganglion cell activity in vitro and in vivo. J Neurosci 27:3981–3986
40. Peirson SN, Oster H, Jones SL, Leitges M, Hankins MW, Foster RG (2007) Microarray analysis and functional genomics identify novel components of melanopsin signaling. Curr Biol 17:1363–1372
41. Hankins MW, Peirson SN, Foster RG (2008) Melanopsin: an exciting photopigment. Trends Neurosci 31:27–36
42. Lucas RJ, Freedman MS, Lupi D, Munoz M, David-Gray ZK, Foster RG (2001) Identifying the photoreceptive inputs to the mammalian circadian system using transgenic and retinally degenerate mice. Behav Brain Res 125:97–102
43. Mrosovsky N, Hattar S (2003) Impaired masking responses to light in melanopsin-knockout mice. Chronobiol Int 20:989–999
44. Thompson S, Foster RG, Stone EM, Sheffield VC, Mrosovsky N (2008) Classical and melanopsin photoreception in irradiance detection: negative masking of locomotor activity by light. Eur J Neurosci 27:1973–1979
45. Thompson S, Lupi D, Hankins MW, Peirson SN, Foster RG (2008) The effects of rod and cone loss on the photic regulation of locomotor activity and heart rate. Eur J Neurosci 28:724–729
46. Lupi D, Oster H, Thompson S, Foster RG (2008) The acute light-induction of sleep is mediated by OPN4-based photoreception. Nat Neurosci 11:1068–1073
47. Lucas RJ, Freedman MS, Munoz M, Garcia-Fernandez JM, Foster RG (1999) Regulation of the mammalian pineal by non-rod, non-cone, ocular photoreceptors. Science 284:505–507
48. Lucas RJ, Douglas RH, Foster RG (2001) Characterization of an ocular photopigment capable of driving pupillary constriction in mice. Nat Neurosci 4:621–626
49. Lucas RJ, Hattar S, Takao M, Berson DM, Foster RG, Yau KW (2003) Diminished pupillary light reflex at high irradiances in melanopsin-knockout mice. Science 299:245–247
50. Mrosovsky N, Lucas R, Foster R (2001) Persistence of masking responses to light in mice lacking rods and cones. J Biol Rhythms 16:585–587

Chapter 5
Circadian Clocks and Metabolism

Henrik Oster

5.1 Introduction

The metabolic system of the human body is a highly sophisticated and carefully tuned machine optimised to maximise the chance of survival under the harsh living conditions of the Palaeolithic. For this purpose, our body has developed mechanisms of efficient energy reduction during times of prolonged starvation as well as means to rapidly utilize large amounts of nutrients by preserving energy in the form of adipose tissue. However, with the advent of industrial food processing and the abundant availability of nutrition during the second half of the twentieth century – at least in our first world societies – this concept has started becoming problematic in several ways: (1) The constant thrive of our metabolic system to store excessive energy in an anticipation of hard times gave rise to an increasing prevalence of obesity throughout all Western cultures, with a plethora of associated diseases such as diabetes, stroke and cancer [1]. (2) Industrially processed food contains various amounts of artificial and chemically modified ingredients that are either difficult to be metabolised (e.g. hydrogenated fats), or (3) can actively interfere with the metabolic and endocrine regulatory pathways of the body, for example steroids sometimes found in meat products and milk.

These days, obesity represents one of the major health challenges worldwide and has grown into an epidemic over the last decades. According to the WHO, the number of obese people in Europe has tripled over the last 20 years [1]. Today, more than 130 million people count as being obese in Europe alone. Another 400 million people are *overweight*. Moreover, obesity is the most common health disorder amongst young people in Europe. According to the WHO, there will be about 150 million obese adults and 15 million obese children and adolescents in the European WHO region by 2010 [1]. Obese people are more likely to develop metabolism-linked pathologies such as type 2 diabetes and coronary heart disease,

H. Oster
Circadian Rhythms Group, Max Planck Institute for Biophysical Chemistry, 37077, Göttingen, Germany
e-mail: henrik.oster@mpibpc.mpg.de

U. Albrecht (ed.), *The Circadian Clock*, Protein Reviews 12,
DOI 10.1007/978-1-4419-1262-6_5, © Springer Science+Business Media, LLC 2010

the leading cause of death in both men and women worldwide, with numbers exceeding those of all cancer deaths together. Obesity not only affects the individual, but has economic implications for the whole of society. About 6% of health costs in the WHO European region are spent on obesity and related diseases [1]. The causes of this pandemic cannot fully be explained by changes in lifestyle factors such as high fat/high carbohydrate diets and low physical activity, but may also reflect circadian rhythm related parameters such as chronic sleep curtailment and the deregulation of endocrine rhythms.

Several lines of evidence link metabolic control with the regulation of the circadian timing system. These include numerous epidemiologic and genetic studies, experimental animal work and mechanistic assays based on cell culture and in vitro approaches. It is the purpose of this review to summarise the current state of knowledge and highlight some of the trends and remaining blind spots of current *chronometabology*. It is our task now to apply the increasing scientific knowledge to the therapy of metabolic disorders – that might well become the Achilles' heel of our modern society.

5.2 Lifestyle and Metabolism

Many aspects of our current lifestyle are believed to directly or indirectly affect the metabolic processes and homeostasis. The major factors comprise chronic sleep curtailment, shift work and the nutritional effects of the so called *Western style* diet. A behaviour that seems to have constantly developed during the past decades and these days has become highly prevalent, particularly amongst the Europeans and North Americans, is sleep curtailment, the continuous accumulation of sleep debts over long periods of time [2]. As per a survey made in 1960, modal sleep duration was found to be 8.0–8.9 h [3], while in 1995 the modal category of a survey conducted by the US National Sleep Foundation had dropped to 7 h [4]. Recent data indicate that a higher percentage of adult Americans report sleeping 6 h or less [5]. In 2005, in the US, more than 30% of adult men and women between the ages of 30 and 64 years reported sleeping on average less than 6 h/night [5]. Interestingly, this decrease in sleep duration has occurred over the same time as the increase in the prevalence of obesity and diabetes [2]. But what is the experimental evidence for a causative link between both phenomena?

In healthy subjects, humoral levels of glucose are tightly regulated in order to avoid hypoglycemic or hyperglycemic blood conditions which both can have adverse, ultimately life threatening, consequences. The term *glucose tolerance* refers to the ability of the system to metabolise exogenous glucose and restore the baseline normoglycemia. In healthy individuals, glucose tolerance varies across the day with plasma glucose responses to exogenous glucose injections being higher in the evening than in the morning. Minimal glucose tolerance, however, is observed in the middle of the night [6]. The reduction in glucose tolerance in the evening is at least partly due to a decrease in insulin sensitivity that goes along with a reduction in the insulin secretion response to elevated glucose blood levels, and might reflect an anticipation of the prolonged fasting period of the nocturnal rest phase. This decrease in glucose tolerance during the night is dependent on the occurrence of sleep, in particular slow wave sleep (SWS) [7]. Repetitive partial sleep deprivation (*sleep debt*) has been shown to inhibit nocturnal glucose tolerance reduction and stimulate appetite on the following day [8].

Sleep loss also activates sympathetic nervous activity which can for example be seen in an increased heart rate variability. Disturbances in the secretory profiles of counter-regulatory hormones such as growth hormone (GH) and cortisol may also contribute to the alterations in glucose metabolism observed during sleep loss. Six days of recurrent partial sleep restriction in young adults were associated with a broadening of the night time GH peak [9] and a marked increase in evening cortisol levels [10]. Elevated GH levels may induce a decrease in muscular glucose uptake which adversely affects glucose regulation. Similarly, high evening cortisol concentrations promote reduced insulin sensitivity on the following morning [6]. Sleep loss is also associated with increased levels of pro-inflammatory cytokines and low grade inflammation, a condition known to predispose to insulin resistance and diabetes [11,12].

Several studies indicate that sleep loss may act on the regulation of human leptin and ghrelin levels, thus directly affecting hunger and appetite control. For example, in a study that subjected healthy young men to 4 h bedtimes for 6 nights followed by 6 nights of recovery sleep, mean leptin levels were 19% lower, the nocturnal acrophase was 2 h earlier and 26% lower, and the amplitude of the diurnal variation was 20% lower during sleep restriction [13]. These changes occurred despite keeping a stable caloric intake and comparable amounts of physical activity [13]. In another large scale study, ghrelin levels were negatively associated with sleep duration. This suggests that by leptin/ghrelin regulation sleep loss could promote an increased appetite which is consistent with reports of elevated food intake in human subjects and in laboratory rodents submitted to total sleep deprivation [14,15].

Closely related to sleep debt effects are the consequences of shift work schedules and *social jet lag*. The latter term, first coined by Till Roenneberg, refers to accumulating sleep debts during the week due to work hours that are incompatible with the natural activity rhythms of the individual [16]. Late risers (so called *owls*) tend to constantly accumulate sleep debts over the working week due to an enforced early wake up time combined with late bedtimes, mainly defined by their internal clockwork. Early risers (*larks*), on the other hand, normally cope well with the standard early work hours, but tend to become sleep deprived during weekends when social events make them stay awake beyond their normal bedtime. Sleep debts can amount up to several hours per day in extreme cases and it has been suggested that social jet lag might be one of the major factors influencing the attention capacities of teenage pupils which tend to have rather late activity rhythms when compared with children or adults [16].

In any urban society an estimated fifth of all people work on alternating shift schedules [17]. Shift workers live much of their lives out of phase with normal local time. However, their circadian rhythms often cannot fully adjust to their activity rhythms due to the changing schedules of the shift work and due to the necessary readjustment to rest days, i.e. during weekends [18]. Thus, shift workers experience regularly re-occurring episodes of jet lag combined with intentional exposures to light at night that, as a potent *Zeitgeber* (= synchroniser), has the potential to disrupt normal circadian physiological and behavioural rhythms. Both mechanisms can evoke a multitude of downstream effects, re-organising the entire physiological state. It has been shown that constant lighting conditions affect the rhythmicity of several hormones including prolactin [19], glucocorticoids [17,20], adrenocorticotropic hormone (ACTH), corticotrophin releasing hormone (CRH) [21], serotonin [22], and melatonin [23]. Exposure to a normal incandescent bulb at night requires only 40 min to suppress melatonin levels by 50% [24]. Such changes in melatonin production and release can affect metabolism, immune function, and endocrine balances via the reproductive, adrenal, and thyroid endocrine axes [25]. Detrimental effects of shift work have been observed for carbohydrate and lipid metabolism, insulin resistance, hypertension, coronary heart disease, and myocardial infarction [26]. They could be the result from either direct physiological effects of light exposure (i.e. via melatonin suppression) or indirect effects associated with a lack of sleep [26]. Melatonin reduces the activity of the sympathetic nervous system and significantly reduces noradrenalin turnover in the heart, thus inhibiting the uptake of low density lipoprotein (LDL) cholesterol [27] which correlates well with the sympathetic activation observed after sleep deprivation. Effects on other hormones are less direct and might reflect a deregulation of circadian timing at the level of the eye, the SCN or downstream endocrine glands such as the pituitary or the adrenal. At least some of the symptoms persist in sleep deprived subjects that receive only low light exposure [28]. Therefore, it remains to be determined to which extend the metabolic disorders present in night shift workers reflect direct effects of light on circadian organisation or downstream processes such as sleep disruption.

Another aspect that only recently appeared as an intervening factor in the interaction between circadian rhythms and metabolic regulation is food. An abnormal timing of food uptake rapidly shifts clock gene expression rhythms in most peripheral organs, un-coupling these clocks from the master pacemaker of the SCN [29]. It appears that the synchronising power of food depends on its nutritive content [30]. Modern lifestyle promotes late dinners of high caloric value, while traditional breakfasts and lunches are often replaced by relatively light *snacks*. It seems likely that this historically young *caloric load shift* might affect peripheral clock entrainment and metabolic regulation. Tightly controlled experiments are still missing, but numerous large scale studies on extreme late dining from Ramadan practitioners point into that direction. In one such study a shift in the onset of cortisol and testosterone secretion, an enhancement of the evening peak of prolactin and a blunting of TSH rhythms was observed [31]. Moreover, a nocturnal increase [32] and a diurnal decrease in cortisol serum levels in Ramadan practitioners suggest that the amplitude of cortisol rhythm is also decreased [32,33]. An increase in afternoon cortisol was also reported in a recent study where blood sampling was carried out at one time point only [34]. In another study, circadian changes in melatonin levels were assessed [31] showing that the amplitude of the melatonin rhythm decreased during Ramadan fasting. Other studies conclude that the circadian rhythms of glucose, insulin, gastrin, gastric pH and calcium are altered [35]. Of note, shifts were also observed in the sleep/wake cycle [36].

Other substances – including various pharmaceutical compounds – are known to affect circadian rhythms and metabolism at the same time. These include numerous neurotransmission interacting drugs such as the anti-depressant olanzapine [37,38] or the anti-psychotic risperidone [39,40], but also non-CNS targeting compounds including anti-hypertensiva such as hydralazine [41,42] and nifedipine [43,44], and immunosuppressants such as prednisone [45,46] and infliximab [47,48]. It is intriguing to speculate that at least some of the metabolic side-effects of these drugs might stem from a disruption of (peripheral) circadian rhythms, and chronopharmacologic approaches to the administration of these drugs will certainly prove beneficial. A well-studied example is the glucocorticoid analogue prednisone. GC levels are tightly circadian clock controlled with peak secretion from the adrenal cortex at the beginning of the activity phase, correlating with the weakened immune responsiveness during early morning hours in humans [49]. Prednisone treatment is most effective during evening hours and metabolic side effects can be minimised by reducing or omitting the morning dosage [50].

Because of the complexity and the network nature of the circadian timing system, it is difficult to determine where exactly these drugs impinge on circadian timekeeping. Only recently, the tools have been developed allowing us to dissect the different components of the mammalian circadian clockwork and study their contribution to physiological and metabolic regulation in the living organism.

5.3 Central and Peripheral Clocks

The discovery of the first murine clock gene, *Clock*, by Martha Hotz Vitaterna and Joe Takahashi in 1994 triggered a revolution in mammalian chronobiology [51]. In the next few years, a whole row of other genes were identified – mostly by sequence homology with *Drosophila* – and we now have a good working model of how cellular circadian timekeeping is organised at the molecular level [52–54]. At the level of the organism, things appeared rather clear until recently: The circadian clock of the suprachiasmatic nuclei (SCN) receives light information via the hypothalamic tract and, by the rhythmic release of humoral factors, synchronises behavioural and physiological rhythms in the periphery [55]. A series of groundbreaking papers in the early 2000s, however, profoundly changed this view, showing that autonomous circadian clocks exist in many – some say all – central and peripheral tissues and even in many cultured cell lines [56–58]. We now believe that the SCN represents rather a conductor than a driver of peripheral circadian rhythms. It relays external time information – perceived by light signalling from a subset of photosensitive retinal ganglion cells (Chap. 4) – via humoral and neuronal signals to thousands of peripheral clocks throughout the body [59].

Like in the SCN, clock genes/proteins in the periphery drive the rhythmic expression of hundreds of clock controlled genes (CCGs). The nature of these CCGs varies across different tissues, allowing for a specific translation of a uniform time signal (clock gene rhythm) into physiologically meaningful commands [60,61]. It is estimated that between five and ten percent of all transcripts are clock controlled in each organ and tissue, many of those encoding key regulatory enzymes of physiologically important pathways [62].

This system is further complicated by the fact that not all rhythmic genes in a tissue are controlled by a local circadian oscillator. Others – and sometimes including even clock genes themselves, such as *Per2* in the liver [63] – can be regulated by external factors such as hormones or variations in temperature that are ultimately controlled by the SCN (e.g. via rhythmic endocrine systems like the hypothalamus-pituitary-adrenal (hpa) axis) [64]. Others, such as adrenal and retinal clocks, are directly light-responsive and, hence, do not depend on a functional SCN pacemaker to track time [65–67]. Only few studies exist that have analysed the division of work between SCN and peripheral clocks in the regulation of specific physiological parameters. Studies from our group have shown that the SCN and a peripheral clock in the adrenal cortex together regulate the circadian rhythm of glucocorticoid (GC) secretion from this endocrine gland (Fig. 5.1) [66,68]. While the SCN controls the rhythmic release of adrenocorticotropic hormone (ACTH) via the hpa axis, the adrenal clock gates the sensitivity of the adrenal cortex to stimulation by ACTH. Only when hpa axis activity and ACTH sensitivity are synchronised, a stable and high-amplitude GC is achieved which itself provides a strong synchronising signal for the other peripheral clocks, e.g. in the liver [66].

A different mechanism has been proposed for the role of the liver clock in the regulation of glucose homeostasis (Fig. 5.1) [69]. In wild-type mice, the SCN drives feeding rhythms and, thus, rhythms of glucose supply from nutrients. These cycles are counteracted by anti-phased rhythms of glucose export from the liver controlled via rhythmic activation of key regulators of glycogen mobilisation such as glycogen synthase (GYS2), glycogen phosphorylase (PYGL), phosphoenolpyruvate carboxykinase (PEPCK), glucokinase (GCK), glucose-6-phosphatase (GLC-6-Pase), and the glucose transporter GLUT2 by hepatocyte clocks [60]. Together, SCN-driven systemic glucose uptake and liver glucose release rhythms result in a nearly constant regulation of blood glucose levels over the whole day [70].

Feedback from peripheral oscillators to the SCN is likely, but effects are difficult to assess because of the dominant role of external *Zeitgebers* (e.g. light) in synchronising SCN activity and only few studies so far have addressed this question. It seems that sex steroids interfere with pacemaker function as shown with gonad-ectomised rodents [71]. A GC feedback has been proposed [72], but SCN neurons do not or only weakly express glucocorticoid receptors (GCRs) [73] and this process likely employs other GCR-positive brain areas that relay GC signals to the SCN by neuronal means such as the paraventricular nucleus (PVN) or the dorso-medial nucleus of the hypothalamus (DMH) [72]. One study has shown that – at least after several weeks – metabolic synchronisation of the periphery by scheduled meals can reset the SCN [74], but other reports argue

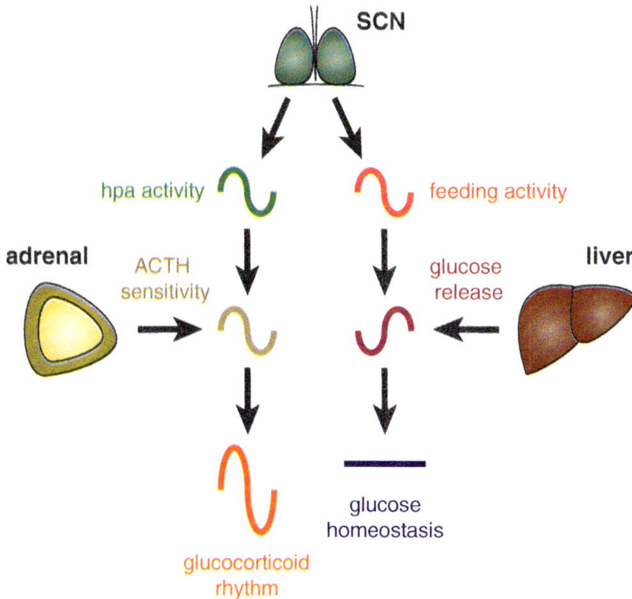

Fig. 5.1 Principles of SCN/periphery interaction in the regulation of physiological rhythms and homeostasis. The SCN clock controls rhythms of hpa axis activity (*via the pituitary*) and feeding behaviour (*via the control of rest/activity cycles*). In the adrenal cortex, a peripheral oscillator gates sensitivity of the adrenal steroidogenic machinery to stimulation by ACTH. Synchrony of SCN and adrenal clock rhythms results in a high amplitude GC output. On the other hand, the liver clock controls glucose release from hepatic glycogen storages with a rhythm anti-phased to that of feeding behaviour and, hence, glucose uptake. During the rest period glucose release from the liver can substitute food glucose in the blood to maintain stable glucose levels throughout the day

against this finding [29,75]. Another candidate for integrating feedback signals to the SCN is the arousal system that regulates the levels of dopamine and noradrenalin in the CNS [76,77], both of which have the capacity to affect circadian clocks in vivo [78,79].

5.4 Circadian Clocks of Metabolic Tissues

Circadian clocks can interact with metabolic function at three different anatomical levels: regulation of feeding behaviour by the SCN, regulation of metabolic control centres within the hypothalamus, and regulation of physiological processes in metabolically active tissues. Neural tracing studies have revealed projections from the SCN to hypothalamic cell clusters that express orexigenic and anorexigenic neuropeptides [80,81] such as the subparaventricular zone (SPZ) and the DMH [77] (Chap. 10). Destruction of the ventral SPZ reduces circadian sleep/wake rhythms but has little effect on body temperature [82]. Opposite effects are seen when targeting the dorsal SPZ [82], demonstrating an anatomical dissociation of circadian regulation of sleep-wakefulness and body temperature [77]. Ablation of the DMH affects circadian regulation of sleep/wakefulness, locomotor activity, hpa axis activity, and feeding behaviour [83]. The DMH also relays SCN outputs to other regions of the brain, including the ventrolateral preoptic nucleus, the PVN, the *locus coeruleus* and the lateral hypothalamus, which regulate sleep, corticosteroid release, and wakefulness/feeding, respectively. Interestingly, the DMH has also been implicated in the ability of organisms to be entrained by time restricted feeding. DMH neurons show robust circadian *Per2* oscillations following food restriction, and ablation of the DMH eliminates the altered activity and feeding rhythms characteristic of food-restricted animals [84,85]. However, controversy remains concerning the precise anatomic localisation of the food-entrainable oscillator (Chap. 10) because Landry et al. [86] have reported that DMH ablation fails to disrupt these phenomena. Destruction of the ventromedial hypothalamic (VMH), PVN, and DMH regions results in obesity [87], whereas ablation of the lateral hypothalamic nucleus promotes anorexia [87]. It has yet to be determined whether these nuclei host their own circadian clocks or whether circadian rhythms in these tissues are driven by rhythmic input from the SCN. Nevertheless, these findings provide an anatomical framework for future studies examining the integration of hormonal and nutrient signals at the level of the hypothalamus.

The fact that the adipocyte satiety signal leptin is expressed in a circadian fashion in addition to fluctuating in response to fasting and feeding [88–92] and that the leptin receptor is strongly expressed in various regions within the hypothalamus, including the arcuate nucleus, the DMH, and the VMH [93], suggests that humoral signals derived from peripheral tissues may transmit the nutritional status of the organism to the hypothalamic centres controlling hunger and satiety in a circadian clock-dependent manner. In the brain, leptin signalling is promoted via the melanocortin system (57, 72). Many of the nuclei involved in feeding behaviour receive input from the SCN and display pronounced circadian rhythms of gene expression (40, 52, 54, 55). These include neurons in the lateral hypothalamic area that produce the hunger-stimulating neuropeptides melanin-concentrating hormone and orexins [94–96]. Orexins A and B display a circadian rhythm of expression. They are strongly induced by fasting [97,98] and are involved in the regulation of sleep-wake rhythms [98–100].

Circadian and energetic centres are intimately linked at both anatomical and endocrine levels. Recent studies suggest that such connections also extend to co-regulation of metabolic and circadian transcription networks within the individual peripheral tissues. The strong impact of circadian rhythms on metabolism is exemplified by microarray studies that have examined gene expression profiles throughout the circadian cycle in the mammalian liver, skeletal muscle, and brown and white adipose tissues (reviewed in [62]). In each tissue, 3–20 % of the transcriptome were shown to be expressed with a distinct circadian rhythm. Many of these CCGs have roles in biosynthetic and metabolic processes, including cholesterol and lipid metabolism, glycolysis and gluconeogenesis, oxidative phosphorylation, and detoxification pathways. Of note, many rate-limiting enzymes are under circadian control [60], suggesting that the clock regulates tissue physiology via the control of key check points of metabolic pathways. Supporting evidence comes from a comprehensive analysis of nuclear receptor mRNA transcription measured throughout the daily cycle in liver, skeletal muscle, brown and white adipose tissues [101]. More than 50 % of the 49 nuclear receptor mRNAs were rhythmic in most tissues. Nuclear receptors interact with dietary lipids and lipophilic hormones, suggestive of a direct link between nutrient-sensing pathways and the circadian control of gene expression.

Many possible signals may control rhythmic gene expression in peripheral tissues, including the central SCN clock, an as yet un-located *food-entrainable oscillator*, local clocks (which themselves are coupled to the SCN), or the absence/presence of food itself. To start to address such issues, Kornmann et al. have generated a conditional transgenic mouse model in which the circadian clock can be reversibly and specifically disabled only in the liver, keeping the rest of the circadian system intact [63]. Microarray analysis revealed that the great majority of (but not all) rhythmic genes stop cycling after liver clock disruption, indicating that expression of these genes is driven by the local liver clock and not by rhythmic systemic signals. This shows that the control of metabolically relevant hepatic gene expression is complex, most likely responding to both systemic and cell-autonomous signals to maintain appropriate temporal coordination of metabolic pathways [102].

Examples of molecular linkers between the metabolic state and peripheral clock function have been described, including PGC-1α, PPARα, RARα/RXRα, and REV-ERBα. PGC-1α stimulates *Bmal1* expression through co-activation of ROR proteins. Liver-specific *knock-down* of PGC-1α disrupts liver rhythms, suggesting that this protein is required for normal clock function in this tissue [103]. *Bmal1* is also transcriptionally regulated by the nuclear receptor PPARα, a key regulator of lipid metabolism [104,105]. In vascular tissues, RARα and RXRα can directly bind to CLOCK and the closely related NPAS2 protein in a hormone-dependent manner, causing inhibition of CLOCK/BMAL1 or NPAS2/BMAL1 activity [106]. Recent studies of REV-ERBα have suggested that the metabolic sensor haem is an endogenous ligand of this core clock protein, raising the possibility that haem may play a central role in coordinating circadian function itself [107,108].

Emerging evidence indicates that clock gene function also plays a crucial role in many other metabolic tissues [109]. Circadian clock gene rhythms have been described in the gut [110], the pancreas [111], white adipose tissue [112], and the endocrine glands including the pituitary [113], the ovaries [114], and the adrenal [66], most of which are responsive to changes in feeding regimens and nutritional status. Clocks within the skeletal muscle may affect the alternation between glycolysis and β-oxidation along the day [115]. Clock function may also be important in adipogenesis (a process that also affects insulin sensitivity) and adipocyte endocrine signalling [116], as well as impact on both adrenal [66] and pancreatic physiology [117]. In the adrenal, clock gene function has been shown to regulate both catecholamine and glucocorticoid secretion [118]. Circadian function at the level of both brain and peripheral organs may also influence the counter-regulatory response to hypoglycemia [66,119].

5.5 Mechanisms of Metabolic Circadian Control

In summary, circadian clocks can impact on metabolic processes via four different routes: (1) the transcriptional control of CCGs and, ultimately, the regulation of the concentration of enzymes available for metabolic pathways; (2) system-wide synchronisation of metabolic state via rhythmic release of humoral factors; (3) neuronal regulation of target organs, e.g. via activation and de-activation of sympathetic and parasympathetic synapses; and (4) routes that require interaction with the external environment and, hence, involve the regulation of activity and behaviour (Fig. 5.2).

Genomic and proteomic studies have shown that a surprisingly large fraction of the transcriptome/proteome of a given tissue is under the control of the circadian clock [62]. It is believed that the majority of these genes/proteins are regulated by the local circadian oscillator as opposed to being responsive to rhythmic signals from outside the cell/tissue. A regulation of tissue physiology by CCGs would not

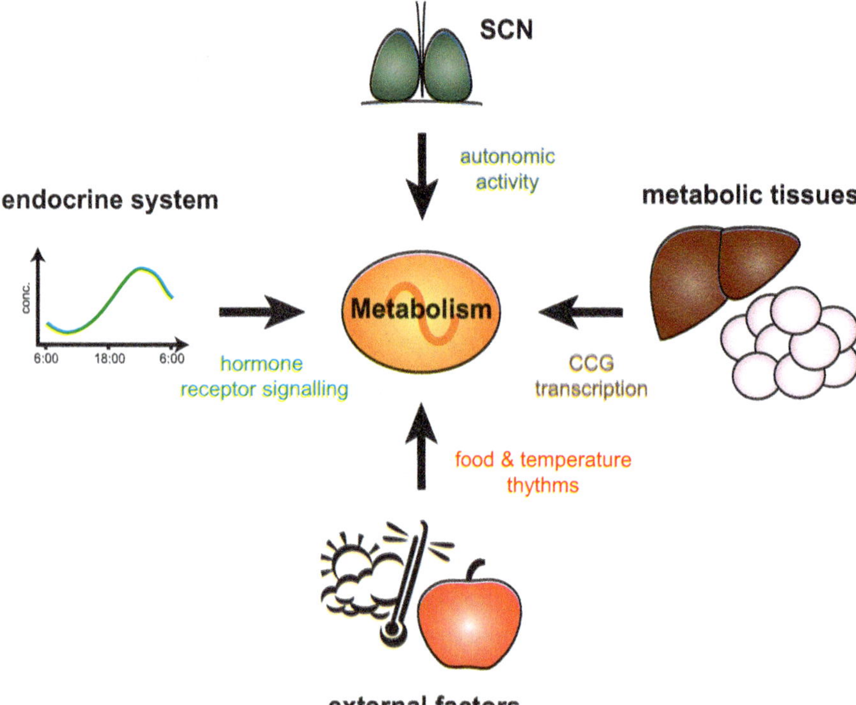

Fig. 5.2 Circadian inputs into metabolic regulation. The circadian timing system can interact with metabolic regulation in four different ways. (a) The SCN controls the activity of the autonomic nervous system. (b) Peripheral clocks control rhythmic expression of clock controlled genes (CCGs). (c) Rhythmically secreted humoral factors affect the physiology of target tissues. (d) Factors that depend on interaction of the organism with its environment such as temperature (via locomotor activity) and food uptake influence tissue function. For details see text

make sense if the same clock targets would exist in each cell of the body. Comparative studies, however, have shown that only very few CCGs overlap between different tissues [60,61]. In fact, each organ contains its own circadian fingerprint that reflects its physiological role for the organism. The mechanisms of tissue specialisation of CCGs are unknown – and only very few circadian regulatory motifs have been identified so far [120] – but may involve the expression of co-factors interacting with the main core clock transcription factors such as CLOCK/BMAL1, REV-ERBα/RORx and DBP/E4BP [53].

The most extensive data set of CCGs has been identified for the hepatocyte clock. Liver CCGs include genes involved in nutrient metabolism such as the glutamate transporter *Glut2*, *pyruvate kinase* and *lipin*, genes important for cholesterol synthesis and processing such as *Hmgcr* and *Npc1*, as well as genes involved in xenobiotic metabolism and detoxification such as the methyltransferase *Nnmt* and the organic cation transporters *Slc22a1* and *2* [60]. CCGs in the adipocyte include key regulators of fatty acid metabolism such as *Fatp1* (*fatty acid transport protein 1*), *Acs1* (*fatty acyl-CoA synthetase 1*), *Adrp* (*adipocyte differentiation-related protein*), and the satiety hormone leptin as well as other adipokines such as adipsin (*Cfd*), resistin (*Rstn*), adiponectin (*Apm1*), and visfatin (*Pbef1*) [121]. Other rhythmic genes have been identified in tissues such as stomach, skeletal muscle and the pancreas [122]. It remains to be shown, however, to which extend these genes are controlled by local circadian clocks or if they are merely responsive to the external time stimuli described below.

Rhythmic hormones that have been shown to influence the metabolic processes derive mainly from hpa and hpt (hypothalamus-pituitary-thyroid) axes activity. Others include melatonin, secreted from the pineal gland, and hypothalamic peptides such as orexin and vasopressin. The adrenal hormones, glucocorticoids (GCs), mineralcorticoids and catecholamines, show circadian rhythms of secretion under non-stressed conditions [118]. The steroidal corticoids signal via nuclear hormone receptors and direct activation of target genes. GCs are potent synchronisers of peripheral circadian clocks and synthetic analogues such as dexamethasone are routinely used to temporally synchronise fibroblast cells in culture [56,57]. In the liver, GCs increase the activities of enzymes involved in fatty acid synthesis and promote the secretion of lipoproteins [123,124]. Since liver lipids appear to be involved in the negative regulation of hepatic insulin sensitivity [125] and are associated with aspects of the Metabolic Syndrome [126–129], GC-induced hepatic fat accumulation is likely to contribute to the pathophysiology of the Metabolic Syndrome. GCs also activate the hepatic gluconeogenic pathway by activation of phosphoenolpyruvate carboxykinase (PEPCK) and glucose-6-phosphatase (G6Pase) transcription, both rate-limiting enzymes of gluconeogenesis [130,131]. The resulting elevation in hepatic glucose output favours hyperglycemia. In adipose tissue, GCs promote the differentiation of pre-adipocytes to adipocytes which could lead to increased body fat [132,133]. However, once differentiated, adipocytes develop insulin resistance in the presence of GCs with decreased insulin-stimulated glucose uptake [134]. The reduced insulin sensitivity is mediated by GCs antagonising the insulin-stimulated translocation of glucose transporters from intracellular compartments to the plasma membrane [135,136]. A similar mechanism might be responsible for the GC-induced insulin resistance in skeletal muscle [137]. GCs also inhibit the insulin secretion by the pancreatic β cells in animals and perturb high-frequency insulin release during fasting in humans [138,139].

Thyroid hormones enter the cells through membrane transporter proteins. There, they stimulate diverse metabolic activities, leading to an increase in basal metabolic rate. One consequence of this activity is to increase body heat production, which might result, at least in part, from increased oxygen consumption and rates of ATP hydrolysis [140]. Elevated thyroid hormone levels stimulate fat mobilisation, leading to increased concentrations of fatty acids in the blood. They also enhance oxidation of fatty acids in many tissues [141]. Finally, plasma concentrations of cholesterol and triglycerides are inversely correlated with thyroid hormone levels – one diagnostic indication of hypothyroidism is an increased blood cholesterol concentration [142]. Thyroid hormones stimulate various aspects of carbohydrate metabolism, including the insulin-dependent entry of glucose into cells, gluconeogenesis and glycogenolysis [137].

Suppression of melatonin secretion has been associated with metabolic disorders like diabetes [143,144], implying a function for melatonin in the regulation of glucose homeostasis [145]. Interestingly, in type 2 diabetes, endogenous glucose production and gluconeogenesis display diurnal rhythms that drive the fasting hyperglycemia typical for these patients and are absent in healthy humans [146]. The melatonin receptors MT1 and MT2 are expressed in human pancreatic tissue [144] and in rat islets [147]. Melatonin stimulates insulin secretion in pancreatic β-cells by IP_3 release and, ultimately, the elevation of intracellular Calcium. Melatonin also affects the lipoprotein exchange in the liver. It ameliorates liver damage in response to high fat feeding or toxin administration in rodents. These effects, however, are difficult to disentangle from its function as antioxidant. Long term effects of melatonin administration on metabolic parameters in rats have been reported, but could as well reflect secondary effects [148].

The orexins (orexin A and B, or hypocretins-1 and -2) are synthesised by specialised neurons in the posterolateral hypothalamus [149]. Both neuropeptides act as potent agonists at the hypocretin-1 and -2 G-protein-coupled receptors, which show different distributions in the brain and differential affinities for the two orexins [150]. Orexin A neurons project to the main arousal centres of the brain stem where they antagonize sleep and muscle atonia and promote wakefulness and arousal [77]. Orexinergic neurons are activated by monoamines, acetylcholine, and metabolic cues, including leptin, glucose, and ghrelin [151]. Other activators are neurotensin, oxytocin, and vasopressin, all of which have functional interactions with hypothalamic feeding systems and monoaminergic and cholinergic centres in the brain [151]. Hence, orexin constitutes a critical link between peripheral energy balance and the central mechanisms coordinating sleep/wakefulness and motivated behaviour (e.g. locomoter activity).

Hormones act broadly throughout the body and obtain their specificity by the expression patterns of their respective receptors, whereas neurons deliver their message to a precisely targeted tissue in the body. The SCN uses both ways to impose its rhythmicity onto the rest of the body. The circadian control of the daily GC rhythm is a good example for how the SCN uses humoral and neuronal cues to create a robust, high amplitude physiological rhythm. A direct input to the neuro-endocrine neurons of the hypothalamus controls the release of ACTH from the pituitary while the autonomic nervous system controls the adrenal sensitivity for ACTH [66]. This dual control mechanism may hold not only for endocrine glands, but also for the other organs. For instance, at the time of awakening, the SCN not only increases insulin sensitivity to cause a physiologically relevant increase in muscle glucose uptake, but at the same time stimulates glucose production by the liver to ensure sufficient glucose availability [152].

5.6 Impact of Clock Disruption on Metabolic Homeostasis

Several lines of evidence suggest that circadian deregulation may impact on metabolic function on a broad scale. At the clinical and epidemiologic level, circadian disruption is associated with cardiovascular and metabolic complications across large segments of the human population (reviewed in [153]). Cross-sectional studies revealed an increased prevalence of metabolic syndrome, obesity, and cardiovascular events in shift workers [154]. These findings raise the possibility that chronic misalignment between the cycles of rest and activity, and those of fasting and feeding, may contribute to the initiation and progression of obesity and metabolic syndrome. With regard to the latter, studies in humans suggest that nocturnal feeding regimens (*night-eating syndrome*) may represent an independent risk for metabolic disease [155].

These findings are in line with recent experimental studies in rodents showing that excess energy uptake during diet-induced obesity is caused by an increased feeding only during the rest period and not during the active period when the animal would normally eat [156]. Thus, it appears that the capacity to defend long-term energy homeostasis is related both to the time of day when food is consumed and to the phase relationship between meal time and the sleep/wake cycle. In animal models, rapid shifting of the external light/dark cycle, a so-called *jet lag* paradigm, has been associated with accelerated cardiomyopathy and premature mortality [157,158].

In view of the fact that metabolic networks are under extensive circadian control at the levels of transcription, translation, and cellular signalling, it could be expected that the integration of circadian and metabolic cycles confers adaptive advantage and optimises energy utilisation. Experiments addressing this hypothesis have been performed in *Cyanobacteria* and *Arabidopsis*, indicating that synchrony of endogenous circadian period length with that of the environment may play a direct role in survival and optimise biological functions ranging from photosynthesis to reproduction [159,160]. The availability of mutant mice with different circadian period lengths now provides the opportunity to formally test the concept that circadian "resonance" (i.e., synchrony between the external light-dark cycle and the internal period) is important in metabolic health and energy balance.

Analyses of mice with genetic lesions that disrupt circadian rhythms have provided insight into the role of several of the major circadian clock genes in physiology and metabolism (Fig. 5.3). Homozygous *Clock* mutant mice display severe alterations in energy balance, with a phenotype showing many characteristics of the Metabolic Syndrome, including obesity, hyperlipidemia, hepatic steatosis, high circulating glucose, and low insulin levels [161]. Feeding rhythms in these mice are dampened, with increased food intake during the day, resulting in significantly increased overall energy consumption. It is likely that this phenotype results, at least in part, from altered rhythms of neuropeptides in the hypothalamus. Ghrelin, CART, and orexin are all expressed at decreased levels in *Clock* homozygous mutants [161]. Whether loss of clock function in peripheral tissues might also contribute to this metabolic phenotype still remains to be seen.

The question was raised whether the phenotypic changes seen in these animals really constitute an effect of circadian clock disruption or whether the mutated

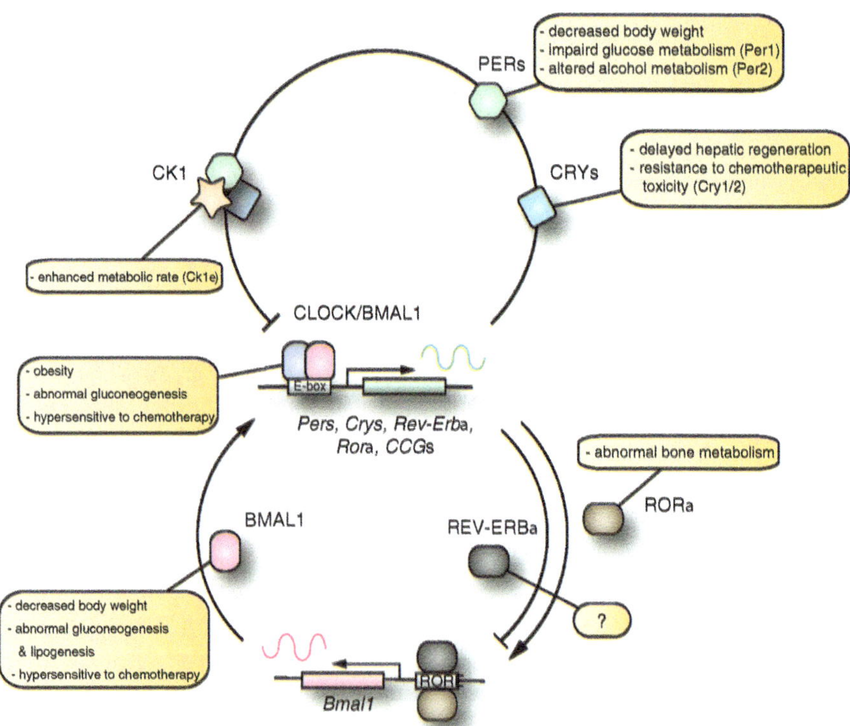

Fig. 5.3 Metabolic consequences of clock gene disruption. Shown are the two main feedback loops of the mammalian circadian timing system. CLOCK and BMAL1 activate E-box-regulated genes including *Per1*, *Per2*, *Cry1*, *Cry2*, *Rev-Erbα* and *Rorα*. PER and CRY proteins form complexes with casein kinase 1 (CK1) and relocate back into the nucleus where they inhibit CLOCK/BMAL1 activity. REV-ERBα and RORα compete for RORE elements on the *Bmal1* promoter, thereby regulating *Bmal1* transcriptional activity. Orange boxes list metabolic phenotypes of the corresponding mouse mutants

genes have functions outside the organisation of circadian timing. This problem will be difficult to solve because of the interconnected regulation of the different clock components via the transcriptional feedback loops that make up the molecular clockwork. In other words, if one modifies expression of one clock gene, it will unavoidably also affect the regulation of several others. Clearly, comparative studies using different clock mutant strains will be needed to address this issue.

5.7 How Metabolism Affects Clock Function

In addition to the effects of circadian disruption on metabolic state, it has become evident that metabolism itself can affect circadian clock function [162]. It is long known that time restricted feeding schedules can entrain peripheral circadian rhythms in rodents. Although the SCN seems resistant to the synchronising effects of food restriction, non-SCN oscillators in the brain and circadian clocks in the periphery are rapidly reset by restricted feeding [29]. Although the molecular pathways by which food restriction entrains these oscillators remain elusive, mechanisms have been proposed for how metabolic state could directly access the clock machinery [163]. It was shown that the oxidative state of the cell (i.e. the NAD(P)H/NAD(P) ratio) modulates CLOCK/BMAL1 or NPAS2/BMAL1 activity by regulating their binding strength to DNA [164]. Moreover, the clock proteins NPAS2 and REV-ERBα can both bind the metabolic sensor haem [107,108,165]. Haem biosynthesis itself is also under circadian regulation which postulates a metabolic feedback on the circadian clockwork [165]. If these results will hold true in vivo, then fluctuations in cellular metabolism could directly influence the transcriptional activity of circadian clock genes and provide a means by which metabolism and circadian timing could regulate each other. Supporting this view, in mice, high fat feeding disrupts circadian organisation on both molecular and behavioural levels [162]. In addition, disruption of *Pgc-1α*, a transcriptional co-activator involved in energy metabolism, is also associated with abnormal circadian rhythms of activity, temperature and metabolic rate in mice [103].

5.8 Conclusion

It seems clear that circadian clocks and metabolic regulation interact at a broad scale – in some cases identical genes are involved in both processes. Given the fact that the prevalence of complex metabolic disorders rises at an alarming rate worldwide, the study of these interactions might provide new therapeutic approaches and ultimately help neutralising the side-effects that came along with two of the greatest achievements of the modern society: the constant availability of food and light. This task is no easy one because we are looking at extremely complex systems that work at the level of the whole organism. New experimental and theoretical tools will be needed to start to disentangle the communication network that links the different circadian oscillators throughout the body and connects them with the metabolic regulatory centres. First attempts have been made with the development of tissue-specific transgenic mouse lines and in vivo reporter gene systems that allow us to monitor the molecular clock function in real-time in the living organism.

Open questions remain, such as the possibility of clock gene function outside of circadian timing. Are there clock-less tissues inside the mammalian body? And are (some of the) clock genes expressed there? Does stopping the clock at different circadian times differentially affect the metabolic homeostasis or is it only important whether the clock is running or not? Experiments in cyanobacteria and plants have beautifully demonstrated the advantage of *resonance* between internal clock rhythms and external light/dark cycles for an organism's energy consumption and, ultimately, its survival. Comparable experiments in mammals are still lacking.

The next big goal – and we have started to take the first steps along this road – will be to apply the findings of molecular chronobiology to humans and the treatment of human diseases. The concepts of chronopharmacology are long known – and widely ignored. A major reason for this is the (im)practicality of circadian measurements, especially in humans. Technical improvements are needed to rapidly and economically assess circadian clock function in human subjects and patients with the final goal of a standardised molecular *chronotyping* protocol. This will be extremely beneficial at all levels of clinical medicine, from diagnosis to therapy. Brown et al. have shown that morningness/eveningness preferences in humans are reflected in the clock gene oscillation profiles of skin-derived fibroblast cultures [166]. Given the tight interaction of clock and metabolic regulation, extending this or similar procedures to multiple clock and clock-controlled genes and other tissue sources could help predicting not only behavioural, but also metabolic circadian characteristics of humans from an in vitro experiment – with the final goal of a personalised, chronotype-adjusted medication.

Acknowledgements I would like to thank Dr. Judit Kovac for helpful comments on the manuscript. H.O.'s work is supported by an Emmy Noether fellowship of the Deutsche Forschungsgemeinschaft.

References

1. International Obesity Task Force (2003) Obesity in Europe - 2: Waiting for a green light for health? Europe at the crossroads for diet and disease. http://www.iotf.org/
2. Knutson KL et al (2007) The metabolic consequences of sleep deprivation. Sleep Med Rev 11(3):163–178
3. Kripke DF et al (1979) Short and long sleep and sleeping pills. Is increased mortality associated? Arch Gen Psychiatry 36(1):103–116
4. The National Sleep Foundation (Ed.) (2006) Sleep in America Poll. http://www.sleepfoundation.org/catalog/sleep-america-poll-reports/
5. National Center for Health Statistics (2005) QuickStats: Percentage of Adults Who Reported an Average of [less than or equal to] 6 Hours of Sleep 24-Hour Period, by Sex and Age Group–United States, 1985 and 2004. Morb Mortal Wkly Rep 54(37):917–948
6. Van Cauter E, Polonsky KS, Scheen AJ (1997) Roles of circadian rhythmicity and sleep in human glucose regulation. Endocr Rev 18(5):716–738
7. Nofzinger EA et al (2002) Human regional cerebral glucose metabolism during non-rapid eye movement sleep in relation to waking. Brain 125(Pt 5):1105–1115

8. Spiegel K et al (2004) Brief communication: sleep curtailment in healthy young men is associated with decreased leptin levels, elevated ghrelin levels, and increased hunger and appetite. Ann Intern Med 141(11):846–850
9. Spiegel K et al (2000) Adaptation of the 24-h growth hormone profile to a state of sleep debt. Am J Physiol Regul Integr Comp Physiol 279(3):R874–R883
10. Spiegel K, Leproult R, Van Cauter E (1999) Impact of sleep debt on metabolic and endocrine function. Lancet 354(9188):1435–1439
11. Vgontzas AN et al (1999) Circadian interleukin-6 secretion and quantity and depth of sleep. J Clin Endocrinol Metab 84(8):2603–2607
12. Vgontzas AN et al (1999) Sleep deprivation effects on the activity of the hypothalamic-pituitary-adrenal and growth axes: potential clinical implications. Clin Endocrinol (Oxf). 51(2):205–215
13. Spiegel K et al (2004) Leptin levels are dependent on sleep duration: relationships with sympathovagal balance, carbohydrate regulation, cortisol, and thyrotropin. J Clin Endocrinol Metab 89(11):5762–5771
14. Dzaja A et al (2004) Sleep enhances nocturnal plasma ghrelin levels in healthy subjects. Am J Physiol Endocrinol Metab 286(6):E963–E967
15. Vioque J, Torres A, Quiles J (2000) Time spent watching television, sleep duration and obesity in adults living in Valencia. Spain. Int J Obes Relat Metab Disord 24(12):1683–1688
16. Wittmann M et al (2006) Social jetlag: misalignment of biological and social time. Chronobiol Int 23(1–2):497–509
17. Scheving LE, Pauly JE (1966) Effect of light on corticosterone levels in plasma of rats. Am J Physiol 210(5):1112–1117
18. Rajaratnam SM, Arendt J (2001) Health in a 24-h society. Lancet 358(9286):999–1005
19. Vaticon MD et al (1980) Effects of constant light on prolactin secretion in adult female rats. Horm Res 12(5):277–288
20. Leproult R et al (2001) Transition from dim to bright light in the morning induces an immediate elevation of cortisol levels. J Clin Endocrinol Metab 86(1):151–157
21. Fischman AJ et al (1988) Constant light and dark affect the circadian rhythm of the hypothalamic-pituitary-adrenal axis. Neuroendocrinology 47(4):309–316
22. Snyder SH et al (1965) Control of the circadian rhythm in serotonin content of the rat pineal gland. Proc Natl Acad Sci U S A 53:301–305
23. Schernhammer ES et al (2006) Urinary 6-sulfatoxymelatonin levels and their correlations with lifestyle factors and steroid hormone levels. J Pineal Res 40(2):116–124
24. Schulmeister K, Weber M, Bogner W (2004) Application of melatonin action spectra on practical lighting issues. In The Fifth International LRO Lighting Research Symposium, Light and Human Health. The Electric Power Research Institute: Palo Alto, CA. p. 103–114
25. Prendergast BJ, Nelson RJ, Zucker I (2002) Mammalian seasonal rhythms: behavior and neuroendocrine substrates. In: Pfaff DW (ed) Hormones Brain and Behavior. Elsevier Science, San Diego, CA
26. Haus E, Smolensky M (2006) Biological clocks and shift work: circadian dysregulation and potential long-term effects. Cancer Causes Control 17(4):489–500
27. Bullough JD, Rea MS, Figueiro MG (2006) Of mice and women: light as a circadian stimulus in breast cancer research. Cancer Causes Control 17(4):375–383
28. Horne JA, Donlon J, Arendt J (1991) Green light attenuates melatonin output and sleepiness during sleep deprivation. Sleep 14(3):233–240
29. Damiola F et al (2000) Restricted feeding uncouples circadian oscillators in peripheral tissues from the central pacemaker in the suprachiasmatic nucleus. Genes Dev 14(23):2950–2961
30. Froy O (2007) The relationship between nutrition and circadian rhythms in mammals. Front Neuroendocrinol 28(2–3):61–71
31. Bogdan A, Bouchareb B, Touitou Y (2001) Ramadan fasting alters endocrine and neuroendocrine circadian patterns. Meal-time as a synchronizer in humans? Life Sci 68(14):1607–1615
32. Al-Hadramy MS, Zawawi TH, Abdelwahab SM (1988) Altered cortisol levels in relation to Ramadan. Eur J Clin Nutr 42(4):359–362

33. Ben Salem L et al (2002) Circadian rhythm of cortisol and its responsiveness to ACTH during Ramadan. Ann Endocrinol (Paris) 63(6 Pt 1):497–501
34. El-Migdadi F et al (2002) Plasma levels of adrenocorticotropic hormone and cortisol in people living in an environment below sea level (Jordan Valley) during fasting in the month of Ramadan. Horm Res 58(6):279–282
35. Aybak M et al (1996) Effect of Ramadan fasting on platelet aggregation in healthy male subjects. Eur J Appl Physiol Occup Physiol 73(6):552–556
36. Roky R et al (2001) Sleep during ramadan intermittent fasting. J Sleep Res 10(4):319–327
37. Goudie AJ, Cooper GD, Halford JC (2005) Antipsychotic-induced weight gain. Diabetes Obes Metab 7(5):478–487
38. Mann K et al (2006) Nocturnal hormone profiles in patients with schizophrenia treated with olanzapine. Psychoneuroendocrinology 31(2):256–264
39. Ayalon L, Hermesh H, Dagan Y (2002) Case study of circadian rhythm sleep disorder following haloperidol treatment: reversal by risperidone and melatonin. Chronobiol Int 19(5):947–959
40. Urichuk L et al (2008) Metabolism of atypical antipsychotics: involvement of cytochrome p450 enzymes and relevance for drug-drug interactions. Curr Drug Metab 9(5):410–418
41. Goyal RK (1999) Hyperinsulinemia and insulin resistance in hypertension: differential effects of antihypertensive agents. Clin Exp Hypertens 21(1–2):167–179
42. Janssen BJ, Tyssen CM, Struyker-Boudier HA (1991) Modification of circadian blood pressure and heart rate variability by five different antihypertensive agents in spontaneously hypertensive rats. J Cardiovasc Pharmacol 17(3):494–503
43. Hermida RC et al (2007) Chronotherapy of hypertension: administration-time-dependent effects of treatment on the circadian pattern of blood pressure. Adv Drug Deliv Rev 59(9–10):923–939
44. O'Byrne S, Feely J (1990) Effects of drugs on glucose tolerance in non-insulin-dependent diabetics (Part I). Drugs 40(1):6–18
45. Challis J et al (1981) Loss of diurnal rhythm in plasma estrone, estradiol, and estriol in women treated with synthetic glucocorticoids at 34 to 35 weeks' gestation. Am J Obstet Gynecol 139(3):338–343
46. Thomson SP et al (2007) Adrenal steroids and the metabolic syndrome. Curr Hypertens Rep 9(6):512–519
47. Charles P et al (1999) Regulation of cytokines, cytokine inhibitors, and acute-phase proteins following anti-TNF-alpha therapy in rheumatoid arthritis. J Immunol 163(3):1521–1528
48. Tam LS et al (2007) Impact of TNF inhibition on insulin resistance and lipids levels in patients with rheumatoid arthritis. Clin Rheumatol 26(9):1495–1498
49. Simpson ER, Waterman MR (1988) Regulation of the synthesis of steroidogenic enzymes in adrenal cortical cells by ACTH. Annu Rev Physiol 50:427–440
50. Xu J et al (2008) Assessment of the impact of dosing time on the pharmacokinetics/pharmacodynamics of prednisolone. AAPS J 10(2):331–341
51. Vitaterna MH et al (1994) Mutagenesis and mapping of a mouse gene, clock, essential for circadian behavior. Science 264(5159):719–725
52. Oster H (2006) The genetic basis of circadian behavior. Genes Brain Behav 5(Suppl 2):73–79
53. Reppert SM, Weaver DR (2002) Coordination of circadian timing in mammals. Nature 418(6901):935–941
54. Takahashi JS et al (2008) The genetics of mammalian circadian order and disorder: implications for physiology and disease. Nat Rev Genet 9(10):764–775
55. Klein DC, Moore RY, Reppert SM (1991) Suprachiasmatic nucleus: the mind's clock. Oxford University Press, New York
56. Balsalobre A, Damiola F, Schibler U (1998) A serum shock induces circadian gene expression in mammalian tissue culture cells. Cell 93(6):929–937
57. Balsalobre A et al (2000) Resetting of circadian time in peripheral tissues by glucocorticoid signaling. Science 289(5488):2344–2347

58. Yoo SH et al (2004) PERIOD2::LUCIFERASE real-time reporting of circadian dynamics reveals persistent circadian oscillations in mouse peripheral tissues. Proc Natl Acad Sci U S A 101(15):5339–5346
59. Davidson AJ, Yamazaki S, Menaker M, (2003) SCN: ringmaster of the circadian circus or conductor of the circadian orchestra? Novartis Found Symp 253:110–121; discussion 121–125, 281–284
60. Panda S et al (2002) Coordinated transcription of key pathways in the mouse by the circadian clock. Cell 109(3):307–320
61. Storch KF et al (2002) Extensive and divergent circadian gene expression in liver and heart. Nature 417(6884):78–83
62. Lowrey PL, Takahashi JS (2004) Mammalian circadian biology: elucidating genome-wide levels of temporal organization. Annu Rev Genomics Hum Genet 5:407–441
63. Kornmann B et al (2007) System-driven and oscillator-dependent circadian transcription in mice with a conditionally active liver clock. PLoS Biol 5(2):e34
64. Oishi K et al (2005) Genome-wide expression analysis reveals 100 adrenal gland-dependent circadian genes in the mouse liver. DNA Res 12(3):191–202
65. Ishida A et al (2005) Light activates the adrenal gland: timing of gene expression and glucocorticoid release. Cell Metab 2(5):297–307
66. Oster H et al (2006) The circadian rhythm of glucocorticoids is regulated by a gating mechanism residing in the adrenal cortical clock. Cell Metab 4(2):163–173
67. Storch KF et al (2007) Intrinsic circadian clock of the mammalian retina: importance for retinal processing of visual information. Cell 130(4):730–741
68. Oster H et al (2006) Transcriptional profiling in the adrenal gland reveals circadian regulation of hormone biosynthesis genes and nucleosome assembly genes. J Biol Rhythms 21(5): 350–361
69. Lamia KA, Storch KF, Weitz CJ (2008) Physiological significance of a peripheral tissue circadian clock. Proc Natl Acad Sci USA 105(39):15172–15177
70. Gatfield D, Schibler U (2008) Circadian glucose homeostasis requires compensatory interference between brain and liver clocks. Proc Natl Acad Sci USA 105(39):14753–14754
71. Iwahana E et al (2008) Gonadectomy reveals sex differences in circadian rhythms and suprachiasmatic nucleus androgen receptors in mice. Horm Behav 53(3):422–430
72. Krout KE et al (2002) CNS inputs to the suprachiasmatic nucleus of the rat. Neuroscience 110(1):73–92
73. Rosenfeld P et al (1993) Ontogeny of corticosteroid receptors in the brain. Cell Mol Neurobiol 13(4):295–319
74. Abe H, Honma S, Honma K (2007) Daily restricted feeding resets the circadian clock in the suprachiasmatic nucleus of CS mice. Am J Physiol Regul Integr Comp Physiol 292(1):R607–R615
75. Mendoza J et al (2008) Daily meal timing is not necessary for resetting the main circadian clock by calorie restriction. J Neuroendocrinol 20(2):251–260
76. Mistlberger RE (2005) Circadian regulation of sleep in mammals: role of the suprachiasmatic nucleus. Brain Res Brain Res Rev 49(3):429–454
77. Saper CB, Scammell TE, Lu J (2005) Hypothalamic regulation of sleep and circadian rhythms. Nature 437(7063):1257–1263
78. Terazono H et al (2003) Adrenergic regulation of clock gene expression in mouse liver. Proc Natl Acad Sci USA 100(11):6795–6800
79. Weaver DR, Rivkees SA, Reppert SM (1992) D1-dopamine receptors activate c-fos expression in the fetal suprachiasmatic nuclei. Proc Natl Acad Sci USA 89(19):9201–9204
80. Swanson LW, Cowan WM (1975) The efferent connections of the suprachiasmatic nucleus of the hypothalamus. J Comp Neurol 160(1):1–12
81. Watts AG, Swanson LW, Sanchez-Watts G (1987) Efferent projections of the suprachiasmatic nucleus: I. Studies using anterograde transport of Phaseolus vulgaris leucoagglutinin in the rat. J Comp Neurol 258(2):204–229

82. Lu J et al (2001) Contrasting effects of ibotenate lesions of the paraventricular nucleus and subparaventricular zone on sleep-wake cycle and temperature regulation. J Neurosci 21(13):4864–4874
83. Chou TC et al (2003) Critical role of dorsomedial hypothalamic nucleus in a wide range of behavioral circadian rhythms. J Neurosci 23(33):10691–10702
84. Gooley JJ, Schomer A, Saper CB (2006) The dorsomedial hypothalamic nucleus is critical for the expression of food-entrainable circadian rhythms. Nat Neurosci 9(3):398–407
85. Mieda M et al (2006) The dorsomedial hypothalamic nucleus as a putative food-entrainable circadian pacemaker. Proc Natl Acad Sci U S A 103(32):12150–12155
86. Landry GJ et al (2006) Persistence of a behavioral food-anticipatory circadian rhythm following dorsomedial hypothalamic ablation in rats. Am J Physiol Regul Integr Comp Physiol 290(6):R1527–R1534
87. Anand BK, Brobeck JR (1951) Localization of a "feeding center" in the hypothalamus of the rat. Proc Soc Exp Biol Med 77(2):323–324
88. Ahren B (2000) Diurnal variation in circulating leptin is dependent on gender, food intake and circulating insulin in mice. Acta Physiol Scand 169(4):325–331
89. Kalsbeek A et al (2001) The suprachiasmatic nucleus generates the diurnal changes in plasma leptin levels. Endocrinology 142(6):2677–2685
90. Licinio J et al (1997) Human leptin levels are pulsatile and inversely related to pituitary-adrenal function. Nat Med 3(5):575–579
91. Licinio J et al (1998) Synchronicity of frequently sampled, 24-h concentrations of circulating leptin, luteinizing hormone, and estradiol in healthy women. Proc Natl Acad Sci USA 95(5):2541–2546
92. Schoeller DA et al (1997) Entrainment of the diurnal rhythm of plasma leptin to meal timing. J Clin Invest 100(7):1882–1887
93. Elmquist JK et al (1998) Distributions of leptin receptor mRNA isoforms in the rat brain. J Comp Neurol 395(4):535–547
94. Flier JS, Maratos-Flier E (1998) Obesity and the hypothalamus: novel peptides for new pathways. Cell 92(4):437–440
95. Friedman JM, Halaas JL (1998) Leptin and the regulation of body weight in mammals. Nature 395(6704):763–770
96. Schwartz MW et al (2000) Central nervous system control of food intake. Nature 404(6778): 661–671
97. Sutcliffe JG, de Lecea L (2000) The hypocretins: excitatory neuromodulatory peptides for multiple homeostatic systems, including sleep and feeding. J Neurosci Res 62(2):161–168
98. Willie JT et al (2001) To eat or to sleep? Orexin in the regulation of feeding and wakefulness. Annu Rev Neurosci 24:429–458
99. Lin L et al (1999) The sleep disorder canine narcolepsy is caused by a mutation in the hypocretin (orexin) receptor 2 gene. Cell 98(3):365–376
100. Nishino S et al (2000) Hypocretin (orexin) deficiency in human narcolepsy. Lancet 355(9197):39–40
101. Yang X et al (2006) Nuclear receptor expression links the circadian clock to metabolism. Cell 126(4):801–810
102. Schibler U, Ripperger J, Brown SA (2003) Peripheral circadian oscillators in mammals: time and food. J Biol Rhythms 18(3):250–260
103. Liu C et al (2007) Transcriptional coactivator PGC-1alpha integrates the mammalian clock and energy metabolism. Nature 447(7143):477–481
104. Canaple L et al (2006) Reciprocal regulation of brain and muscle Arnt-like protein 1 and peroxisome proliferator-activated receptor alpha defines a novel positive feedback loop in the rodent liver circadian clock. Mol Endocrinol 20(8):1715–1727
105. Inoue I et al (2005) CLOCK/BMAL1 is involved in lipid metabolism via transactivation of the peroxisome proliferator-activated receptor (PPAR) response element. J Atheroscler Thromb 12(3):169–174

106. McNamara P et al (2001) Regulation of CLOCK and MOP4 by nuclear hormone receptors in the vasculature: a humoral mechanism to reset a peripheral clock. Cell 105(7):877–889
107. Raghuram S et al (2007) Identification of heme as the ligand for the orphan nuclear receptors REV-ERBalpha and REV-ERBbeta. Nat Struct Mol Biol 14(12):1207–1213
108. Yin L et al (2007) Rev-erbalpha, a heme sensor that coordinates metabolic and circadian pathways. Science 318(5857):1786–1789
109. Zvonic S et al (2007) Circadian rhythms and the regulation of metabolic tissue function and energy homeostasis. Obesity (Silver Spring) 15(3):539–543
110. Hoogerwerf WA et al (2007) Clock gene expression in the murine gastrointestinal tract: endogenous rhythmicity and effects of a feeding regimen. Gastroenterology 133(4): 1250–1260
111. Mühlbauer E et al (2004) Indication of circadian oscillations in the rat pancreas. FEBS Lett 564(1–2):91–96
112. Ando H et al (2005) Rhythmic messenger ribonucleic acid expression of clock genes and adipocytokines in mouse visceral adipose tissue. Endocrinology 146(12):5631–5636
113. von Gall C et al (2002) Rhythmic gene expression in pituitary depends on heterologous sensitization by the neurohormone melatonin. Nat Neurosci 5(3):234–238
114. Fahrenkrug J et al (2006) Diurnal rhythmicity of the clock genes Per1 and Per2 in the rat ovary. Endocrinology 147(8):3769–3776
115. Pessacq MT, Gagliardino JJ (1975) Glycogen metabolism in muscle: its circadian and seasonal variations. Metabolism 24(6):737–743
116. Shimba S et al (2005) Brain and muscle Arnt-like protein-1 (BMAL1), a component of the molecular clock, regulates adipogenesis. Proc Natl Acad Sci USA 102(34):12071–12076
117. Allaman-Pillet N et al (2004) Circadian regulation of islet genes involved in insulin production and secretion. Mol Cell Endocrinol 226(1–2):59–66
118. Kemppainen RJ, Behrend EN (1997) Adrenal physiology. Vet Clin North Am Small Anim Pract 27(2):173–186
119. Bartness TJ, Song CK, Demas GE (2001) SCN efferents to peripheral tissues: implications for biological rhythms. J Biol Rhythms 16(3):196–204
120. Ukai-Tadenuma M, Kasukawa T, Ueda HR (2008) Proof-by-synthesis of the transcriptional logic of mammalian circadian clocks. Nat Cell Biol 10(10):1154–1163
121. Bray MS, Young ME (2007) Circadian rhythms in the development of obesity: potential role for the circadian clock within the adipocyte. Obes Rev 8(2):169–181
122. Ramsey KM et al (2007) The clockwork of metabolism. Annu Rev Nutr 27:219–240
123. Diamant S, Shafrir E (1975) Modulation of the activity of insulin-dependent enzymes of lipogenesis by glucocorticoids. Eur J Biochem 53(2):541–546
124. Wang CN et al (1995) Effects of dexamethasone on the synthesis, degradation, and secretion of apolipoprotein B in cultured rat hepatocytes. Arterioscler Thromb Vasc Biol 15(9): 1481–1491
125. Samuel VT et al (2004) Mechanism of hepatic insulin resistance in non-alcoholic fatty liver disease. J Biol Chem 279(31):32345–32353
126. Marchesini G et al (1999) Association of nonalcoholic fatty liver disease with insulin resistance. Am J Med 107(5):450–455
127. Nguyen-Duy TB et al (2003) Visceral fat and liver fat are independent predictors of metabolic risk factors in men. Am J Physiol Endocrinol Metab 284(6):E1065–E1071
128. Seppala-Lindroos A et al (2002) Fat accumulation in the liver is associated with defects in insulin suppression of glucose production and serum free fatty acids independent of obesity in normal men. J Clin Endocrinol Metab 87(7):3023–3028
129. Tiikkainen M et al (2002) Liver-fat accumulation and insulin resistance in obese women with previous gestational diabetes. Obes Res 10(9):859–867
130. Argaud D et al (1996) Regulation of rat liver glucose-6-phosphatase gene expression in different nutritional and hormonal states: gene structure and 5'-flanking sequence. Diabetes 45(11):1563–1571

131. Friedman JE et al (1993) Glucocorticoids regulate the induction of phosphoenolpyruvate carboxykinase (GTP) gene transcription during diabetes. J Biol Chem 268(17):12952–12957
132. Hauner H, Schmid P, Pfeiffer EF (1987) Glucocorticoids and insulin promote the differentiation of human adipocyte precursor cells into fat cells. J Clin Endocrinol Metab 64(4):832–835
133. Hauner H et al (1989) Promoting effect of glucocorticoids on the differentiation of human adipocyte precursor cells cultured in a chemically defined medium. J Clin Invest 84(5):1663–1670
134. Olefsky JM (1975) Effect of dexamethasone on insulin binding, glucose transport, and glucose oxidation of isolated rat adipocytes. J Clin Invest 56(6):1499–1508
135. Carter-Su C, Okamoto K (1987) Effect of insulin and glucocorticoids on glucose transporters in rat adipocytes. Am J Physiol 252(4 Pt 1):E441–E453
136. Horner HC, Munck A, Lienhard GE (1987) Dexamethasone causes translocation of glucose transporters from the plasma membrane to an intracellular site in human fibroblasts. J Biol Chem 262(36):17696–17702
137. Dimitriadis G et al (1997) Effects of glucocorticoid excess on the sensitivity of glucose transport and metabolism to insulin in rat skeletal muscle. Biochem J 321(Pt 3):707–712
138. Hollingdal M et al (2002) Glucocorticoid induced insulin resistance impairs basal but not glucose entrained high-frequency insulin pulsatility in humans. Diabetologia 45(1):49–55
139. Lambillotte C, Gilon P, Henquin JC (1997) Direct glucocorticoid inhibition of insulin secretion. An in vitro study of dexamethasone effects in mouse islets. J Clin Invest 99(3):414–423
140. Harper ME, Seifert EL (2008) Thyroid hormone effects on mitochondrial energetics. Thyroid 18(2):145–156
141. Ribeiro MO (2008) Effects of thyroid hormone analogs on lipid metabolism and thermogenesis. Thyroid 18(2):197–203
142. Papi G et al (2007) Subclinical hypothyroidism. Curr Opin Endocrinol Diabetes Obes 14(3):197–208
143. Peschke E et al (2006) Diabetic Goto Kakizaki rats as well as type 2 diabetic patients show a decreased diurnal serum melatonin level and an increased pancreatic melatonin-receptor status. J Pineal Res 40(2):135–143
144. Peschke E et al (2007) Melatonin and type 2 diabetes - a possible link? J Pineal Res 42(4):350–358
145. Claustrat B, Brun J, Chazot G (2005) The basic physiology and pathophysiology of melatonin. Sleep Med Rev 9(1):11–24
146. Radziuk J, Pye S (2006) Diurnal rhythm in endogenous glucose production is a major contributor to fasting hyperglycaemia in type 2 diabetes. Suprachiasmatic deficit or limit cycle behaviour? Diabetologia 49(7):1619–1628
147. Mühlbauer E, Peschke E (2007) Evidence for the expression of both the MT1- and in addition, the MT2-melatonin receptor, in the rat pancreas, islet and beta-cell. J Pineal Res 42(1):105–106
148. Peschke E (2008) Melatonin, endocrine pancreas and diabetes. J Pineal Res 44(1):26–40
149. Sakurai T et al. (1998) Orexins and orexin receptors: a family of hypothalamic neuropeptides and G protein-coupled receptors that regulate feeding behavior. Cell 92(5):1 page following 696
150. Marcus JN et al (2001) Differential expression of orexin receptors 1 and 2 in the rat brain. J Comp Neurol 435(1):6–25
151. Sakurai T (2006) Roles of orexins and orexin receptors in central regulation of feeding behavior and energy homeostasis. CNS Neurol Disord Drug Targets 5(3):313–325
152. Buijs RM, Kalsbeek A (2001) Hypothalamic integration of central and peripheral clocks. Nat Rev Neurosci 2(7):521–526
153. Laposky AD et al (2008) Sleep and circadian rhythms: key components in the regulation of energy metabolism. FEBS Lett 582(1):142–151
154. Karlsson B, Knutsson A, Lindahl B (2001) Is there an association between shift work and having a metabolic syndrome? Results from a population based study of 27, 485 people. Occup Environ Med 58(11):747–752

155. Allison KC et al (2007) Binge eating disorder and night eating syndrome in adults with type 2 diabetes. Obesity (Silver Spring) 15(5):1287–1293
156. Kohsaka A, Bass J (2007) A sense of time: how molecular clocks organize metabolism. Trends Endocrinol Metab 18(1):4–11
157. Davidson AJ et al (2006) Chronic jet-lag increases mortality in aged mice. Curr Biol 16(21):R914–R916
158. Penev PD et al (1998) Chronic circadian desynchronization decreases the survival of animals with cardiomyopathic heart disease. Am J Physiol 275(6 Pt 2):H2334–H2337
159. Dodd AN et al (2005) Plant circadian clocks increase photosynthesis, growth, survival, and competitive advantage. Science 309(5734):630–633
160. Ouyang Y et al (1998) Resonating circadian clocks enhance fitness in cyanobacteria. Proc Natl Acad Sci USA 95(15):8660–8664
161. Turek FW et al (2005) Obesity and metabolic syndrome in circadian clock mutant mice. Science 308(5724):1043–1045
162. Kohsaka A et al (2007) High-fat diet disrupts behavioral and molecular circadian rhythms in mice. Cell Metab 6(5):414–421
163. Tu BP, McKnight SL (2006) Metabolic cycles as an underlying basis of biological oscillations. Nat Rev Mol Cell Biol 7(9):696–701
164. Rutter J, Reick M, McKnight SL (2002) Metabolism and the control of circadian rhythms. Annu Rev Biochem 71:307–331
165. Kaasik K, Lee CC (2004) Reciprocal regulation of haem biosynthesis and the circadian clock in mammals. Nature 430(6998):467–471
166. Brown SA et al (2008) Molecular insights into human daily behavior. Proc Natl Acad Sci USA 105(5):1602–1607

Chapter 6
Circadian Clock, Cell Cycle and Cancer

Circadian Rhythm and Cell Growth Regulation

Zhaoyang Zhao and Cheng Chi Lee

Abbreviations

AML	Acute myeloid leukemia
APC	Anaphase-promoting complex
ATM	Ataxia telangiectasia mutated
ATR	Ataxia telangiectasia Rad-3-related
ATRIP	ATR interacting protein
Bmal1	Brain and muscle ARNT-like 1 gene, also known as aryl hydrocarbon receptor nuclear translocator-like, Arntl
BRCA1	Breast cancer 1
Chk1/2	Checkpoint kinase 1/2
Cdk	Cyclin dependent-kinases
Cry1	Cryptochrome1 gene
DDR	DNA damage response
DNA-PK	DNA-dependent protein kinase
DSBs	Double-strand breaks
Frq	Frequency gene
G0 phase	Quiescent cells stay in G0 phase, outside the cell cycle
G1 phase	The first gap in the cell cycle
G2 phase	The second gap in the cell cycle
HR	Homologous recombination
IR	Ionizing radiation
M phase	The mitotic phase in the cell cycle
mPer2	Mouse period2 gene
Npas2	Neuronal PAS domain protein2, a paralogue of Clock gene
Rb	Retinoblastoma gene
ROS	Reactive oxygen species

Z. Zhao and C.C. Lee (✉)
Department of Biochemistry and Molecular Biology, University of Texas Health Science Center at Houston, 6431 Fannin Street, MSB 6.200, Houston, TX, 7030, USA
e-mail: Cheng.C.Lee@uth.tmc.edu

U. Albrecht (ed.), *The Circadian Clock*, Protein Reviews 12,
DOI 10.1007/978-1-4419-1262-6_6, © Springer Science+Business Media, LLC 2010

S phase	The DNA sysnthesis phase of the cell cycle
SCN	Suprachiasmatic nuclei
SSBs	Single-strand breaks
Tim	Timeless gene
UV	ultraviolet light

6.1 Introduction

Epidemiology studies have revealed that individuals in professions with activity patterns that disrupt normal circadian rhythm, bear an increased risk for a number of disorders [1, 2]. For example, flight attendants and night shift workers have an increased risk of developing breast cancer. The level of risk is proportional to the cumulative length of time an individual is exposed to this stressor [3, 4]. Recently, Kubo and colleagues reported that compared with day workers, rotating-shift workers had a significantly higher incidence of prostate cancer [5]. The prevailing view is that the majority of tumorigenic processes start from dysfunction of the cell cycle that leading to abnormal and unregulated growth. With advancements in our understanding of the cell cycle, tumorigenesis and the circadian clock at the molecular level, the mechanistic inter-relationship among these biological processes are now being unraveled.

The genetic and molecular basis of the circadian clock was pioneered by genetic studies in Drosophila and Neurospora that led to the discoveries of essential clock regulators in mammals [6, 7]. The discovery of the first two essential genes that control the mammalian circadian clock, mouse Clock [8, 9] and mouse Period1 (mPer1, also known as Rigui) [10] in 1997, marked the beginning of an explosion of informative investigations into the molecular basis of the mammalian circadian clock [11]. The findings that clock regulators are highly conserved between Drosophila and mouse indicated that this mechanism is evolutionarily conserved.

Since the first study linking the loss of circadian function to an abnormal DNA damage response (DDR) and tumorigenesis in mouse period2 (mPer2) deficient mice in 2002 [12], many independent studies have further expanded our insight into the molecular ties between regulators of the circadian clock and those involved in the cell cycle and the DDR. In addition, abnormal levels of key clock proteins have been observed in many different types of human cancers suggesting a more prevalent role for circadian clock regulators in tumorigenesis [13, 14].

6.2 The Circadian Clock

The activities of many living organisms on planet Earth follow a cyclic active and rest behavioral pattern in a period of about 24 hours – known as a circadian rhythm. Circadian is a word derived from Latin, with *circa* means "about", and *dies* means “day”. The daily rhythmic behavioral pattern is ultimately driven by an endogenous biological clock mechanism, simply called the circadian clock. The circadian clock exerts temporal regulation of biological process at the molecular, cellular, and physiological levels.

Studies reported in the early 1970s established the suprachiasmatic nuclei (SCN) in the hypothalamus as the master neural structure, the clock, that drives the circadian rhythm in mammals [15]. It was later demonstrated that bilateral SCN lesions in rats resulted in a complete loss of circadian rhythmic behavior [16]. Further, the circadian rhythm could be restored in animals with ablated SCN when they receive healthy SCN tissue grafts in their SCN regions [17]. The SCN circadian clock is entrained by the light signal received by retina and transmitted through the retino-hypothalamic track. In its dominant role as the central clock, SCN serves as a pacemaker in the coordination of the peripheral clock that regulates circadian rhythmic activity patterns [18].

The current understanding of the molecular clock is that it is composed of a group of nuclear proteins that regulates each other's expression in an auto-feedback loop. The key clock regulators will be described only briefly here, since a more detailed insight into their functions is described in Chaps. 2 and 3. Central clock genes that have been identified include Clock, Npas2 (Neuronal PAS domain protein2, a paralogue of Clock gene), Bmal1 (also known as aryl hydrocarbon receptor nuclear translocator-like, Arntl), Per1 (homologue of Drosophila Period), Per2, Cryptochrome1 (Cry1) and Cry2. The protein products of these genes regulate each other's expression in a primary clock feedback loop. The basic functional unit of the clock transcription complex is the heterodimer of CLOCK/BMAL1 or NPAS2/BMAL1, depending on the cell type. Typically, the CLOCK/BMAL1 transcriptional complex binds to the E-box enhancers of the promoter region of Per1, Per2, Cry1 and Cry2 genes, leading to the activation of the expression of PER1, PER2, CRY1 and CRY2. In the nucleus, CRYs directly interact with CLOCK/BMAL1 heterodimer and suppresses its mediated transcription activation, thus inhibiting the transcription of target genes, including that of Cry and Per. Mouse genetics studies indicate that PER2 acts as a transcriptional coactivator [19, 20], but the exact mechanistic role played by PER2 remains unclear. Both PER and CRY proteins are phosphorylated by casein kinase 1 delta (CSNK1δ) and casein kinase1 epsilon (CSNK1ε), thus marking them for poly ubiquitylation and degradation through a 26S proteosomal pathway. Removal of the inhibition of Cry at the end of the circadian cycle allows CLOCK/BMAL1 to restart a new transcription activation cycle [19, 21]. NPAS2, a redundant analog of CLOCK, is highly abundant in the central nervous system [22] and many peripheral tissues. Unlike constitutively expressed CLOCK, expression of NPAS2 is under robust temporal control. The NPAS2/BMAL1 heterodimer plays the same role as CLOCK/BMAL1 where they are expressed [23]. In addition to the core feedback loop described above, the CLOCK/BMAL1 complex also acts as an activator for Rev-erbα [24, 25] and RORα genes [26]. RORα activates the BMAL1 promoter and forms a positive feedback loop, while Rev-erbα represses the BMAL1 promoter and forms a negative feedback loop [27]. Other studies have found that two other basic helix-loop-helix transcription factors, Dec1 and Dec2, repress Per1 expression by direct protein–protein interaction with Bmal1 and/or competition for the E-box elements in the Per1 promoter [28].

All of the identified mammalian core circadian clock regulators are expressed in SCN [8-10, 29, 30]. They are also abundantly expressed in peripheral tissues and in most cell types. It is now clear that there is an endogenous circadian clock in most cells [8-10, 31]. A still unanswered question is how the central clock in the SCN coordinates the peripheral clocks.

6.3 The Cell Cycle and Cancer

The cell cycle is the process by which a single cell grows and divides into two daughter cells. Since early 1970s, the cell cycle has been described as a series of sequential events composed of four distinct phases: G1, the first gap, when gene expression and protein synthesis proceed to prepare for DNA duplication and cell growth; S, referred to as the DNA synthesis phase, when the cell undergoes DNA replication to double its genetic material; G2, the second gap, when the cell produces additional protein and metabolites to prepare for cell division; and M, the mitotic phase, when cell divides into two daughter cells. Cells that are not actively involved in cell cycling, but remain in a quiescent state are considered to be in the G0 phase. Cells can exit the cell cycle at the beginning of the G1 phase to enter the G0 phase when they are deprived of growth factors and nutrients. They can reenter the cell cycle and return to the G1 in response to specific signals depending on the cell type. In the past two decades, much insightful progress has been made in identifying the molecular components and pathways that contribute to the regulation of the cell cycle. Two key families of proteins, the cyclins and the cyclins dependent-kinases (Cdk), play key roles in controlling the cell cycles. The completion of a cell cycle requires specific spatial and temporal coordinated activities of all Cdks, and the functional activity of each Cdk depends on its association with a specific cyclin. Most Cdks are constantly expressed throughout the cell cycle, while expression of each cyclin is controlled in a temporal manner. Thus, the activities and specific functional roles of Cdks throughout the cell division cycle are also under temporal control, governed by the availability of specific cyclins.

The fidelity of genetic material transfer from parental cell to the daughter cells is vitally important for an organism. At checkpoints during the cell cycle, the surveillance of DNA damage serves to protect the genomic integrity of an organism [32]. The G1/S, S, and G2/M checkpoints trigger DNA damage response (DDR) upon detecting protein complexes that sense DNA damage and engage in DNA repair, which signal that DNA repair has not been completed [33-35, 36]; The S phase checkpoint can also detect protein complexes engaged in DNA replication, which signal that DNA duplication is not complete; The M phase checkpoint activates DDR when detecting incorrectly attached chromatides to the bipolar spindle, which could result in improper distribution of chromatides into the daughter cells [36]. The activation of any of the checkpoints would set off DDR pathways, including activation of the DNA repair mechanism, and triggers downstream pathways that either blocks the cell cycle progression until these repairs or DNA replication are completed, or trigger apoptosis – programmed cell death. Thus, normal functions of the DNA damaged surveillance and response mechanism would prevent damaged DNA or mutated genetic material from being passed on to the next generation of cells.

The normal growth of the organism through cell division depends on the accurate coordinated performance of every player in the cell cycle mechanism. Deregulation and dysfunction of any of these cell cycle regulatory components would result in abnormal cell growth. When these mistakes are not corrected in time or are terminated by apoptosis, they are passed on and amplified in future generations of cells, ultimately resulting in pathological conditions, including mal-development, premature aging, or tumorigenesis.

Defects in oncogenes or tumor suppressors involved in DNA damage response pathways and cell cycle regulation contribute to tumorigenesis. A gene, mutated or overexpressed, that helps to convert a normal cell into a cancer cell is an oncogene [37]. A gene with a function that protects a cell from tumorigenesis is a tumor suppressor gene [38]. Any reduction in the function of a tumor suppressor gene usually makes the cell susceptible to cancer-inducing damages. Thus the deregulation of a tumor suppressor generally promotes tumorigenesis only in combination with another genetic change, the second hit. The list of identified oncogenes and suppressors continues to expand as our understanding of tumorigenesis develops.

The relationship between the cell cycle and cancer can be appreciated by looking at some common and relatively well-studied cancer contributors and their associated pathways. The cyclin D-dependant kinase and Retinoblastoma (Rb) are key players that control G1 phase DNA damage checkpoint. Defects in either the cyclin D-dependant kinase, or Rb, or both, are found in a large number of tumors. The oncogene Ras, tumor suppressors PTEN and p16^{ink4}, as well as CDK4 and CKI p27^{KIP1} are involved in the regulation of Rb phosphorylation, and thus their functional defects lead to mis-regulation of the G1 checkpoint. Their genetic alterations are commonly found in many types of human cancers [39, 40]. The tumor suppressor p53, known as the guardian of the genome, regulates multiple DNA damage checkpoints. Defects in p53 are found in more than half of the sporadic human cancers [35, 41]. In response to DNA damages, p53 activates multiple pathways that engage in cell cycle arrest, DNA damage repair, and apoptosis. Other commonly seen cancer associated genes, p21^{CIP1}, Cdk2, cyclin E, and Cdk4, are involved in the cell cycle arrest mechanism [42]. Oncogene cdc25A phosphatase promotes cell cycle progression impacting checkpoints from G1 to S phase [35, 43]. Similarly, cdc25B and cdc25C are key regulators of the G2/M and M phase checkpoints [44]. The coordinated actions of Wee1 and cdc25C phosphatase constitute the focal DNA damage response mechanism at the G2/M checkpoint [45].

Oncogenes Aurora2 kinase, Aurora1 kinase[46], and Securin[47] are involved in sister chromatid attachment to the spindle and their defects are seen in many types of human tumors [35]. A few other M-phase checkpoint regulators, Mad1[48], MAD2[49], Bub1[50] and BubR1 have been frequently associated with tumorigenesis in colon and colorectal cancers [35].

Mutations in the genes encoding two phosphatidylinositol 3-kinase-like kinases, ataxia telangiectasia mutated (ATM) and ataxia telangiectasia and Rad-3-related (ATR), cause ataxia telangiectasia disorders. Individuals with ataxia telangiectasia manifest hypersensitivity to DNA damage-causing environmental factors such as ionizing radiation (IR) and ultraviolet light (UV), suffer immunodeficiency and are predisposed to tumorigenesis [51-53]. Studies in the past decade revealed that both ATM and ATR are DNA damage signal transducers, and both are key players in initiating DDR [35, 54-56]. ATM is activated at double-stranded DNA damage sites, and is more responsive to IR. ATM triggers the checkpoint mechanism and inhibits cell cycle progression mainly by phosphorylating Chk2, which sets off the Chk2 activated pathways [55, 57, 58]. ATM also activates p53 pathways through activation of Chk2 [59]. In contrast, ATR is activated by factors that compromise DNA replication, such as hydroxyurea and UV, which cause single-strand breaks in DNA [60, 61]. ATR switches on the checkpoint mechanism primarily by phosphorylating Chk1 and activating its downstream pathways, such as cell cycle regulatory pathways of cdc25A and cdc25C [62]. Recent studies have shown that there are multiple cross-talks between ATM, ATR and their downstream pathways. Any DNA damage signal eventually activate both ATM and ATR signaling systems through their crosstalk, impacting checkpoints throughout the cell cycle, from G1 phase to M phase [54].

6.4 The Circadian Clock and the Cell Cycle

Observations from prokaryotic to eukaryotic organisms have suggested that the cell division cycle is linked to the regulation of the circadian clock. In the unicellular organism Gonyaulax Polyedra, 85% of all cell division happens during the 5 hour time period spanning the transition from dark to light cycle, exhibiting circadian temporal control [63]. In Cyanobacteria, strains with the period lengths similar to environmental light–dark cycle cues demonstrate survival advantages against strains with different period lengths or defective circadian rhythms, illustrating circadian regulation as a key adaptive mechanism in cyclic environments [64]. In mice, DNA synthesis of dividing cells from many different tissues followed a 24-hour rhythm with peak activity occurring at the end of the dark cycle and the beginning of the light cycle [65, 66]. It is thought that this evolutionarily conserved timing of the cell cycle S phase before sunrise protects replicating DNA from the higher level of radiation damage that may occur during the day.

In recent years, independent studies have begun to elucidate the mechanistic links between behavioral observations, physiological processes and the underlining molecular pathways. Notably, circadian rhythm regulates the vast majority of physiological processes in an organism via numerous molecular pathways. Gene expression profiling studies have shown that up to 10% of RNA transcripts in a tissue oscillate with a circadian rhythm. A majority of these oscillating genes are tissue specific, and involved in rate-limiting steps of essential functional pathways

[67, 68]. In liver and muscle tissues, about 10–15% of the oscillating genes are involved in the cell cycle and its regulation [69].

One of the genes exhibits an oscillating expression is $p21^{CIP1}$, an important DDR gene involved in cell cycle arrest. The rhythmicity of $p21^{CIP1}$ expression is abolished in BMAL1 knockout mice. The expression of $p21^{CIP1}$ appears to be under the control of the negative feedback loop in the circadian clock comprised of BMAL1/CLOCK and Rev-erb α/β [70].

Okamura and colleagues observed that regeneration of an injured liver in mice is initiated by re-entry into the cell cycle at a specific circadian time, unrelated to the time of injury [71]. The same study also demonstrated that the temporal expression of Wee1 is regulated by the CLOCK/BMAL1 complex. The promoter of the Wee1 gene has a signature E-box for the CLOCK/BMAL1 complex. Wee1 is a G2-M phase check-point phosphatase that counteracts another phosphatase, cdc25C. The counter-balanced actions of Wee1 and cdc25C determine the phosphorylation state of cdc2/cyclin B1, which in turn controls the timing of G2-M transition [71, 72].

Several studies have linked a clock gene, Timeless (tim), to cell cycle regulation. Tim is essential for circadian rhythm in Drosophila. The Timeless orthologue in mammals (Mammalian Tim) is associated with PER proteins in the clock mechanism and is required for the maintenance of rhythmicity [73]. The human Tim protein has also been found to interact with the cell cycle checkpoint proteins Chk1 and the ATR-ATRIP complex, which are both involved DDR [74]. In addition, immunoprecipitation studies have revealed that the ATR-ATRIP complex is associated with CRY2 and that loss of TIM decreases PER2 expression. In the cells depleted of Tim, the replication fork progresses at a rate of 52% of the control rate. Tim is believed to be necessary to maintain DNA replication fork progression in the absence of DNA damage [75].

The connection between the cell cycle and the circadian clock is observed in less complex organisms. For example, mutation of the Prd-4 gene in Neurospora shortens the circadian period length and abolishes the phase shift response to DNA damage. Prd-4 is the Neurospora ortholog of mammalian Chk2. At the transcriptional level, the expression of Prd-4 is regulated by the circadian clock. At the posttranscriptional level, PRD-4 physically interacts with the clock component frequency (FRQ), promoting its phosphorylation. Presumably, through PRD-4 and FRQ interaction, DNA-damaging agents can phase shift the Neurospora clock in a temporal manner [76].

Circadian clocks have been found in dividing cell and quiescent cell types. There is ample evidence to support the notion that the cell cycle is regulated by the circadian clock, but little data to substantiate the converse notion that circadian rhythm is regulated by the cell cycle. Using micro-array profiling and functional analysis, Gery and colleagues reported that two components of the core circadian clock mechanism, Per2 and Rev-erbα, were target genes of C/EBPs. Specifically, Per2 is up-regulated by C/EBPα and C/EBPε [77]. C/EBPs are a family of transcription factors that regulate target genes via binding to CCAAT/enhancers. They regulate cell growth and differentiation, and are found in many types of cells. C/EBPα has been found to be a p53 induced gene active in the G1 DNA damage checkpoint [78], and thus Per2 expression is not only regulated by the CLOCK/BMAL1 heterodimer, but also by the cell cycle regulator p53 via C/EBPα. This study presents an example of how the circadian clock may respond to feedback from developmental and DDR pathways.

Currently, the majority of the intersection points between the circadian clock mechanism and the cell cycle mechanism involve various DDR pathways triggered by DNA damage checkpoints, as well as at replication (G1/S phase transition) checkpoint, and spindle (M phase) checkpoint. Taken together, these findings suggest that many cell cycle checkpoint regulators are downstream targets of the circadian clock mechanism. In their studies in yeast, Chen and colleagues observed that the metabolic cycle-directed restriction on cell cycle was obstructed in cell cycle mutants, which resulted in an increased rate of spontaneous mutation [79].. The yeast cell division cycle is coupled to the circadian cycle by a DNA checkpoint kinase. The blockade of the expression of the checkpoint kinase desynchronizes not only the cell cycle and the circadian cycle, but also the metabolic and the cell cycles. Chen and Mcknight advocated that DNA damage signaling is the ultimate force coordinating the three fundamental life cycles: the circadian cycle, the cell cycle, and the metabolic cycle [80].

The homeostasis of important physiological functions is usually safeguarded by primary feedback loops and supporting feedback loops, with multiple input and output pathways involving other basic functions. We are in the early phase of recognizing the complexity of these functions, such as the cell cycle and the circadian rhythm. Current studies are unraveling the deeper level of mechanistic complexity of cell cycle regulation, the circadian clock, and their web of interactions.

6.5 The Circadian Clock and Cancer

Long before the molecular biology era, there has been deep interest in the connections between circadian rhythm and physiological functions in normal and disease conditions, including cancer, [81]. In early 1970s, significant differences between the circadian rhythm of skin temperatures of cancerous and normal breast, were documented [82]. Klevecz and colleague found that ovarian cancer tumor cells predominantly displayed a 12-hour shift of circadian gating of the cell cycle in contrast to the normal cells [83].

Following the elucidation of the core molecular mechanism of the mammalian circadian clock, one of the first genetic studies to investigate whether compromised circadian clock function would predispose an individual to tumorigenesis was carried out with mice deficient in mPER2. These mice carry a deletion mutation in the PAS domain of the mPer2 gene. When kept in constant darkness, homozygous mPer2 mutant mice exhibit a shorter circadian period initially, and then a complete loss of circadian rhythmic behavior [20]. Mice deficient in both mPER1 and mPER2 also display apparent signs of premature aging, with an early decline in soft tissues, kyphosis and the loss of fertility in female mice after 6–8 months of age [84]. The observation of enlarged, hyperplastic salivary glands in older mPer2 mutant mice provided the first indication of mPer2's involvement in growth control regulation. Further investigation using sublethal IR treatments showed that the mPer2 mutant mice developed spontaneous lymphomas at a higher rate than the wild type control. These IR treated mPer2 mutant mice also exhibited rapid coat graying, suggesting that they are hypersensitive in their response to genotoxic stress from IR exposure [12]. Interestingly, one of the phenotypes observed in the mouse model of ATM is, accelerated coat graying in heterozygous mice after sub-lethal IR exposure [85]. Hypersensitivity to IR and predisposition to cancer are typical characteristics of ataxia telangiectesia patients. Thus, it is possible that PER2's role in DDR is to couple the circadian clock regulation to the ATM-dependent DNA damage response pathway [12].

The same study also revealed that several genes involved in cell cycle regulation, such as Cyclin D1, Cyclin A, Mdm-2, Gadd45α and c-Myc, were circadian controlled and their expression was deregulated in mPer2 mutant mice. While Cyclin D1 and Cyclin A are core regulators of the cell division cycle, both Mdm2 and Gadd45α are key players of p53 regulated DDR pathways. Mdm2 is an E3 ubiquitin ligase that negatively regulates p53 at two levels, inhibiting transcriptional activation of p53 and targeting p53 for degradation [86]. Gadd45α is a p53 target gene that involved in G2/M checkpoint and can be activated in multiple stress conditions. Deletion of Gadd45α results in centrosome amplification and leads to anneupliody, and thus, genomic instability [87]. C-Myc is a classic proto-oncogene. Over expression of c-Myc is often associated with the abnormal proliferation observed in various human tumors [88]. Several studies have directly linked over-expression of c-Myc to DNA damage, mutation, and hence, genomic instability [89]. The expression of c-Myc was shown to be controlled by the NPAS2/BMAL1 complex, through one of the two signature E-box motifs in its promoter. The overall c-Myc expression level is significantly elevated in mPer2 mutant mice, suggesting that PER2 normally suppresses c-Myc expression. Recently, Yang and colleagues observed an increase in Cyclin D and Cyclin E levels correlating with down regulation of Per2, suggesting that a Per2 defect can also cause deregulation of these core cell cycle regulators [90]. Consistent with the above studies, over-expression of Per2 suppresses the expression of proto-oncogenes including c-Myc, Bcl-X(L) and Bcl-2, while up-regulate tumor suppressors such as p53 and Bax. As a result, over-expression of Per2 in cancer cells inhibited cell proliferation and promotes their apoptosis, but interestingly over expression of PER2 in non-cancerous cells did not enhanced their apoptotic rate [91].

Other studies have also reported deregulation of Per2 in tumor models and human cancers, as well as Per2's action as a tumor suppressor. Gery and colleagues found Per2 expression levels were lower in lymphoma cell lines and in acute myeloid leukemia (AML) patient samples, in comparison to normal controls. When Per2 was over expressed in K562 tumor cells, inhibition of proliferation, cell cycle arrest, apoptosis, and loss of clonogenic ability were observed. These authors proposed that Per2 deficiency or loss of function contributes to the development of

AML [77]. In a grafted breast cancer mouse model, when Per2 expression was down-regulated with siRNA and shRNA, an increased in vivo tumor growth and a doubling of breast cancer cell proliferation rate in vitro were observed. Interestingly, down regulation of Per2 doubles the daily amplitude of the tumor growth rhythm [90]. Further, a study of tumor development in intestine and colon using a cell culture model and mutant mPer2 mice suggested that inactivation of PER2 elevated the expression of β-catenin, cyclin D, and enhanced the cell proliferation of colon cell lines (HCT116 and SW480) [92]. Beta-catenin is a component of the Wnt signaling pathway. Defects in Wnt signaling have been associated with human breast carcinoma. It was reported recently that increased Wnt signaling switches on DDR, and compromises p53 and Rb function [93]. Increased β-catenin production has also been directly linked to cell proliferation and carcinogenesis [94]. It was suggested that an mPer2 regulated pathway suppresses tumorigenesis by down regulation of β-catenin and its targeted genes. Another antitumor function of Per2 is illustrated by a recent study showing that mPer2 protects cells from γ-ray radiation. Cells with higher mPer2 expression display a lower level of DNA damage [95]. Together, these studies suggest that Per2 regulated pathways protect the cells from environmental damage.

In addition to Per2, deregulation of its homologs, Per1 and Per3 has also been associated with tumorigenesis in several studies. Per1 has been found to interact with the checkpoint proteins ATM and Chk2 [96]. Over-expression of Per1 in human cancer cells makes the cells more sensitive to DNA damage - induced apoptosis. Conversely, apoptosis was suppressed by inhibition of Per1 in similarly treated cells. Another study revealed that the expression of Per1, Per2 and Per3 genes is in disarray in most (>95%) of breast cancer cells in comparison with the noncancerous cells from the same tissue sample [97]. This study further indicated that genetic mutation is not the cause of the deregulation, but rather the abnormal methylation of the CpG islands in the promoters of Per genes. The study suggested that an aberrant circadian rhythm resulted from the abnormal Per expression, and this in turn benefited the survival and growth of the tumor cells. In addition, it was also observed that the expression level of c-erbB2 strongly correlated with the methylation levels of Per promoters. The further linkage of C-erbB2 to DNA damage checkpoints and cancer, is revealed by observations that C-erbB2 induces MDM2 dependent p53 degradation, found in many human cancers, and is a target of clinical intervention [98].

Winter and colleagues identified a putative interaction between breast cancer 1 (BRCA1), Per1 and Per2, in a yeast two-hybrid system [99]. BRCA1 has been suggested to be regulated by ATM and ATR and is involved in regulation of the G2/M checkpoint in response to DNA damage [100]. A recent study shows BRCA1 is regulated by Chk2 in response to spindle damage [101]. The demonstrated interaction with BRCA1, brings Per1 and Per2's actions directly into the DDR pathways. Using human breast tumor samples, Winter and colleagues observed down regulation of Per1 gene expression in both sporadic and familial breast tumors and down regulation of Per2 gene expression in familial breast tumors.

Other recent studies have revealed that several other circadian genes are involved in tumor suppression or tumorigenesis. Gene expression profiles of pancreatic cancer tissues suggest that many circadian genes, such as Per1, Dec1 and Tim are under expressed in comparison with normal pancreatic tissue. As expected, many circadian target genes were also down regulated in tumor tissues when compared to normal pancreatic tissues [102]. NPAS2, the paralog of the Clock gene, has been characterized as a possible tumor suppressor. Using RNA interference-mediated depletion of NPAS2, the expected cell cycle delay in response to mutagen treatment was abolished in NPAS2 deficient cells. These cells also show impaired DNA repair capacity, as measured by the comet assay. In addition, the expression of a number of DNA damage repairs and cell cycle genes were repressed in cells depleted of NPAS2 [103]. Therefore, NPAS2 appears to be involved in a DNA damage check-point role similar to that of a tumor suppressor. Clinically, polymorphic variants of NPAS2 are significantly associated with distinct risks for cancers. For example, Zhu and colleagues found that a heterozygous Ala394Thr genotype of NPAS2 is found in women carriers with significantly higher breast cancer risk compared with women with the common homozygous Ala394Ala genotype [104]. In a separate study, they found that the variant Thr genotypes (Ala/Thr and Thr/Thr) of the NPAS2 gene are associated with a reduced risk for non-Hodgkin's lymphoma [105].

While the majority of the above studies clearly established an involvement of circadian clock regulators with cell cycle regulation and tumorigenesis, the mechanistic details remain poorly understood.

6.6 Discussion

It appears that almost all tumor-causing cell cycle deregulation events occur in cell cycle checkpoint pathways: the genetic material damage (DNA breakage, stalled replicons, or erratic chromotids distribution) surveillance pathways, and the damage activated cell cycle arrest, DNA repair, or apoptosis pathways.

The Circadian clock casts a web of control over every physiological process in a living organism, including the most fundamental development and growth process – the cell cycle. Early studies have revealed many pathways through which the circadian clock regulates the cell cycle. Increasing evidence also suggest that the clock mechanism receives feedback from DNA damage checkpoints. It is now clear that the circadian clock system of a living organism is the focal point that integrates inputs from environmental signals including the light/dark cycle, nutrient availability, cellular-damaging agents, as well as endogenous signals for homeostasis. As its output, the circadian clock exerts coordinated temporal control that fine-tunes physiological processes to occur at a precise temporal pace, so that the organism maximally benefits from available resources and keeps exposure to adverse agents to a minimum. From an evolutionary point-of-view, organisms that grow into the dominant surviving species are those with an inheritable circadian mechanism that coordinate their physiology to maximize benefit and minimize harm from the environment. At least in Neurospora, such view has been directly demonstrated.

This temporal control is carried out by core clock control proteins that regulate output pathways that in turn control specific pathways involved in every known physiological function. We have reviewed some of the pathways that link the circadian clock to the cell cycle and the DDR, as well as deregulation of these pathways that results in tumorigenesis.

Proper function of our circadian clock sets the rhythmic regulation that coordinates each physiological process to occur at the proper temporal cycle. Dysfunction of circadian clock impacts the organism from two levels: disrupted optimal rhythm and compromised specific clock controlled pathways. Firstly, deregulation of the optimal rhythmic control could expose sensitive vital physiological processes to

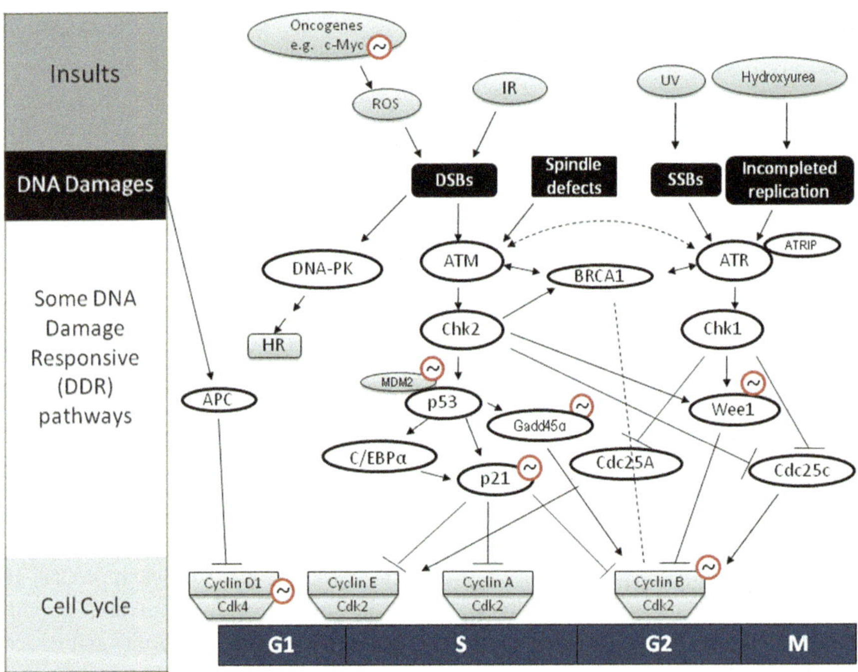

Fig. 6.1 Identified links between the mammalian circadian clock regulators and the cell cycle regulatory pathways: The top portion of the diagram shows the common insulting agents that cause damages to genomic DNA. The blocked boxes depict the commonly seen damages that would trigger cell cycle check points. The middle portion of the diagram shows several common DDR pathways and the key regulators in these pathways. The bottom portion of the diagram shows the key checkpoints and related core regulators in the cell cycle. The flows of the pathways are depicted by lines: the *arrow lines* indicate activation interactions and the *blocked lines* indicate inhibitory interactions. The *red circles* with a sine-like drawing mark the genes whose expression has been identified to be under the circadian clock control

environmental threats. It is believed that optimal circadian rhythmic control instructs DNA synthesis to be initiated before the onset of the light cycle when there is minimum environmental radiation that can cause DNA damage. Exposure of DNA synthesis to radiation, in case of circadian clock deregulation, is likely to increase the rate of DNA damage, elevate the risk of carrying mutations to future generations, and increase the possibility of developing tumors. Meanwhile, deregulated circadian rhythm can impair metabolic processes from harvesting sufficient nutrients from the environment, thus placing the organism at a disadvantage for survival. At the molecular level, defective clock factors can directly disrupt the normal function of regulatory pathways they participate in, resulting in imbalances in the physiological processes under regulation. In the case of the cell cycle, imbalanced regulation elevates the risk for tumorigenesis.

6.7 Summary

Advancements in our understanding of the cell division cycle and the circadian clock at the molecular level have revealed common elements between these fundamental pathways. The extent of the integration of the cell cycle, the DDR, and the circadian clock mechanism remains to be investigated. A review of the genetic, molecular and clinical studies published thus far, suggests that these fundamental biological processes are likely to be highly integrated. An important question is whether the key regulators of the cell cycle and the DDR are also key players in the mammalian clock mechanism. Further molecular and genetic studies will lead to a better understanding of the circadian clock's role in tumorigenesis and conceivably result in better therapeutic strategies in cancer treatment.

References

1. Akerstedt T et al (1984) Shift work and cardiovascular disease. Scand J Work Environ Health 10(6 Spec No):409–414
2. Stevens RG et al (1992) Electric power, pineal function, and the risk of breast cancer. FASEB J 6(3):853–860
3. Davis S, Mirick DK, Stevens RG (2001) Night shift work, light at night, and risk of breast cancer. J Natl Cancer Inst 93(20):1557–1562
4. Rafnsson V et al (2001) Risk of breast cancer in female flight attendants: a population-based study (Iceland). Cancer Causes Control 12(2):95–101
5. Kubo T et al (2006) Prospective cohort study of the risk of prostate cancer among rotating-shift workers: findings from the Japan collaborative cohort study. Am J Epidemiol 164(6):549–555
6. Rosbash M (1995) Molecular control of circadian rhythms. Curr Opin Genet Dev 5(5):662–668
7. Iwasaki K, Thomas JH (1997) Genetics in rhythm. Trends Genet 13(3):111–115

8. Antoch MP et al (1997) Functional identification of the mouse circadian Clock gene by transgenic BAC rescue. Cell 89(4):655–667
9. King DP et al (1997) Positional cloning of the mouse circadian clock gene. Cell 89(4): 641–653
10. Sun ZS et al (1997) RIGUI, a putative mammalian ortholog of the Drosophila period gene. Cell 90(6):1003–1011
11. Hastings M, Maywood ES (2000) Circadian clocks in the mammalian brain. Bioessays 22(1):23–31
12. Fu L et al (2002) The circadian gene Period2 plays an important role in tumor suppression and DNA damage response in vivo. Cell 111(1):41–50
13. Chen-Goodspeed M, Lee CC (2007) Tumor suppression and circadian function. J Biol Rhythms 22(4):291–298
14. Gery S, Koeffler HP (2007) The role of circadian regulation in cancer. Cold Spring Harb Symp Quant Biol 72:459–464
15. Schwartz WJ, Gainer H (1977) Suprachiasmatic nucleus: use of 14C-labeled deoxyglucose uptake as a functional marker. Science 197(4308):1089–1091
16. Stephan FK, Zucker I (1972) Circadian rhythms in drinking behavior and locomotor activity of rats are eliminated by hypothalamic lesions. Proc Natl Acad Sci U S A 69(6):1583–1586
17. Ralph MR et al (1990) Transplanted suprachiasmatic nucleus determines circadian period. Science 247(4945):975–978
18. Buijs RM, Kalsbeek A (2001) Hypothalamic integration of central and peripheral clocks. Nat Rev Neurosci 2(7):521–526
19. Shearman LP et al (2000) Interacting molecular loops in the mammalian circadian clock. Science 288(5468):1013–1019
20. Zheng B et al (1999) The mPer2 gene encodes a functional component of the mammalian circadian clock. Nature 400(6740):169–173
21. Ko CH, Takahashi JS (2006) Molecular components of the mammalian circadian clock. Hum Mol Genet. 15 Spec No. 2:R271–R277
22. Reick M et al (2001) NPAS2: an analog of clock operative in the mammalian forebrain. Science 293(5529):506–509
23. Bertolucci C et al (2008) Evidence for an overlapping role of CLOCK and NPAS2 transcription factors in liver circadian oscillators. Mol Cell Biol 28(9):3070–3075
24. Preitner N et al (2002) The orphan nuclear receptor REV-ERBalpha controls circadian transcription within the positive limb of the mammalian circadian oscillator. Cell 110(2):251–260
25. Triqueneaux G et al (2004) The orphan receptor Rev-erbalpha gene is a target of the circadian clock pacemaker. J Mol Endocrinol 33(3):585–608
26. Sato TK et al (2004) A functional genomics strategy reveals Rora as a component of the mammalian circadian clock. Neuron 43(4):527–537
27. Guillaumond F et al (2005) Differential control of Bmal1 circadian transcription by REV-ERB and ROR nuclear receptors. J Biol Rhythms 20(5):391–403
28. Honma S et al (2002) Dec1 and Dec2 are regulators of the mammalian molecular clock. Nature 419(6909):841–844
29. Albrecht U et al (1997) A differential response of two putative mammalian circadian regulators, mper1 and mper2, to light. Cell 91(7):1055–1064
30. DeBruyne JP, Weaver DR, Reppert SM (2007) CLOCK and NPAS2 have overlapping roles in the suprachiasmatic circadian clock. Nat Neurosci 10(5):543–545
31. Yamamoto T et al (2004) Transcriptional oscillation of canonical clock genes in mouse peripheral tissues. BMC Mol Biol 5:18
32. Hartwell LH, Weinert TA (1989) Checkpoints: controls that ensure the order of cell cycle events. Science 246(4930):629–634
33. Lim DS et al (2000) ATM phosphorylates p95/nbs1 in an S-phase checkpoint pathway. Nature 404(6778):613–617
34. Kastan MB, Lim DS (2000) The many substrates and functions of ATM. Nat Rev Mol Cell Biol 1(3):179–186

35. Niida H, Nakanishi M (2006) DNA damage checkpoints in mammals. Mutagenesis 21(1):3–9
36. Garrett MD (2001) Cell cycle control and cancer. Curr Sci 81(5):8
37. Todd R, Wong DT (1999) Oncogenes. Anticancer Res 19(6A):4729–4746
38. Sherr CJ (2004) Principles of tumor suppression. Cell 116(2):235–246
39. Stambolic V, Mak TW, Woodgett JR (1999) Modulation of cellular apoptotic potential: contributions to oncogenesis. Oncogene 18(45):6094–6103
40. Marshall CJ (1988) The ras oncogenes. J Cell Sci Suppl 10:157–169
41. Carson DA, Lois A (1995) Cancer progression and p53. Lancet 346(8981):1009–1011
42. Boulaire J, Fotedar A, Fotedar R (2000) The functions of the cdk-cyclin kinase inhibitor p21WAF1. Pathol Biol (Paris) 48(3):190–202
43. Mailand N et al (2000) Rapid destruction of human Cdc25A in response to DNA damage. Science 288(5470):1425–1429
44. Nilsson I, Hoffmann I (2000) Cell cycle regulation by the Cdc25 phosphatase family. Prog Cell Cycle Res 4:107–114
45. Perry JA, Kornbluth S (2007) Cdc25 and Wee1: analogous opposites? Cell Div 2:12
46. Bischoff JR, Plowman GD (1999) The Aurora/Ipl1p kinase family: regulators of chromosome segregation and cytokinesis. Trends Cell Biol 9(11):454–459
47. Zou H et al (1999) Identification of a vertebrate sister-chromatid separation inhibitor involved in transformation and tumorigenesis. Science 285(5426):418–422
48. Jin DY, Spencer F, Jeang KT (1998) Human T cell leukemia virus type 1 oncoprotein Tax targets the human mitotic checkpoint protein MAD1. Cell 93(1):81–91
49. Michel LS et al (2001) MAD2 haplo-insufficiency causes premature anaphase and chromosome instability in mammalian cells. Nature 409(6818):355–359
50. Cahill DP et al (1998) Mutations of mitotic checkpoint genes in human cancers. Nature 392(6673):300–303
51. Meyn MS (1995) Ataxia-telangiectasia and cellular responses to DNA damage. Cancer Res 55(24):5991–6001
52. Liu A et al (2005) Alterations of DNA damage-response genes ATM and ATR in pyothorax-associated lymphoma. Lab Invest 85(3):436–446
53. Shiloh Y (2003) ATM and related protein kinases: safeguarding genome integrity. Nat Rev Cancer 3(3):155–168
54. Hurley PJ, Bunz F (2007) ATM and ATR: components of an integrated circuit. Cell Cycle 6(4):414–417
55. Matsuoka S et al (2000) Ataxia telangiectasia-mutated phosphorylates Chk2 in vivo and in vitro. Proc Natl Acad Sci U S A 97(19):10389–10394
56. Bakkenist CJ, Kastan MB (2004) Initiating cellular stress responses. Cell 118(1):9–17
57. Shiloh Y (2006) The ATM-mediated DNA-damage response: taking shape. Trends Biochem Sci 31(7):402–410
58. Matsuoka S, Huang M, Elledge SJ (1998) Linkage of ATM to cell cycle regulation by the Chk2 protein kinase. Science 282(5395):1893–1897
59. Giaccia AJ, Kastan MB (1998) The complexity of p53 modulation: emerging patterns from divergent signals. Genes Dev 12(19):2973–2983
60. Osborn AJ, Elledge SJ, Zou L (2002) Checking on the fork: the DNA-replication stress-response pathway. Trends Cell Biol 12(11):509–516
61. Kumagai A, Dunphy WG (2006) How cells activate ATR. Cell Cycle 5(12):1265–1268
62. Syljuasen RG et al (2005) Inhibition of human Chk1 causes increased initiation of DNA replication, phosphorylation of ATR targets, and DNA breakage. Mol Cell Biol 25(9): 3553–3562
63. Sweeney BMAH (1958) Woodland, rhythmic cell division in populations of gonyaulax polyedra. J Protozool 5:217–224
64. Woelfle MA et al (2004) The adaptive value of circadian clocks: an experimental assessment in cyanobacteria. Curr Biol 14(16):1481–1486
65. Scheving LE et al (1978) Circadian variation in cell division of the mouse alimentary tract, bone marrow and corneal epithelium. Anat Rec 191(4):479–486

66. Scheving LE (1981) Circadian rhythms in cell proliferation: their importance when investigating the basic mechanism of normal versus abnormal growth. Prog Clin Biol Res 59C(00):39–79
67. Panda S et al (2002) Coordinated transcription of key pathways in the mouse by the circadian clock. Cell 109(3):307–320
68. Storch KF et al (2002) Extensive and divergent circadian gene expression in liver and heart. Nature 417(6884):78–83
69. Miller BH et al (2007) Circadian and CLOCK-controlled regulation of the mouse transcriptome and cell proliferation. Proc Natl Acad Sci U S A 104(9):3342–3347
70. Grechez-Cassiau A et al (2008) The circadian clock component BMAL1 is a critical regulator of p21WAF1/CIP1 expression and hepatocyte proliferation. J Biol Chem 283(8):4535–4542
71. Matsuo T et al (2003) Control mechanism of the circadian clock for timing of cell division in vivo. Science 302(5643):255–259
72. Cardone L, Sassone-Corsi P (2003) Timing the cell cycle. Nat Cell Biol 5(10):859–861
73. Barnes JW et al (2003) Requirement of mammalian Timeless for circadian rhythmicity. Science 302(5644):439–442
74. Unsal-Kacmaz K et al (2005) Coupling of human circadian and cell cycles by the timeless protein. Mol Cell Biol 25(8):3109–3116
75. Unsal-Kacmaz K et al (2007) The human Tim/Tipin complex coordinates an Intra-S checkpoint response to UV that slows replication fork displacement. Mol Cell Biol 27(8):3131–3142
76. Pregueiro AM et al (2006) The Neurospora checkpoint kinase 2: a regulatory link between the circadian and cell cycles. Science 313(5787):644–649
77. Gery S et al (2005) Transcription profiling of C/EBP targets identifies Per2 as a gene implicated in myeloid leukemia. Blood 106(8):2827–2836
78. Yoon K, Smart RC (2004) C/EBPalpha is a DNA damage-inducible p53-regulated mediator of the G1 checkpoint in keratinocytes. Mol Cell Biol 24(24):10650–10660
79. Chen Z et al (2007) Restriction of DNA replication to the reductive phase of the metabolic cycle protects genome integrity. Science 316(5833):1916–1919
80. Chen Z, McKnight SL (2007) A conserved DNA damage response pathway responsible for coupling the cell division cycle to the circadian and metabolic cycles. Cell Cycle 6(23):2906–2912
81. Reinberg A (1975) Circadian changes in the temperature of human beings. Bibl Radiol 6:128–139
82. Mansfield CM et al (1973) Circadian rhythm in the skin temperature of normal and cancerous breasts. Int J Chronobiol 1(3):235–243
83. Klevecz RR et al (1987) Circadian gating of S phase in human ovarian cancer. Cancer Res 47(23):6267–6271
84. Lee CC (2006) Tumor suppression by the mammalian period genes. Cancer Causes Control 17(4):525–530
85. Barlow C et al (1999) Atm haploinsufficiency results in increased sensitivity to sublethal doses of ionizing radiation in mice. Nat Genet 21(4):359–360
86. Vassilev LT et al (2004) In vivo activation of the p53 pathway by small-molecule antagonists of MDM2. Science 303(5659):844–848
87. Hollander MC, Fornace AJ Jr (2002) Genomic instability, centrosome amplification, cell cycle checkpoints and Gadd45a. Oncogene 21(40):6228–6233
88. Ramsay G, Evan GI, Bishop JM (1984) The protein encoded by the human proto-oncogene c-myc. Proc Natl Acad Sci U S A 81(24):7742–7746
89. Prochownik EV, Li Y (2007) The ever expanding role for c-Myc in promoting genomic instability. Cell Cycle 6(9):1024–1029
90. Yang X et al (2009) Down regulation of circadian clock gene Period 2 accelerates breast cancer growth by altering its daily growth rhythm. Breast Cancer Res Treat 117(2): 423–431
91. Hua H et al (2006) Circadian gene mPer2 overexpression induces cancer cell apoptosis. Cancer Sci 97(7):589–596
92. Wood PA et al (2008) Period 2 mutation accelerates ApcMin/+ tumorigenesis. Mol Cancer Res 6(11):1786–1793

93. Ayyanan A et al (2006) Increased Wnt signaling triggers oncogenic conversion of human breast epithelial cells by a Notch-dependent mechanism. Proc Natl Acad Sci U S A 103(10):3799–3804
94. Saldanha G et al (2004) Nuclear beta-catenin in basal cell carcinoma correlates with increased proliferation. Br J Dermatol 151(1):157–164
95. Zhang J et al (2008) High expression of circadian gene mPer2 diminishes radiosensitivity of tumor cells. Cancer Biother Radiopharm 23(5):561–570
96. Gery S et al (2006) The circadian gene per1 plays an important role in cell growth and DNA damage control in human cancer cells. Mol Cell 22(3):375–382
97. Chen ST et al (2005) Deregulated expression of the PER1, PER2 and PER3 genes in breast cancers. Carcinogenesis 26(7):1241–1246
98. Zhou BP et al (2001) HER-2/neu induces p53 ubiquitination via Akt-mediated MDM2 phosphorylation. Nat Cell Biol 3(11):973–982
99. Winter SL et al (2007) Expression of the circadian clock genes Per1 and Per2 in sporadic and familial breast tumors. Neoplasia 9(10):797–800
100. Yarden RI et al (2002) BRCA1 regulates the G2/M checkpoint by activating Chk1 kinase upon DNA damage. Nat Genet 30(3):285–289
101. Chabalier-Taste C et al (2008) BRCA1 is regulated by Chk2 in response to spindle damage. Biochim Biophys Acta 1783(12):2223–2233
102. Pogue-Geile KL, Lyons-Weiler J, Whitcomb DC (2006) Molecular overlap of fly circadian rhythms and human pancreatic cancer. Cancer Lett 243(1):55–57
103. Hoffman AE et al (2008) The circadian gene NPAS2, a putative tumor suppressor, is involved in DNA damage response. Mol Cancer Res 6(9):1461–1468
104. Zhu Y et al (2008) Non-synonymous polymorphisms in the circadian gene NPAS2 and breast cancer risk. Breast Cancer Res Treat 107(3):421–425
105. Zhu Y et al (2007) Ala394Thr polymorphism in the clock gene NPAS2: a circadian modifier for the risk of non-Hodgkin's lymphoma. Int J Cancer 120(2):432–435

Chapter 7
Comparative Clocks

Martha Merrow, David Lenssen, and Till Roenneberg

7.1 Introduction

Most of the chapters in this book deals with state of the art research on the circadian system in animals. Although this is fascinating for *us*, especially as we begin to understand the impact that the circadian system has on our health and well-being, non-animal and non-mammal clocks deserve some attention. We are truly surrounded by biological clocks – in plants, fungi, bacteria and other animals. It is through a comparative approach that the field of chronobiology has progressed as far and as fast as it has. The clock in the plant follows most of the same rules as those in animals, fungi and bacteria, and vice versa. The circadian properties that we think of most often are a self-sustained free running period, entrainability (and specific entrainment characteristics) and temperature and metabolic compensation, the phenomenon whereby the free running period remains stable over a wide range of temperatures or nutritional states. These properties have been used to tease apart clock mechanisms across the biological world. The comparative approach has led to a simple concept concerning the organization of a circadian system (Fig. 7.1). It stipulates that each clock has an input pathway to transmit signals from zeitgeber to the rhythm generator or central oscillator. The oscillator in turn transmits signals to output pathways, leading to a phenotype that is used for monitoring the circadian rhythm. In general, the dominant zeitgeber is light (more specifically, light/dark cycles) but there are many environmental cycles that derive from the light cycle, such as temperature and humidity. Given that these cycles are also somewhat predictable, they are typically also used as zeitgebers. Concepts of central oscillators are in a state of flux, with an undeniable and important role for transcriptional feedback loops but with additional post-transcriptional feedbacks acting either as zeitnehmers [1–3] or as part of the oscillator mechanism.

This chapter briefly reviews the circadian behaviours, circadian organization and molecular mechanisms of circadian rhythms in plants, fungi, animals and cyanobacteria.

M. Merrow (✉)
The University of Groningen, Haren, The Netherlands
e-mail: m.merrow@rug.nl

U. Albrecht (ed.), *The Circadian Clock*, Protein Reviews 12,
DOI 10.1007/978-1-4419-1262-6_7, © Springer Science+Business Media, LLC 2010

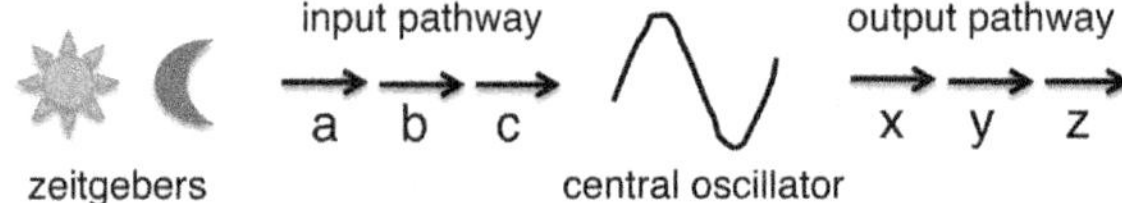

Fig. 7.1 A simple scheme of a circadian system. Light (the sun and the moon) is the predominant zeitgeber for most circadian clocks. An input pathway transduces a signal to a central oscillator which in turn directs oscillations in clock controlled outputs

Despite the huge taxonomic divergence, at least within the eukaryotes, there are many shared features between the systems despite that the molecules and the circadian behaviours that they regulate are different.

7.2 The Plant Clock

7.2.1 The First Observation in "Modern" Times

The story of the Frenchman, Jean Jacques d'Ortous de Mairan, is something of a legend in circadian biology. De Mairan was an astronomer who kept a tamarind plant on his desk. He noted that the leaves folded themselves downward each night, prompting him to perform an experiment whereby he placed the plant in a closed, darkened cupboard in his desk and monitored the state of the leaves by checking them periodically. He was impressed to see that – even without a light signal – the plants remained rhythmic, unfurling their leaves during daytime and folding them down at night. He published a *circa* 350 word paper, speculating on circadian rhythms even in humans and anticipating circadian rhythms research and the slow progress that it was likely to make [4]. He was right on just about everything that he wrote, including the possible flaw in his experimental setup: it is probable – given the time in history and the state of central heating – that the plant experienced temperature cycles, which are fully capable of entraining cyanobacterial, plant, fungal and animal clocks [5–9], and possibly all others too.

7.2.2 The Clock in the Plant: Arabidopsis thaliana

Circadian timing in plants has long since been solidly established. In addition to the leaf movement-rhythm, plants open and close the stomatal openings on the underside of their leaves according to a circadian rhythm, they change their growth rate over the day, they organize the expression of proteins used in photosynthesis to coincide with the appearance of light, and they do this even in changing photoperiods, showing seasonally adjusted entrainment. Plants are among the most robust practitioners of photoperiodism, with some plants setting seeds following long days, and others doing so after short days. Mutations in clock genes in plants often lead to disrupted photoperiodic responses [10]. They also alter the general fitness of the plant, in that short or long period mutants made more chlorophyll and biomass in cycle lengths that approximately matched the mutant period [11]. Conversely, wild type plants (with a free running period very close to 24 h) were the most robust in cycles of 24 h versus 20 h and 28 h.

Regarding the organization of the clock in the plant on the structural level, the clock in the roots functions differently, and as a slave to, the clock in the shoots [12]. The molecular components of the root clock are also abbreviated when compared with the clock above the ground. On the level of what could be called an organ, specifically, in the leaves, the clock in the plant cells is essentially an individual, precise, self-sustained oscillator. That the cells remain (mostly) un-coupled was shown by clever experiments involving plants, clock-regulated LUCIFERASE expression and a bit of tin foil [13]. The same leaf could be entrained such that gene expression was anti-phase on the proximal and distal portions of a leaf, a condition facilitated by switching protective foil to these respective regions at opposite times of day whilst the culture container was constantly lit. (Subsequent experiments showed that there is in fact very weak coupling between cells in the leaf [14]; perhaps coupling is condition dependent).

Referring the input side of the clock mechanism (Fig. 7.1), light reaches the plant clock via many photoreceptors. Plants are experts at harvesting light and using it for signaling and environmental information as well as for energy. The collection of plant photoreceptors has demonstrated some basic properties that turn out to be common features of most circadian systems: light input pathways in plants are clock regulated [3, 15], a finding that could explain aspects of the shape of the phase response curve (PRC) [16] (see also Carl Johnson's PRC Atlas at http://www.cas.vanderbilt.edu/johnsonlab/prcatlas/index.html). Photoreceptor (e.g., CRYPTOCHROMES 1 and 2, PHYTOCHROMES A-E) mutants in plants are also often period mutants, albeit subtle ones, showing how photoreceptors can effect the basic clock properties – not just entrainment but also free running period [17, 18]. This observation suggests a model whereby photoreceptors are part of the central oscillator of the plant clock (see Fig. 7.1), but this is a somewhat radical interpretation.

Regarding the central oscillator mechanism, the clock genes that have been identified in plants form a transcriptional feedback loop, similar in structure and function to that described for animals (Fig. 7.2, panels A, C). Specifically, it is a transcriptionally regulated set of coupled feedbacks; two are shown in (Fig. 7.2), but it is clear that there are several more loops in the plant clock formed by genes encoding homologs of TOC1 (the PRR family) [19]. The loops were recently "closed" by identification of a missing component (GIGANTEA or GI) in a model-driven experiment [20]. On the level of sequence, the plant clock proteins are not obviously related to those from fungi or animals, even though the molecular feedback loops that they form work in a similar fashion. They utilize the MYB-domain DNA binding proteins (LHY, CCA1) as transcription factors [21, 22]. In plants, like animals and fungi, some clock-involved proteins have PAS/LOV domains [23],

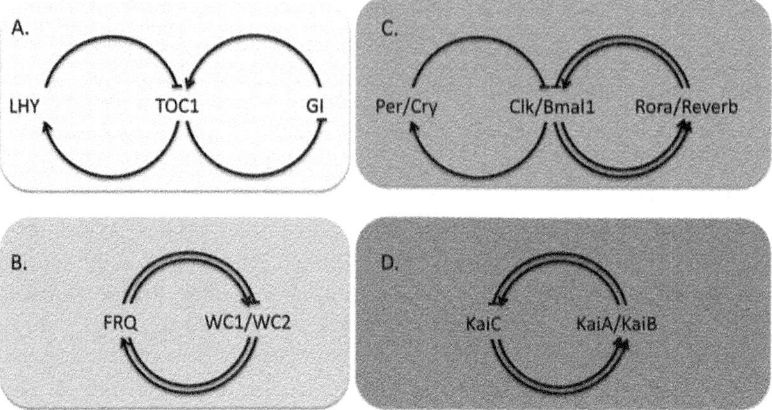

Fig. 7.2 The minimal, central, genetic clock networks in plants, fungi, animals and prokaryotes are formulaic, in that they all use negative feedback and transcriptional regulation. The schemes shown here are meant to be the minimal representations of genetic networks that function as "central oscillators." (**A**) The *Arabidopsis* clock gene network; (**B**) the *Neurospora* clock gene network; (**C**) the mouse clock gene network; (**D**) the *Synechococcus* clock gene network. The *arrows* indicate a positive effect (as in transcriptional activator or promotion of phosphorylation in the case of KaiA on KaiC) and the *perpendicular lines* indicate a negative effect (as in transcriptional repression or promoting dephosphorylation in the case of KaiB on KaiC)

sequences which are likely involved in binding, whether it be another clock protein or a co-factor.

A novel circadian regulatory feedback loop was recently identified in *Arabidopsis* [24]. The cyclic adenosine di-phosphate ribose (cADPR) protein regulates the oscillations in calcium. By controlling expression of cADPR, the levels of clock gene expression (and thus presumably those of clock-regulated genes downstream of the clock loop) were altered, showing how cell signaling molecules are integrated into the clock mechanism as feedbacks.

The central transcriptional feedback loop directs the expression of downstream targets (outputs). For instance, the rhythm in growth rate has been linked to rhythmic expression of the gene encoding phytohormone [25]. A microarray survey of gene expression over a variety of photo- and thermo-periods indicates that, although rhythms in gene expression are modest in a given condition, across the checkerboard of light and temperature cycles that are experienced throughout the year, much of the genome is rhythmic in at least one of the conditions [9].

Ultimately, the clock does not just regulate the gene expression but reaching into metabolism. If the purpose of a functioning clock is to enhance fitness (see cyanobacteria below also), then ill-timed metabolism would be predicted to decrease fitness. An indication of metabolic in-efficiency can be seen in the GI clock mutant. Whereas, a normal plant produces starch granules during the day, metabolizing and dissipating them as a source of energy through the night, the GI mutant fails to fully use them up overnight [26].

7.2.3 Emerging (and Submerged) Plant Model Systems

The small, weedy *Arabidopsis* is a model organism for higher plants, serving as a practical lab tool for experiments that will eventually lead to better understanding of crop plant production and yields. But even this little plant is bulky for many high throughput approaches. There are several photosynthetic unicells that serve as potentially interesting model systems for the plants and have illustrated notable clock regulatory mechanisms. One of the first unicells that was recruited for circadian experiments was the marine diatom, *Gonyaulax polyedra* (now called *Lingulodinium polyedrum*). It has been characterized extensively for its circadian properties (e.g., the first PRC) by virtue of its natural bioluminescense. It has all the benefits of clock-regulated reporter genes but as part of its own repertoire [16]! (In the "modern" era, the main drawback to working with *Gonyaulax* is that it carries up to hundreds of copies of its genome). The circadian program of this organism is a stunning example of how the clock is used to adapt to a spatial niche by forming a temporal one [27]. The diatom propels itself to the surface of the ocean during the day to photosynthesize and accumulate carbon. At night, it pushes itself to the depths, traveling phenomenal distances, to the nitrogen-rich ocean bottom. It emits bioluminescent glows and flashes at night, which though never experimentally tested, are thought to function to ward off predator copepods. When *Gonyaulax* sits at the surface during the day, they will form swarms (during a population explosion, this causes the red tides), an apparent social mechanism.

Gonyaulax was used to characterize light input pathways to the clock, showing multiple, clock-regulated pathways even in this ancient unicell [28, 29]. With the advent of the molecular era, a certain amount of effort went into identification of molecular mechanisms of clock (i.e., central oscillator) regulation. This turned into a fascinating demonstration of a translationally regulated clock. The regulation of luminescence was shown to depend on binding of the 3′end of the luciferin RNA transcript [30, 31]. No transcriptional regulation has ever been demonstrated in this model system.

The so-called green yeast, *Chlamydomonas reinhardtii*, has been studied for circadian rhythm for years and it also uses translational regulation as part of its circadian program, also with an mRNA binding protein [32]. *Chlamydomonas* shows time-of-day differences in chemotaxis towards ammonium [33] and it gates cell division to certain windows during the day [34]. It was used to demonstrate how the clock contributes to increased fitness: pulses of UV light were more lethal in the early subjective night than during the morning hours [35] (once again, suggesting that a light input pathway could be regulated by the circadian clock, but this time, not the input that entrains the cells).

Although clock mutant strains were generated in recent years, no candidate clock genes were uncovered until a recent breakthrough study implemented a high-throughput mutant screen (similar to that used for cyanobacteria, see below) to reveal a panel of clock genes in *Chlamydomonas* [36, 37]. Since the genome of this strange microbe shares gene homologs with animals and plants, it was not obvious what clock genes would look like. A sequence comparison of the putative clock genes from the mutant screen shows a strong resemblance to plant clock genes, in that MYB-like DNA binding domains appear.

Ostreococcus is a unicellular alga with a sequenced genome that shows transcriptionally regulated circadian rhythms [38]. Although little is so far known concerning circadian rhythms at the level of physiology and behavior, it is clear that there is a transcriptional program under circadian control. This system seems ripe for rapid development as an efficient model system for the plants.

7.3 The Fungal Clock

7.3.1 Neurospora crassa

Pittendrigh was one of the first researchers to work on the rhythms in fungi. He was amazed by the rhythms in spore shooting by *Pilobolus*, which was first described some years earlier by Ueblmesser [39, 40]. Due to the foul smell of the media required for this fungus (the primary ingredient was dung), he switched and started experimenting with *Neurospora crassa*, an ascomycete – the division/phylum within the fungal kingdom that includes brewer's yeast, *Saccharomyces cerevisiae*. *Neurospora* makes tufts of asexual spores once per 22 h in constant darkness [41, 42]. Experiments using different photoperiods showed that *Neurospora* makes more asexual spores, carotenoids and pre-sexual structures (protoperithecia) in certain photoperiods, indicating the capacity for seasonal regulation [43].

Concerning the organization of the clock in *Neurospora*, there is little known. Although it is generally considered a simple cellular system, it grows as syncitia, in which the nuclei stream from front to back. The older tissue plugs the pores in the syncitia so that the nuclei tend to be restricted to a limited portion of the hyphae. The hyphae can also fuse with the other hyphae (in the absence of incompatibility factors, so mostly with hyphae from within an individual), theoretically creating possibilities for synchronization.

Historically, *Neurospora* was one of the powerhouse genetic model organisms, so it was also an early entry in the race to find clock genes. As in plants and animals, a transcriptional feedback loop was characterized (Fig. 7.2, panel B). Indeed, the feedback loop hypothesis was originally demonstrated in *Neurospora* [44]. The components of the clock feedback loop in *Neurospora* include the negative feedback protein, FREQUENCY (FRQ), and it's activators, WHITE COLLAR1 (WC1) and WC2 [45]. It is notable that WC1 is a blue light photoreceptor as well as a DNA binding transcriptional activator. Whereas in plants, the photoreceptors have subtle effects on the circadian system, in *Neurospora*, a photoreceptor is one of the central transcription factors in the feedback loop. Without WC1, almost all light responses are lost, for instance, concerning light-induced carotenogenesis in hyphae. However, the *wc-1* knockout mutant can still be entrained for circadian conidiation using light, but with high light levels, indicating a substantial decrease in sensitivity in the mutant [46, 47]. A recent publication extends the observation of residual light responsiveness in this mutant using a high throughput approach (microarrays) [48].

In addition to the dominant feedback loop, which already contains multiple positive and negative regulatory functions, there is an auxiliary loop in *Neurospora* containing an additional blue light photoreceptor, VIVID (VVD), which regulates photoadaptation. As a result, the VVD mutant is recognized by its bright orange conidia. Concerning the clock, the only deficit in the *vvd* mutant is the circadian phase of entrainment. Free running period is un-changed [49, 50]. Additional clock-affecting the components include the FRQ RNA helicase, the COP9 signalasome and numerous kinases and phosphatases [51–55]. Furthermore, a completely independent oscillator has been hypothesized based on many different types of data. The so-called FRQ-less oscillator (FLO) is surmised because of the residual circadian entrainment of clock mutants in Tcycles [5–7] and residual circadian and non-circadian rhythmicity [56–58] in mutant strains. How these loops are coupled together to create a circadian system that is a clock for all seasons is not yet clear.

Neurospora has been a facile organism for testing various ideas concerning mechanisms. Beyond the original mapping out of a transcriptional loop(s), there are additional concepts that have been worked out in *Neurospora*. The molecular basis of temperature sensing (at the level of clock gene expression) was shown to lie in the regulation of FRQ transcription and translation [59, 60].

By visual (or microscopic) inspection, there is little else besides spore formation that has been observed as a circadian output in *Neurospora*, but of course supporting this massive developmental program each day must be numerous molecular oscillations. This was confirmed in pioneer experiments hypothesizing that clock-controlled gene (*ccg*) transcription would be present [61]. In terms of the clock control of metabolism, oscillations in NAD/NADH and NADP/NADPH levels were first shown decades ago for the *Neurospora* clock [62], establishing a link between the clock and metabolism.

We have been able to do a lot of things in *Neurospora* that we were not practical to study using animals. Transformation, regulated gene expression and the simplicity of bulk cultures are among the strengths of the fungal system. The animal model systems were considered substantially more complex. The distinction now is not as great, since even in the animal systems, cellular clocks are easily captured either at tissue culture cell lines of as organ explants.

7.3.2 Beyond Neurospora

Besides *Pilobolus* and *Neurospora*, only a handful of experiments have shown evidence of a circadian program in other fungi. Rhythmic production of asexual spores was also seen in *Aspergillus flavus* [63]. The genome sequence of this organism shows homologs of the WC proteins but no evidence of FRQ. Thus, FRQ looks more and more like a species-restricted facilitator of spore formation. The comparable molecule in *Aspergillus* has yet to be discovered.

There has been a single report of rhythms in *S. cerevisiae*, showing that by slowing down the cell division to once per day (growth at 12°C), the cells would entrain to divide at night in a light/dark cycle [64]. On release to constant darkness, a non-precise rhythm continued for several days. Most of the "litmus tests" for clocks (e.g., entrainment in T-cycles, temperature compensation) have not been performed in this preliminary, non-peer reviewed experiment, thus it is difficult to definitively label the observed phenomenon as "clock-like." Furthermore, inspection of the yeast genome shows no homologs to known clock genes. There are, however, several PAS domain containing proteins [65].

Interestingly, there are numerous reports describing the ultradian rhythms in yeast, with periods ranging from 40 min to several hours [66, 67]. These oscillations are read out of chemostat cultures as oscillations in dissolved oxygen in minimal media. There is some speculation that these metabolic cycles are the underlying basis [68] of circadian rhythms.

Unfortunately, none of these isolated reports for circadian rhythms in non-*Neurospora* fungi have been followed up. The implications are however that clocks will be widely found throughout the fungal kingdom, using a variety of molecular mechanisms to facilitate daily timing.

7.4 The Animal Clock

7.4.1 The Clock in the Fly: Drosophila melanogaster

The human species certainly owes a debt to the fruit fly. As an experimental tool, it has revealed the intricacies of development and the causes of disease in humans. Some of this work was jump-started by the pioneering studies initiated by Seymour Benzer [69], as he dared to look for genes involved in complex behaviours, already in the 1960s. One of the traits he studied was circadian behavior, which had been dissected in *Drosophila*, circadian property by circadian property, in elegant experiments by Colin Pittendrigh [70] as well as by many others. Adults show circadian rhythms of activity in constant darkness, which become almost bimodal in entraining light cycles. Eclosion from pupae is gated to a distinct phase. Already in his landmark Cold Spring Harbor paper, Pittendrigh surmised the existence of two oscillators in the fruit fly by submitting the animals to conflicting temperature and light cycles, a phenomenon now understood in terms of a handful of neurons [71, 72]. The pattern of activity changes its distribution over the day in long versus short photoperiods, which in *D. melanogaster* has few consequences on reproduction, but in other *Drosophila* species, day length regulates diapause resulting in pronounced seasonality [73].

Regarding structural organization of the fly clock, the sophisticated toolbox that the Drosophilists use has identified a finite number of neurons in the brain that are involved in regulating the circadian behavior [71, 74]. In addition to the central brain clock, the fly has peripheral clocks, for instance in the Malpighian tubules, the equivalent of the fly kidney [75]. All sorts of fly bits keep oscillating even after they are separated from the head [76]. The transparency of the fly allows for direct synchronization of these peripheral oscillators with the environmental light/dark cycles, unlike those in higher organisms. Although one might expect that in the intact animal the peripheral organs are coupled to the brain clock, when the Malpighian tubules are transplanted, they maintain their own phase [77]. Furthermore, when the Malpighian oscillations are followed in flies with and without decapitation, the amplitude of the PER oscillations is the same [75]. Both of these experiments indicate that, in flies, the peripheral oscillators are un-coupled from the brain.

The major photoreceptor that *Drosophila* uses is CRYPTOCHROME (dCRY), a blue light photoreceptor with ancestral roots in DNA photolyase [78, 79]. Similar to what is understood in plants and their circadian photoreceptors, mis-expression of or mutation of dCRY leads to an aberrant clock phenotype [80]. Exactly, how the photoreceptor interacts with the clock proteins is well understood, in that dCRY complexes directly with dTIMELESS (dTIM), promoting its light-dependent degradation.

The reference to dTIM leads us to the central oscillator with its clock genes. Although *Drosophila* shows circadian rhythms in locomotor activity, the trait that was used to find the clock genes was eclosion from pupae. The dPeriod (dPer) locus was implicated in a collection of mutants that were either arrhythmic, early or late in their phase of eclosion [69]. The Rosbash lab hypothesized that dPER was participating in a negative feedback loop [81], leading to the key experiments that demonstrated the principle in *Neurospora* [44]. At the time, it was not obvious what other genes participated in this genetic mechanism of the *Drosophila* clock, but now the feedback loop model is filled in with activating and inhibiting heterodimers (dCYCLE and dCLOCK, and dPER and dTIM, respectively) [82]. Like in plants, a multiple loop construction is posited, well supported with genetic experiments [83].

In addition to the well-studied clock in the fruit fly, many and various insect species have been used, especially for their photoperiodism, but also with respect to ecology [84, 85]. Some of these (e.g., the house fly) even have sequenced genomes, but progress concerning mechanisms is somewhat limited due to the (lack of) availability of reverse genetic tools [86–88].

7.4.2 *The Clock in the Mouse*

By now, the reader of this book has gotten a substantial dose of mouse clock behavior and genetics. The most commonly studied phenotype is that of activity, but with the permeation of circadian regulation through many molecular levels, there are surely hundreds of markers that could be assayed for oscillations over the day. The similarities between the organization of the mouse clock and the *Drosophila* clock are notable, including evidence for multiple oscillators in the brain, peripheral oscillators and a homologous set of clock genes [83]. While the fly uses just a few cells to regulate the circadian behavior, the mouse uses a small, discrete brain region called the suprachiasmatic nucleus (SCN) which itself has subregions that apparently perform different tasks with respect to controlling phase in different photoperiods [89]. Even within the SCN, there is evidence for a morning and an evening (M/E) oscillator [89, 90]. Although the SCN is the pacemaker, coordinating oscillations in peripheral tissues, it is clear that most cells have the capacity to oscillate with a circadian rhythm. They may be coupled into the system by the SCN or they can oscillate according to the other stimuli (such as oscillations in gene expression in the liver and feeding behavior) [91].

Circadian photoreception and the interlocked feedback loop structure of the clock gene network (Fig. 7.2, panel C) have been covered in the other chapters and will not be further elaborated here (see Chaps. 2 and 3). Of note, however, are recent reports linking feedbacks from metabolism to the circadian transcriptional machinery [92, 93]. The connection was elaborated through the discovery of orthologs of yeast SIR2 – SIRT1 – acting to modulate CLK/BMAL1 activity through de-acetylation or BMAL1 and PER2 [94, 95]. This basic mechanism introduces the possibility that clock control of gene expression is facilitated by a combination of transcriptional regulation as well as the genome wide chromatin re-modeling. The sirtuins are regulated by NAD^+ levels. Thus, nicotinamide metabolism was investigated, showing daily oscillations in NAD^+ (primarily via clock control of the NAD^+ salvage pathway) and feedbacks on clock gene expression according to NAD^+ expression. These data should make us re-think how the transcriptional feedback loop model regulates the clock.

The SCN clock directs a web of oscillating tissues in the periphery. The liver, kidney and heart, for instance, show oscillations in gene expression, including clock genes and tissue specific clock outputs [96–98]. In the liver it is the expression of RNAs for liver enzymes that are oscillating, whilst in the kidney, the genes for ion channels are rhythmically expressed. Somewhere between 5 and 10% of transcripts are estimated to be rhythmically expressed in a given tissue. Moving to the next level, an analysis of the liver proteome showed about 20% of proteins expressed differently over the day [99]. Furthermore, within the rhythmic proteome, about half of these proteins were constitutively expressed at the RNA level. Clearly, the clock is engaging post-transcriptional regulatory mechanisms. Thus, regarding the sheer volume of molecular oscillations at numerous levels, it is possible that moving through the various levels from transcriptome to metabolome, most genetic and metabolic pathways will be rhythmic (Fig. 7.3).

7.4.3 *The Clock in Us*

One of the big questions hovering over chronobiology research is how is it related to the clock in humans. In everyday life, we see obvious comparisons between our activities and those of mice as they jump onto the running wheel and we jump into action. Yet, in contrast to mice, we think of our activity as plastic and un-fixed (the alarm clock can work wonders) despite that our physiology is oscillating just as that

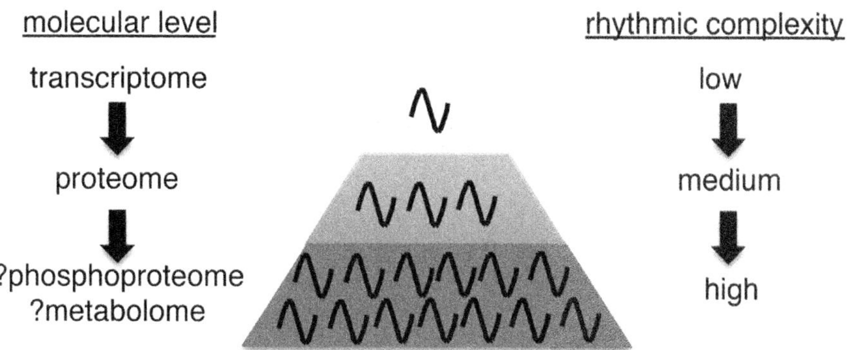

Fig. 7.3 The number of rhythmic components increases moving from transcriptome to proteome [99]. If this trend continues as we move to phosphoproteome, metabolome and the other functions regulated by the clock, then the rhythmic landscape in the cell becomes much more complex than reported by the transcriptome

of the other eukaryotes. On one hand, it turns out that mice are not so fixed to function as purely nocturnal animals, in that when they are submitted to some non-standard lab conditions, they can be induced to become diurnal [100]. On the other hand, we now know that humans are so tuned into the environmental light cycle that population-wide behavior is seen across a single time zone, as the sun rises progressively later or earlier relative to the social clock [101].

Human circadian behavior has been extensively cataloged. Aschoff spent decades collecting data on behavior and physiology from humans experiencing a month in temporal isolation [102, 103]. Hundreds if not thousands of parameters, from metabolism to performances and cognition, are changing over the course of the day. Although the acrophase and nadir of a given hormone, activity or physiological rhythm will have it's own characteristic timing, there are also inter-individual differences. The highly individualized timing reflects the properties of synchronization of the circadian clock with the external zeitgeber cycles. The term chronotype generally refers to this entrained phase. Chronotype was determined for thousands of individuals using a simple questionnaire (the MCTQ, go to http://www.euclock.org/) based on habitual sleep times on work and on free days [104–107]. (This is analogous to analyzing the timing of activity rhythms in other animals, except that it relies on subjective reporting.) Analysis of a large database revealed age and gender differences: on a population level, at the age of 20–21 years, the latest chronotype is attained, and for much of their adult lives, women are slightly earlier chronotypes than are men [107].

The anatomical organization of the human circadian clock shows parallels with respect to the clock in mice. The human SCN was implicated early on when a patient with a tumor in that region showed symptoms of lack of consolidated sleep and a disturbed circadian temperature rhythm [108].

Inspection of the human genome sequence reveals a full set of genes that are homologous to clock genes characterized in mice. Until recently, there was little proof that these were indeed acting as clock genes in humans. This changed with the presentation of two large pedigrees with extremely early chronotypes segregating according to simple Mendelian inheritance. The affected genes were hPeriod2 and hCasein Kinase I [109, 110]. A second route of confirmation comes with the development of human tissue culture cell lines as model systems for cellular clock function. Extensive characterization of one such cell line – U2OS cells, an osteosarcoma cell line – by mis-expressing putative clock genes [111] clearly shows that orthologous genes regulate rhythms in humans cells as they are in mice.

7.5 The Cyanobacteria Clock

Regarding chronobiology research, the only model genetic system from the prokaryotic kingdom is the cyanobacterium, *Synechococcus elongatus*. This prolific alga is photosynthetic, preferring a fresh water habitat. It generates whopping amounts of oxygen, and uses its circadian system to sequester nitrogen-fixation to nighttime, so as to avoid the toxic effects of mixing nitrogen-fixation with oxygen [112]. In this way, they need not form organelles to keep these compounds separate, as many microbes do.

Although their circadian clock is less elaborated than that of plants, animals and fungi with respect to formalisms (especially concerning entrainment), this microbe has delivered stunning insights into the ecology and the molecular mechanism of the clock. It is from cyanobacteria that we first learned that the clock is adaptive, conferring a selective advantage that is apparent easily within a 100 generations [113]. But the majority of what we know about the cyanobacteria clock stems from molecular and genetic analyses.

Light input to the molecular clock occurs via at least two kinases, CikA and CikR [114, 115]. This pathway, in turn, acts via the clock protein, KaiA.

Cyanobacteria may be the only model organism for which a saturating screen has been performed to look for clock genes. This involved a clever, high-throughput, automated setup, and a strain expressing a clock-regulated luciferase gene [116]. The results yielded hundreds of clock mutants, most of them mapping to three genes, KaiA, B and C.

The feedback loop model was initially invoked, with the stunning demonstration that virtually all transcripts in the cell are rhythmically expressed (thus, the entire genome is ccg's) [117]. However, the Kai genes work somewhat differently when compared with the transcription factors in the eukaryotic world. KaiC is a member of the RecA/DNAB superfamily of ATPases. It autophosphorylates and then dephosphorylates itself on at least two residues, and in an ordered fashion [118]. In addition, it forms hexamers. The roles of KaiA and KaiB are to promote the autophosphorylation and dephosphorylation, respectively.

One phenotype of the clock that can be followed as a circadian rhythm is the phosphorylation state of KaiC proteins. When the Kondo lab suppressed transcription on a global scale by growing the microbes in darkness, they observed that the phosphorylation continued, a seeming contradiction of the role of the transcriptional negative feedback loop in the generation of circadian rhythm [119]. They combined the purified clock proteins, KaiA, B and C with ATP and were able to follow self-sustained circadian rhythms for days [120]. These oscillations furthermore are temperature compensated and can be phase shifted with a temperature pulse. The conclusion is that the phosphorylation loop – a completely post-translational mechanism – is the core functional element in the cyanobacterial clock. There has been speculation that phosphorylation could be performing a similar role in the eukaryotic clocks [121], but so far there is no evidence to support this hypothesis. The cyanobacterial clock is thought to regulate transcription on a global scale by directing oscillations in chromatin structure [122].

We do not understand why there are not more examples of circadian clocks in prokaryotes beyond the cyanobacteria or indeed beyond *Synechococcus* [123]. Prokaryotic organisms are subject to the same daily challenges that are created by a changing landscape concerning light, temperature and nutrition as the systems that have been described above, meaning that they would be more likely to thrive if they had a mechanism that creates a temporal structure. Furthermore, they generally have a rapid division time – like cyanobacteria – providing ample substrate on which selection can act. It is interesting to note that many of the model organisms that we study in the lab would not show circadian rhythms under typical lab growth conditions. For instance, the wild type *Neurospora* strain is typically grown at 30°C on defined but rich media. This strain almost never shows regulated sporulation in this condition. Conversely, there are many reports of light dependent sporulation of the appearance of growth rings on fungi and bacteria (e.g., the corn smut *Ustilago maydis*, *Bacillus subtilis*, and *Streptomyces*). A careful investigation of some of these might open up a much more diverse set of model systems. More comparison can only mean more insights into the rules for the wiring of a circadian clock.

7.6 Clocks in Comparison – A Summary

We have briefly discussed some basic characteristics of circadian clocks in plants, animals, fungi and the cyanobacteria. In so doing, we have noted the aspects of circadian behavior, the anatomy of the circadian system and the molecular mechanism of rhythms in representative model systems. Already in 1960, Colin Pittindrigh summarized the commonalities between the circadian systems, and that list still stands as essentially complete [70, 124]. He noted that the circadian rhythms share these properties:

- They are ubiquitous, endogenous and innate
- Their period (*tau*) is near 24 h and is usually – but not always – self-sustaining
- They are precise and robust
- The *tau* is plastic and temperature compensated
- They are entrainable and can be phase shifted, despite their compensation to some zeitgebers and general resistance to noise

All of the model systems that have been studied to date generally conform to these characteristics, however not all of the characteristics have been systematically investigated. For instance, entrainment has not been demonstrated in cyanobacteria yet the experiments showing increased fitness in certain entraining regimes certainly implies it.

In closing, we emphasize several issues. Concerning molecular mechanisms of clocks, it is puzzling that the different phyla all use different clock genes. (A case has been made for the relationship of fungal WC-1 and animal BMAL1 [125], but this is not overwhelmingly convincing). Especially within the eukaryotes, circadian clocks have implemented very similar strategies, namely transcriptional feedback loops that apparently incorporate metabolic feedbacks (this has been best demonstrated in animals and plants). Yet, the proteins that act within these transcriptional loops are completely distinct, with the exception of PAS/LOV domains. Given what we know from cyanobacteria concerning the relatively rapid selection pressure exerted by the circadian clock [113], it is unlikely that the clock emerged only after plants, animals and fungi diverged from their common ancestors. Rather, we propose that the feedback loops that are hard-wired into the phyla- and species-specific clocks were add-ons to a more primitive clock. The evolution of these feedback loops permitted a niche-appropriate, clock regulated, zeitgeber input pathway. Some of these might be considered zeitnehmer pathways [1], clock-regulated cellular pathways that take endogenous cell signals and push them to the clock, resulting in qualities like robustness.

As to the nature of the pre-transcriptional feedback clock machinery, experiments that reveal the residual rhythmicity or entrainment in clock mutant strains suggest that remnants of the primitive system can be resurrected. A possible pre-feedback loop mechanism may lie in protein kinases [121]. In this chapter, we have concentrated on the clock genes that function as transcription factors, but in each of these regulatory loops, protein kinases play an essential role. The double-time mutant in *Drosophila* and the *tau* mutant in hamsters were identified as mutations in casein kinase I (CKI), now implicated as a key clock kinase in plants, animals and fungi [83, 126, 127]. Many additional kinases have been identified [128–131], beyond CKI, not surprising, given the dependence of transcription factors on phosphorylation state for function. Phosphorylation was one of the primitive forms of regulation, using molecular energy to alter the biochemical properties of molecules in a directed fashion. Specifically regarding CKI, it regulates itself by autophosphorylation, which in turn is controlled with input from PROTEIN PHOSPHATASE I (PPI). PPI is regulated according to binding of CRY1 and CRY2. Since these proteins are targets of CKI, a provisional, regulatory phosphorylation loop is suggested. The broad dependence of cell functions on phosphorylation has probably been one of the limiting factors, in further pinpointing how this process is invoked in circadian regulation: on one hand, deletion of kinases is often lethal, and on the other, their activity is somewhat promiscuous in vitro.

As discussed above, there is now evidence from plants and animals that calcium and NAD^+ metabolism also form clock feedback loops, both regulated by and feeding back onto the transcriptional mechanism (again, somewhat like a zeitnehmer).

The regulation of metabolism by the clock has been understood for decades. A simple demonstration is the oscillation in core body temperature in humans that persists even in the absence of activity. The concept of metabolism feeding into the clock is less well established even though it has been discussed over the last decade [132, 133]. Using the example of body temperature, increased activity generally does not disrupt the circadian timing despite that it greatly modifies body temperature. So, missing information still concerns the question of metabolic compensation, which is likely related to temperature and nutritional compensation. How is it that the fluctuations in metabolism do not alter circadian timing?

We have provisionally added the metabolic feedbacks to the clock mechanism (Fig. 7.4). The clock mechanism becomes so extensive that it is difficult to understand where the so-called central oscillator ends and where outputs begin. Here, the comparative approach could be useful. Clock regulated processes which are purely species specific might be hypothesized to be outputs, whereas, processes that are regulated by all clocks will be more likely to be part of a primordial clock mechanism. Candidates for this latter classification might be phosphate, nicotinamide or calcium metabolism, which have been recently elucidated with respect to the animal and plant clocks. Experiments should be undertaken to determine how widespread these properties are in the circadian clocks. If they are broadly utilized, then they are likely to be part of the primordial clock. If they are species specific, then they could be part of the clock mechanism that includes the transcriptional feedback loop or they could be part of the input or output circuits.

This last point brings us back to the ubiquity of circadian clocks, which in some ways is the point of this chapter. We speculate that the daily clocks are present in practically all organisms on earth. Over the next decade, a concerted effort should

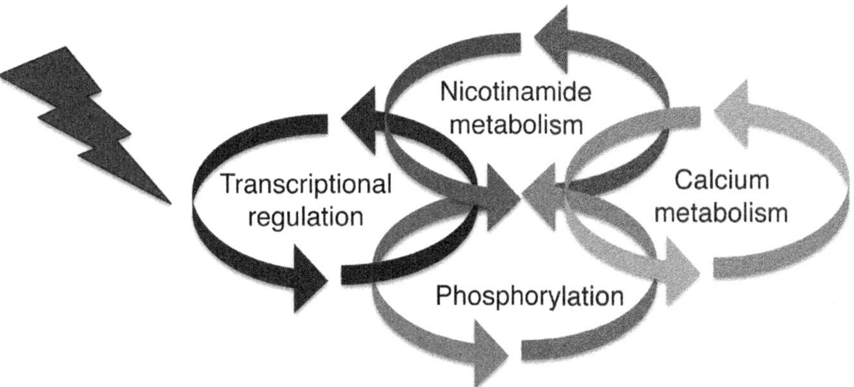

Fig. 7.4 Models of circadian clock mechanisms are inherently complex, involving impingement of multiple regulatory mechanisms into the central oscillator. The transcriptional feedback loop is co-regulated by feedback loops from nicotinamide metabolism (NAD^+ metabolism via regulation of acetylation state by SIRT1) and calcium and phosphorylation metabolism and signaling. Additional epigenetic mechanisms such as DNA methylation have yet to be systematically investigated, and could constitute additional regulatory loops that are part of the central oscillator mechanism

go into characterizing the circadian phenotypes especially in microbes. In these diverse, simple and genetically amenable systems, we stand to gain substantial insights into the variety of regulatory mechanisms that can shape the list of characteristics that Pittendrigh enumerated. These in turn will help us understand not only the species that are potentially, commercially and ecologically interesting, but also to understand the clock in us.

Acknowledgments Our work is supported by the Dutch Science Foundation (the NWO), the German Science Foundation (DFG), The Hersenen Stichting, EUCLOCK, a 6th Framework Program of the European Union and the Rosalind Franklin Fellowships of the University of Groningen.

References

1. Roenneberg T, Merrow M (1999) Circadian clocks and metabolism. J Biol Rhythms 14:449–459
2. Roenneberg T, Merrow M (1998) Molecular circadian oscillators – an alternative hypothesis. J Biol Rhythms 13:167–179
3. McWatters HG, Bastow RM, Hall A, Millar AJ (2000) The *ELF3 zeitnehmer* regulates light signalling to the circadian clock. Nature 408:716–720
4. De Mairan JJDO (1729) Observation botanique. Histoir de l'Academie Royale des Science, Paris, pp 35–36
5. Pregueiro A, Price-Lloyd N, Bell-Pedersen D, Heintzen C, Loros JJ, Dunlap JC (2005) Assignment of an essential role for the *Neurospora frequency* gene in circadian entrainment to temperature cycles. Proc Natl Acad Sci USA 102:2210–2215
6. Roenneberg T, Dragovic Z, Merrow M (2005) Demasking biological oscillators: properties and principles of entrainment exemplified by the *Neurospora* circadian clock. Proc Natl Acad Sci USA 102:7742–7747
7. Merrow M, Brunner M, Roenneberg T (1999) Assignment of circadian function for the *Neurospora* clock gene *frequency*. Nature 399:584–586
8. Brown SA, Zumbrunn G, Fleury-Olela F, Preitner N, Schibler U (2002) Rhythms of mammalian body temperature can sustain peripheral circadian clocks. Curr Biol 12:1574–1583
9. Michael TP, Mockler TC, Breton G, McEntee C, Byer A, Trout JD, Hazen SP, Shen R, Priest HD, Sullivan CM et al (2008) Network discovery pipeline elucidates conserved time-of-day-specific cis-regulatory modules. PLoS Genet 4:e14
10. Hicks KA, Millar AJ, Carré IA, Somers DE, Straume M, Meeks-Wagner R, Kay SA (1996) Conditional circadian dysfuncion of the *Arabidopsis early-flowering* 3 mutant. Science 274:790–792
11. Dodd AN, Salathia N, Hall A, Kevei E, Toth R, Nagy F, Hibberd JM, Millar AJ, Webb AA (2005) Plant circadian clocks increase photosynthesis, growth, survival, and competitive advantage. Science 309:630–633
12. James AB, Monreal JA, Nimmo GA, Kelly CL, Herzyk P, Jenkins GI, Nimmo HG (2008) The circadian clock in *Arabidopsis* roots is a simplified slave version of the clock in shoots. Science 322:1832–1835
13. Thain SC, Hall A, Millar AJ (2000) Functional independence of circadian clocks that regulate plant gene expression. Curr Biol 10:951–956
14. Rascher U, Hutt MT, Siebke K, Osmond B, Beck F, Luttge U (2001) Spatiotemporal variation of metabolism in a plant circadian rhythm: the biological clock as an assembly of coupled individual oscillators. Proc Natl Acad Sci USA 98:11801–11805
15. Anderson SL, Somers-DE DE, Millar AJ, Hanson K, Chory J, Kay SA (1997) Attenuation of *phytochrome A* and *B* signaling pathways by the *Arabidopsis* circadian clock. Plant Cell 9: 1727–1743

16. Hastings JW, Sweeney BM (1958) A persistent diurnal rhythm of luminescence in *Gonyaulax polyedra*. Biol Bull 115:440–458
17. Millar AJ, Straume M, Chory J, Chua NH, Kay SA (1995) The regulation of circadian period by phototransduction pathways in *Arabidopsis*. Science 267:1163–1166
18. Somers DE, Devlin PF, Kay SA (1998) Phytochromes and cryptochromes in the entrainment of the *Arabidopsis* circadian clock. Science 282:1488–1490
19. Nakamichi N, Matsushika A, Yamashino T, Mizuno T (2003) Cell autonomous circadian waves of the APRR1/TOC1 Quintet in an established cell line of *Arabidopsis thaliana*. Plant Cell Physiol 44:360–365
20. Locke JC, Southern MM, Kozma-Bognar L, Hibberd V, Brown PE, Turner MS, Millar AJ (2005) Extension of a genetic network model by iterative experimentation and mathematical analysis. Mol Syst Biol 1(2005):0013
21. Alabadi D, Oyama T, Yanovsky MJ, Harmon FG, Mas P, Kay SA (2001) Reciprocal regulation between TOC1 and LHY/CCA1 within the *Arabidopsis* circadian clock. Science 293:880–883
22. Wang ZY, Tobin EM (1998) Constitutive expression of the *CIRCADIAN CLOCK ASSOCIATED 1* (*CCA1*) gene disrupts circadian rhythms and suppresses its own expression. Cell 93:1207–1217
23. Somers DE, Schultz TF, Milnamow M, Kay SA (2000) *ZEITLUPE* encodes a novel clock associated PAS protein from *Arabidobsis*. Cell 101:319–329
24. Dodd AN, Gardner MJ, Hotta CT, Hubbard KE, Dalchau N, Love J, Assie JM, Robertson FC, Jakobsen MK, Goncalves J et al (2007) The *Arabidopsis* circadian clock incorporates a cADPR-based feedback loop. Science 318:1789–1792
25. Michael TP, Breton G, Hazen SP, Priest H, Mockler TC, Kay SA, Chory J (2008) A morning-specific phytohormone gene expression program underlying rhythmic plant growth. PLoS Biol 6:e225
26. Eimert K, Wang SM, Lue WI, Chen J (1995) Monogenic recessive mutations causing both late floral initiation and excess starch accumulation in *Arabidopsis*. Plant Cell 7:1703–1712
27. Roenneberg T (1996) The complex circadian system of *Gonyaulax polyedra*. Physiol Plant 96:733–737
28. Deng T-S, Roenneberg T (1997) Photobiology of the *Gonyaulax* circadian system: II Allopurinol inhibits blue light effects. Planta 202:502–509
29. Roenneberg T, Deng T-S (1997) Photobiology of the *Gonyaulax* circadian system: I different phase response curves for red and blue light. Planta 202:494–501
30. Mittag M, Lee D-H, Hastings JW (1994) Circadian expression of the luciferin-binding protein correlates with the binding of a protein to the 3′ untranslated region of its mRNA. Proc Natl Acad Sci USA 91:5257–5261
31. Lee D-H, Mittag M, Sczekan S, Morse D, Hastings JW (1993) Molecular cloning and genomic organization of a gene for luciferin-binding protein from the dinoflagellate *Gonyaulax polyedra*. J Biol Chem 268:8842–8850
32. Mittag M (1996) Conserved circadian elements in phylogenetically diverse algae. Proc Natl Acad Sci USA 93:14401–14404
33. Byrne TE, Wells MR, Johnson CH (1992) Circadian rhythms of chemotaxis to ammonium and of methylammonium uptake in *Chlamydomonas*. Plant Physiol 98:879–886
34. Mori T, Binder B, Johnson CH (1996) Circadian gating of cell division in cyanobacteria growing with average doubling times of less than 24 hours. Proc Natl Acad Sci USA 93:10183–10188
35. Nikaido SS, Johnson CH (2000) Daily and circadian variation in survival from ultraviolet radiation in *Chlamydomonas reinhardtii*. Photochem Photobiol 71:758–765
36. Brunner M, Merrow M (2008) The green yeast uses its plant-like clock to regulate its animal-like tail. Genes Dev 22:825–831
37. Matsuo T, Okamoto K, Onai K, Niwa Y, Shimogawara K, Ishiura M (2008) A systematic forward genetic analysis identified components of the *Chlamydomonas* circadian system. Genes Dev 22:918–930
38. Moulager M, Monnier A, Jesson B, Bouvet R, Mosser J, Schwartz C, Garnier L, Corellou F, Bouget FY (2007) Light-dependent regulation of cell division in *Ostreococcus*: evidence for a major transcriptional input. Plant Physiol 144:1360–1369

39. Bruce VC, Weight F, Pittendrigh CS (1960) Resetting the sporulation rhythm in *Pilobolus* with short light flashes of high intensity. Science 131:728–730
40. Uebelmesser E-R (1954) Über den endogenen Tagesrhythmus der Sporangienbildung von *Pilobolus*. Arch Mikrobiol 20:1–33
41. Pittendrigh CS, Bruce VG, Rosensweig NS, Rubin ML (1959) Growth patterns in *Neurospora crassa*. Nature 184:169–170
42. Merrow M, Roenneberg T, Macino G, Franchi L (2001) A fungus among us: the *Neurospora crassa* circadian system. Semin Cell Dev Biol 12:279–285
43. Tan Y, Merrow M, Roenneberg T (2004) Photoperiodism in *Neurospora crassa*. J Biol Rhythms 19:135–143
44. Aronson BD, Johnson KA, Loros JJ, Dunlap JC (1994) Negative feedback defining a circadian clock: autoregulation of the clock gene *frequency*. Science 263:1578–1584
45. Crosthwaite SK, Dunlap JC, Loros JJ (1997) *Neurospora wc-1* and *wc-2*: transcription, photoresponses, and the origin of circadian rhythmicity. Science 276:763–769
46. Dragovic Z, Tan Y, Görl M, Roenneberg T, Merrow M (2002) Light reception and circadian behavior in "blind" and "clock-less" mutants of *Neurospora crassa*. EMBO J 21:3643–3651
47. Ballario P, Vittorioso P, Magrelli A, Talora C, Cabibbo A, Macino G (1996) *White collar-1*, a central regulator of blue light responses in *Neurospora*, is a zinc finger protein. EMBO J 15:1650–1657
48. Chen CH, Ringelberg CS, Gross RH, Dunlap JC, Loros JJ (2009) Genome-wide analysis of light-inducible responses reveals hierarchical light signalling in *Neurospora*. EMBO J 28:1029–1042
49. Heintzen C, Loros JJ, Dunlap JC (2001) The PAS protein VIVID defines a clock-associated feedback loop that represses light input, modulates gating, and regulates clock resetting. Cell 104:453–464
50. Shrode LB, Lewis ZA, White LD, Bell-Pedersen D, Ebbole DJ (2001) *vvd* is required for light adaptation of conidiation-specific genes of *Neurospora crassa*, but not circadian conidiation. Fungal Genet Biol 32:169–181
51. Querfurth C, Diernfellner A, Heise F, Lauinger L, Neiss A, Tataroglu O, Brunner M, Schafmeier T (2007) Posttranslational regulation of *Neurospora* circadian clock by CK1a-dependent phosphorylation. Cold Spring Harb Symp Quant Biol 72:177–183
52. Yang Y, He Q, Cheng P, Wrage P, Yarden O, Liu Y (2004) Distinct roles for PP1 and PP2A in the *Neurospora* circadian clock. Genes Dev 18:255–260
53. Yang Y, Cheng P, Liu Y (2002) Regulation of the *Neurospora* circadian clock by casein kinase II. Genes Dev 16:994–1006
54. He Q, Cheng P, He Q, Liu Y (2005) The COP9 signalosome regulates the *Neurospora* circadian clock by controlling the stability of the SCF^{FWD-1} complex. Genes Dev 19:1518–1531
55. Schaffmeier T, Haase A, Kaldi K, Scholz J, Fuchs M, Brunner M (2005) Transcriptional feedback of *Neurospora* circadian clock gene by phosphorylation-dependent inactivation of its transcription factor. Cell 122:235–246
56. Aronson BD, Johnson KA, Dunlap JC (1994) The circadian clock locus *frequency*: a single ORF defines period length and temperature compensation. Proc Natl Acad Sci USA 91:7683–7687
57. Correa A, Lewis ZA, Greene AV, March IJ, Gomer RH, Bell-Pedersen D (2003) Multiple oscillators regulate circadian gene expression in *Neurospora*. Proc Natl Acad Sci USA 100:13597–13602
58. Christensen MK, Falkeid G, Loros JJ, Dunlap JC, Lillo C, Ruoff P (2004) A nitrate-induced frq-less oscillator in *Neurospora crassa*. J Biol Rhythms 19:280–286
59. Diernfellner A, Colot HV, Dintsis O, Loros JJ, Dunlap JC, Brunner M (2007) Long and short isoforms of *Neurospora* clock protein FRQ support temperature-compensated circadian rhythms. FEBS Lett 581:5759–5764
60. Diernfellner AC, Schafmeier T, Merrow MW, Brunner M (2005) Molecular mechanism of temperature sensing by the circadian clock of *Neurospora crassa*. Genes Dev 19:1968–1973
61. Loros JJ, Denome SA, Dunlap JC (1989) Molecular cloning of genes under control of the circadian clock in *Neurospora*. Science 243:385–388

62. Brody S, Harris S (1973) Circadian rhythms in *Neurospora*: spatial differences in pyridine nucleotide levels. Science 180:498–500
63. Greene AV, Keller N, Haas H, Bell-Pedersen D (2003) A circadian oscillator in *Aspergillus* spp. regulates daily development and gene expression. Eukaryot Cell 2:231–237
64. Edmunds LN Jr (1988) Cellular and molecular bases of biological clocks: models and mechanisms of circadian time keeping. Springer, New York
65. Rutter J, Probst BL, McKnight SL (2002) Coordinate regulation of sugar flux and translation by PAS kinase. Cell 111:17–28
66. Klevecz RR, Bolen J, Forrest G, Murray DB (2004) A genomewide oscillation in transcription gates DNA replication and cell cycle. Proc Natl Acad Sci USA 101:1200–1205
67. Tu BP, Kudlicki A, Rowicka M, McKnight SL (2005) Logic of the yeast metabolic cycle: temporal compartmentalization of cellular processes. Science 310:1152–1158
68. Tu BP, McKnight SL (2006) Metabolic cycles as an underlying basis of biological oscillations. Nat Rev Mol Cell Biol 7:696–701
69. Konopka R, Benzer S (1971) Clock mutants of *Drosophila melanogaster*. Proc Natl Acad Sci USA 68:2112–2116
70. Pittendrigh CS (1960) Circadian rhythms and the circadian organization of living systems. Cold Spring Harb Symp Quant Biol 25:159–184
71. Stoleru D, Peng Y, Agosto J, Rosbash M (2004) Coupled oscillators control morning and evening locomotor behaviour of *Drosophila*. Nature 431:862–868
72. Grima B, Chélot E, Xia R, Rouyer F (2004) Morning and evening peaks of activity rely on different clock neurons of the *Drosophila* brain. Nature 431:869–873
73. Emerson KJ, Bradshaw WE, Holzapfel CM (2009) Complications of complexity: integrating environmental, genetic and hormonal control of insect diapause. Trends Genet 25(5): 217–225
74. Veleri S, Brandes C, Helfrich-Förster C, Hall JC, Stanewsky R (2003) A self-sustaining, light-entrainable circadian oscillator in the *Drosophila* brain. Curr Biol 13:1758–1767
75. Hege DM, Stanewsky R, Hall JC, Giebultowicz JM (1997) Rhythmic expression of a PER-reporter in the Malpighian tubules of decapitated *Drosophila*: evidence for a brain-independent circadian clock. J Biol Rhythms 12:300–308
76. Plautz JD, Kaneko M, Hall JC, Kay SA (1997) Independent photoreceptive circadian clocks throughout *Drosophila*. Science 278:1632–1635
77. Giebultowicz JM, Stanewsky R, Hall JC, Hege DM (2000) Transplanted *Drosophila* excretory tubules maintain circadian clock cycling out of phase with the host. Curr Biol 10:107–110
78. Stanewsky R, Kaneko M, Emery P, Beretta B, Wagner-Smith K, Kay SA, Rosbash M, Hall JC (1998) The cry^{b} mutation identifies *cryptochrome* as a circadian photoreceptor in *Drosophila*. Cell 95:681–692
79. Helfrich-Förster C, Winter C, Hofbauer A, Hall JC, Stanewsky R (2001) The circadian clock of fruit flies is blind after elimination of all known photoreceptors. Neuron 30:249–261
80. Dissel S, Codd V, Fedic R, Garner KJ, Costa R, Kyriacou CP, Rosato E (2004) A constitutively active *cryptochrome* in *Drosophila melanogaster*. Nat Neurosci 7:834–840
81. Hardin PE, Hall JC, Rosbash M (1990) Feedback of the *Drosophila period* gene product on circadian cycling of its messenger RNA levels. Nature 343:536–540
82. Glossop NRG, Lyons LC, Hardin PE (1999) Interlocked feedback loops within the *Drosophila* circadian oscillator. Science 286:766–778
83. Young MW, Kay SA (2001) Time zones: a comparative genetics of circadian clocks. Nat Rev Genet 2:702–715
84. Saunders DS (ed) (1982) Insect clocks . Pergamon, Oxford, UK
85. Meinertzhagen IA, Pyza E (1996) Daily rhythms in cells of the flys optic lobe: taking time out from the circadian clock. Trends Neurosci 19:285–291
86. Codd V, Dolezel D, Stehlik J, Piccin A, Garner KJ, Racey SN, Straatman KR, Louis EJ, Costa R, Sauman I et al (2007) Circadian rhythm gene regulation in the housefly *Musca domestica*. Genetics 177:1539–1551
87. Mazzotta GM, Sandrelli F, Zordan MA, Mason M, Benna C, Cisotto P, Rosato E, Kyriacou CP, Costa R (2005) The clock gene period in the medfly *Ceratitis capitata*. Genet Res 86:13–30

88. Kenny NAP, Saunders DS (1991) Adult locomotor rhythmicity as "hands" of the maternal photoperiodic clock regulating larval diapause in the Blowfly, *Calliphora vicina*. J Biol Rhythms 6:217–235
89. Tournier BB, Dardente H, Simonneaux V, Vivien-Roels B, Pevet P, Masson-Pevet M, Vuillez P (2007) Seasonal variations of clock gene expression in the suprachiasmatic nuclei and pars tuberalis of the European hamster (*Cricetus cricetus*). Eur J NeuroSci 25:1529–1536
90. Hamada T, Antle MC, Silver R (2004) Temporal and spatial expression patterns of canonical clock genes and clock-controlled genes in the suprachiasmatic nucleus. Eur J NeuroSci 19:1741–1748
91. Damiola F, Minh NL, Preitner N, Kornmann B, Fleury-Olela F, Schibler U (2000) Restricted feeding uncouples circadian oscillators in peripheral tissues from the central pacemaker in the suprachiasmatic nucleus. Genes Dev 14:2950–2961
92. Nakahata Y, Sahar S, Astarita G, Kaluzova M, Sassone-Corsi P (2009) Circadian control of the NAD^+ salvage pathway by CLOCK-SIRT1. Science 324:654–657
93. Ramsey KM, Yoshino J, Brace CS, Abrassart D, Kobayashi Y, Marcheva B, Hong HK, Chong JL, Buhr ED, Lee C et al (2009) Circadian clock feedback cycle through NAMPT-mediated NAD^+ biosynthesis. Science 324:651–654
94. Hirayama J, Sahar S, Grimaldi B, Tamaru T, Takamatsu K, Nakahata Y, Sassone-Corsi P (2007) CLOCK-mediated acetylation of BMAL1 controls circadian function. Nature 450:1086–1090
95. Asher G, Gatfield D, Stratmann M, Reinke H, Dibner C, Kreppel F, Mostoslavsky R, Alt FW, Schibler U (2008) SIRT1 regulates circadian clock gene expression through PER2 deacetylation. Cell 134:317–328
96. Panda S, Antoch MP, Millar BH, Su AI, Schook AB, Straume M, Schultz PG, Kay SA, Takahashi JS, Hogenesch JB (2002) Coordinated transcription of key pathways in the mouse by the circadian clock. Cell 109:307–320
97. Storch KF, Lipan O, Leykin I, Viswanathan N, Davis FC, Wong WH, Weitz CJ (2002) Extensive and divergent circadian gene expression in liver and heart. Nature 417:78–83
98. Akhtar RA, Reddy AB, Maywood ES, Clayton JD, King VM, Smoth AG, Gant TW, Hastings MH, Kyriacou CP (2001) Circadian cycling of the mouse liver transcriptome, as revealed by cDNA microarray, is driven by the suprachiasmatic nucleus. Curr Biol 12:540–550
99. Reddy AB, Karp NA, Maywood ES, Sage EA, Deery M, O'Neill JS, Wong GK, Chesham J, Odell M, Lilley KS et al (2006) Circadian orchestration of the hepatic proteome. Curr Biol 16:1107–1115
100. Doyle SE, Castrucci AM, McCall M, Provencio I, Menaker M (2006) Nonvisual light responses in the Rpe65 knockout mouse: rod loss restores sensitivity to the melanopsin system. Proc Natl Acad Sci USA 103:10432–10437
101. Roenneberg T, Kumar CJ, Merrow M (2007) The human circadian clock entrains to sun time. Curr Biol 17:R44–R45
102. Aschoff J (1967) Human circadian rhythms in activity, body temperature and other functions. Life Sci Space Res 5:159–173
103. Aschoff J (1965) Circadian rhythms in man. Science 148:1427–1432
104. Roenneberg T, Wirz-Justice A, Merrow M (2003) Life between clocks – daily temporal patterns of human chronotypes. J Biol Rhythms 18:80–90
105. Wittmann M, Dinich J, Merrow M, Roenneberg T (2006) Social jetlag: misalignment of biological and social time. Chronobiol Int 23:497–509
106. Kanterman T, Juda M, Merrow M, Roenneberg T (2007) The human circadian clock's seasonal adjustment is disrupted by daylight saving time. Curr Biol 17:1996–2000
107. Roenneberg T, Kuehnle T, Pramstaller PP, Ricken J, Havel M, Guth A, Merrow M (2004) A marker for the end of adolescence. Curr Biol 14:R1038–R1039
108. Schwartz WJ, Busis NA, Hedley-Whyte ET (1986) A discrete lesion of ventral hypothalamus and optic chiasm that disturbed the daily temperature rhythm. J Neurol 233:1–4
109. Toh KL, Jones CR, He Y, Eide EJ, Hinz WA, Virshup DM, Ptacek LJ, Fu YH (2001) An *hPer2* phosphorylation site mutation in familial advanced sleep phase syndrome. Science 291:1040–1043

110. Xu Y, Padiath QS, Shapiro RE, Jones CR, Wu SC, Saigoh N, Saigoh K, Ptacek LJ, Fu Y-H (2005) Functional consequences of a CKIδ mutation causing familial advanced sleep phase syndrome. Nature 434:640–644
111. Baggs JE, Price TS, DiTacchio L, Panda S, Fitzgerald GA, Hogenesch JB (2009) Network features of the mammalian circadian clock. PLoS Biol 7:e52
112. Mitsui A, Kumazawa S, Takahashi A, Ikemoto H, Cao S, Arai T (1986) Strategy by which nitrogen-fixing unicellular cyanobacteria grow phototrophically. Nature 323:720–733
113. Yan OY, Andersson CR, Kondo T, Golden SS, Johnson CH, Ishiura M (1998) Resonating circadian clocks enhance fitness in cyanobacteria. Proc Natl Acad Sci USA 95:8660–8664
114. Schmitz O, Katayama M, Williams SB, Kondo T, Golden SS (2000) *CikA*, a bacteriophytochrome that resets the cyanobacterial circadian clock. Science 289:765–768
115. Golden SS (2007) Integrating the circadian oscillator into the life of the cyanobacterial cell. Cold Spring Harb Symp Quant Biol 72:331–338
116. Kondo T, Tsinoremas NF, Golden SS, Johnson CH, Kutsuna S, Ishiura M (1994) Circadian clock mutants of cyanobacteria. Science 266:1233–1236
117. Liu Y, Tsinoremas NF, Johnson CH, Lebedeva NV, Golden SS, Ishiura M, Kondo T (1995) Circadian orchestration of gene expression in cyanobacteria. Genes Dev 9:1469–1478
118. Rust MJ, Markson JS, Lane WS, Fisher DS, O'Shea EK (2007) Ordered phosphorylation governs oscillation of a three-protein circadian clock. Science 318:809–812
119. Tomita J, Nakajima M, Kondo T, Iwasaki H (2005) No transcription-translation feedback in circadian rhythm of KaiC phosphorylation. Science 307:251–254
120. Nakajima M, Imai K, Ito H, Nishiwaki T, Murayama Y, Iwasaki H, Oyama T, Kondo T (2005) Reconstitution of circadian oscillation of cyanobacterial KaiC phosphorylation in vitro. Science 308:414–415
121. Merrow M, Mazzotta G, Chen Z, Roenneberg T (2006) The right place at the right time: regulation of daily timing by phosphorylation. Genes Dev 20:2629–2633
122. Johnson CH (2007) Bacterial circadian programs. Cold Spring Harb Symp Quant Biol 72:395–404
123. Grobbelaar N, Huang T-C, Lin HY, Chow TC (1986) Dinitrogen fixation endogenous rhythm in *Synechococcus* RF-1. FEMS Microbiol Lett 37:173–177
124. Roenneberg T, Merrow M (2005) Circadian clocks – the fall and rise of physiology. Nat Rev Mol Cell Biol 6:965–971
125. Lee K, Loros JJ, Dunlap JC (2000) Interconnected feedback loops in the *Neurospora* circadian system. Science 289:107–110
126. Price JL, Blau J, Rothenfluh A, Abodeely M, Kloss B, Young MW (1998) *Double-time* is a novel *Drosophila* clock gene that regulates PERIOD protein accumulation. Cell 94:83–95
127. Lowrey PL, Shimomura K, Antoch MP, Yamazaki S, Zemenides PD, Ralph MR, Menaker M, Takahashi JS (2000) Positional syntenic cloning and functional characterization of the mammalian circadian mutation tau. Science 288:483–491
128. Sugano S, Andronis C, Ong MS, Green RM, Tobin EM (1999) The protein kinase CK2 is involved in regulation of circadian rhythms in *Arabidopsis*. Proc Natl Acad Sci USA 96:12362–12366
129. Maier B, Wendt S, Vanselow JT, Wallach T, Reischl S, Oehmke S, Schlosser A, Kramer A (2009) A large-scale functional RNAi screen reveals a role for CK2 in the mammalian circadian clock. Genes Dev 23:708–718
130. Lin JM, Kilman VL, Keegan K, Paddock B, Emery-Le M, Rosbash M, Allada R (2002) A role for casein kinase 2alpha in the *Drosophila* circadian clock. Nature 420:816–820
131. Yin L, Wang J, Klein PS, Lazar MA (2006) Nuclear receptor Rev-erbalpha is a critical lithium-sensitive component of the circadian clock. Science 311:1002–1005
132. Merrow M, Roenneberg T (2001) Circadian clocks: running on redox. Cell 106:141–143
133. Rutter J, Reick M, Wu LC, McKnight SL (2001) Regulation of clock and NPAS2 DNA binding by the Redox state of NAD cofactors. Science 293:510–514

Chapter 8
Circadian Neural Networks

Erik D. Herzog and Paul H. Taghert

8.1 Introduction

That we wake, breakfast, lunch, dinner, and sleep on a regular schedule are clues to the existence of an underlying network of biological oscillators. We anticipate important daily events with changes in neural activity, metabolism and blood flow under the command of a circadian timing system. The circadian timing system drives daily rhythms in physiology and behavior. This chapter focuses on the elements in the nervous system which generate and coordinate these circadian rhythms.

8.2 Definition of Terms

The study of biological rhythms is rapidly integrating results from disparate fields such as molecular biology and nonlinear dynamics. As a result, terms that derive from individual fields can be very useful, but have definitions not universally understood in the larger chronobiological context. To overcome such potential intersectional confusion, we hereby introduce terms and offer definitions for the analysis of a network of circadian neurons.

A circadian system represents a system of coupled oscillators (Fig. 8.1). The *oscillators* intrinsically generate near-24 h time. Without exception, they entrain (directly or indirectly) to environmental cycles that display a 24-h period. Circadian oscillators have periods ranging from approximately 18–30 h – an outcome that depends on some specific genetic combinations, but not on ambient temperature (i.e. their period is temperature-compensated). *Coupling* is a circadian signal from one oscillator to another which shifts the rhythm of the recipient (directly or indirectly)

E.D. Herzog (✉)
Department of Biology, Washington University, St. Louis, MO 63130, USA
e-mail: herzog@wustl.edu

U. Albrecht (ed.), *The Circadian Clock*, Protein Reviews 12,
DOI 10.1007/978-1-4419-1262-6_8,

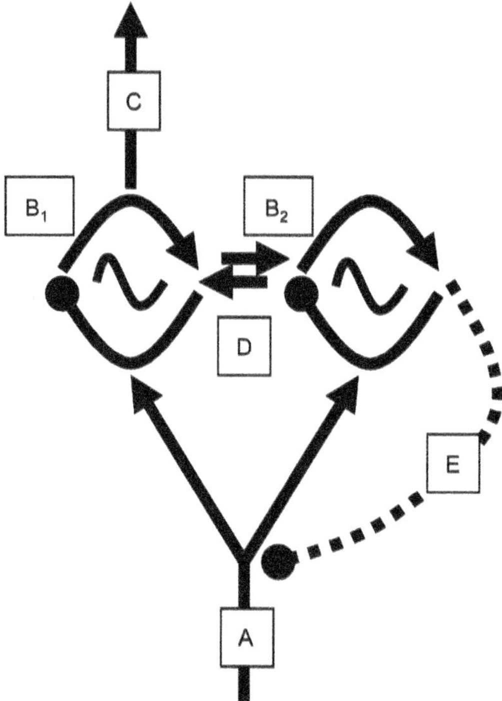

Fig. 8.1 Schematic of components of a circadian system. Input ((**A**); e.g. environmental light or food cues) to the system acts on multiple circadian oscillators ((**B**1) and (**B**2)) to synchronize (entrain) their daily rhythms to local time. The oscillators interact through a coupling pathway (**D**) to adjust their phase relationships to each other and shape rhythmic output (**C**). Some outputs feedback (**E**) to gate inputs to the oscillators and further shape the behavior of this timing system. Interactions can be excitatory (*arrowheads*) or inhibitory (*filled circles*). Dual-oscillator systems like this can entrain to and anticipate environmental cycles, sustain rhythmicity in the absence of timing cues, and adapt to changes in daily cycles like seasonal changes in dawn and dusk or food availability

and leads to the oscillators expressing the same period. *Inputs* from sensory pathways provide environmental timing information to the network of oscillators. It is important to recognize that the coupling signals must be rhythmic to successfully transmit timing information to and synchronize other oscillators. Signals which provide no timing information, but are required for synchrony (e.g. like the air which carries the daily radio signal from the atomic clock to the world's clocks), can be treated as part of the coupling mechanism. *Outputs* from the network of oscillators act on target structures to ultimately drive daily rhythms in behavior. An output which feeds back onto an input is called a *Zeitnehmer* (from German, "Time taker"). In practice, a Zeitnehmer functions as a *gate* which passes input information to the oscillator at some times, but not at others.

8.3 Circadian Clocks: Network Nodes

The breakthrough discovery by Hajime Tei and Chen C. Lee and their colleagues of the mammalian homologs of the *Drosophila Period* (*Per*) gene in 1997 reinvigorated the hypothesis that circadian pacemakers, at least in animals, share some common mechanisms and functions [1, 2]. Similarities between the circadian neurons in flies and rodents include the expression of homologs of the *Per*, *Cryptochrome*, *Bmal1*, and *Clock* genes [3, 4]. These genes have been classified as "clock genes" because of their central importance in generating circadian rhythms. In addition, circadian neurons often secrete neuropeptides including pigment dispersing factor (PDF) and NPLP1 in flies and vasoactive intestinal polypeptide (VIP), prokineticin2 and vasopressin (AVP) in mammals [5–12]. As with other circadian pacemakers, these cells express rhythms in physiology (e.g. firing rate or transcription) with periods near 24 h that change little with temperature, but that entrain to daily stimulation. Thus, the past decade of discoveries has led us to the fundamental belief that the circadian networks start with single cells which are self-sustained, entrainable circadian pacemakers.

8.3.1 Circadian Neural Clocks

In 1993, the laboratory of Gene Block isolated neurons from the retina of a marine snail to test the hypothesis that each is a competent circadian pacemaker [13]. They placed basal retinal neurons individually into the wells of a microtiter plate. After about a day, they began recording membrane conductance (the inverse of resistance) by passing a known current into each neuron. They found that, on average, the isolated neurons had a lower conductance during the day than at night. This provided the most rigorous evidence that single cells in a multicellular organism can generate circadian rhythms.

To date, no other circadian pacemaking neurons have been tested when physically isolated from each other. There is an indirect evidence, however, that the machinery for circadian rhythmogenesis is intrinsic to individual neurons in a variety of animals. Evidence generally comes from other experiments which show that the rhythmicity persists when the cells are cultured far from each other, or from results of genetic manipulations which create mosaic populations by enabling or disabling the oscillator properties in subsets of pacemakers.

In *Drosophila melanogaster*, the expression pattern of the essential clock gene, *Per* has focused attention on a population of about 150 neurons (out of the approximately 100,000) in the brain [14–16]. Several experiments have tested the hypothesis that different subsets of the 150 pacemakers have distinct functions using genetic mosaic experiments to differentially permit or repress locomotor activity during the day and not at night. Those observations strongly support a critical role for the eight small ventral lateral neurons (s-LNv; these express neuropeptide PDF) to drive the morning activity peak in light cycles and the single persistent peak in constant darkness. As a result, these cells have been termed the "M oscillator" for "morning" [115, 17]. In addition, on each side of the brain, the six dorsal lateral neurons (dLNs) and a fifth small LNv (the 5th cell does not express PDF) perform a specific role in driving the evening activity peak in light cycles; they represent the core of the "E oscillator" [17]. The E oscillator has a long intrinsic period and is accelerated by light; the M oscillator group has a short intrinsic period and is normally decelerated by light [93]. Finally, recent evidence implicates subsets of the dorsal neurons 1 group (DN1) in supporting circadian activity during the light phase [18, 19]. Other circadianly-active neurons in flies include chemosensory neurons in the antennae which regulate daily rhythms in olfactory sensitivity and attraction to odorants [20, 21]. Numerous other organs and appendages produce rhythmic *per* transcriptional activity (including legs and wings) suggesting a similar or identical molecular clockwork machinery helps regulating physiology in diverse tissues [22].

In mammals, the three *Period* gene homologs appear to be widely and circadianly expressed in the brain [23–25]. Since 1972, research has focused on the suprachiasmatic nucleus (SCN) of the ventral hypothalamus as the master circadian pacemaker. In that year, the labs of Robert Moore and Irving Zucker independently showed that the loss of the SCN abolished circadian rhythms in hormone secretion, body temperature and locomotion [26, 27]. Daily rhythms in locomotion, feeding and sleep-wake can be restored by transplanting fetal or neonatal SCN into a SCN-lesioned rodent [10, 11, 28–30]. Critically, the period of the donor SCN determines the period of the restored rhythm indicating that the SCN carry what is needed to pattern daily behaviors [31]. When plated at low density SCN neurons continue to express daily rhythms in firing rate, often with periods that differ from their neighbors [32]. Addition of tetrodotoxin (TTX) blocks voltage gated sodium channels and firing in these cells, but doesn't appear to change their periodicity [32, 33]. TTX also leads cells within a cultured SCN slice to gradually drift out of phase with each other and lose rhythm amplitude [34]. Thus, the SCN is comprised of multiple circadian oscillators. Taken together, these results indicate, but do not prove, that individual SCN neurons are circadian pacemakers. It is not known which, if any, SCN neurons are intrinsically circadian.

As seen in *Drosophila*, circadian neurons are distributed throughout the intact mammalian brain. For example, neurons from the ex vivo olfactory bulb and hippocampus express circadian rhythms [35, 36].

8.3.2 Heterogeneity Among Circadian Neurons

Growing evidence indicates that circadian neurons vary in their amplitude, period, phasing and even their ability to sustain rhythmicity (Fig. 8.2). For example, in fruit flies, the PDF-containing pacemakers differ in their behavior during constant conditions: the molecular oscillations in large LNvs damp out, whereas they are durably expressed in small LNv [37, 38]. Likewise, loss of PDF leads to a gradual loss of daily locomotor rhythms in constant darkness and a concomitant gradual loss or desynchronization of PERIOD expression in numerous pacemaker neurons, including the ones that secrete PDF [39–44]. Recent work taking advantage of conditions which lead to over-expression of PDF has revealed differences among clock neurons in their response to PDF [45]. There is increasing evidence in flies that some pacemaker neurons are "dominant" oscillators while others are "damped": for example, only some pacemakers are durable oscillators under constant conditions [18, 19, 38, 40, 46].

8.3.3 Heterogeneity Among Circadian Neurons

Growing evidence indicates that circadian neurons vary in their amplitude, period, phasing and even their ability to sustain rhythmicity (Fig. 8.2). For example, in fruit flies, the PDF-containing pacemakers differ in their behavior during constant conditions: the molecular oscillations in large LNvs damp out, whereas they are durably expressed in small LNv [116, 38].

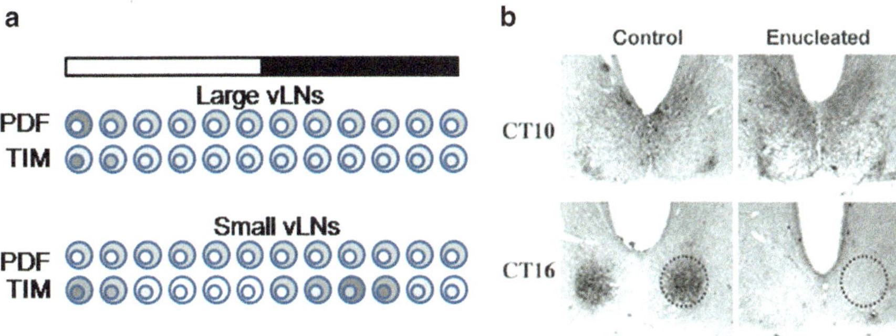

Fig. 8.2 Evidence for diverse types of circadian neurons. (**a**) In flies kept in constant darkness, some clock neurons sustain rhythmicity while others damp out. Expression of the Timeless (TIM) and pigment dispersing factor (PDF) proteins in the large ventral lateral neurons (vLNs) at roughly 2-h intervals over the first day in total darkness. Note that TIM is in the nucleus initially and then remains at low levels in the cytoplasm and nucleus. In contrast, TIM expression in the small vLNs disappears from the nucleus and then appears in cytoplasm and, later, in the nucleus again. (**b**) In hamsters kept in constant darkness, some clock neurons require input from the eyes for circadian rhythmicity. Immunostaining for phosphorylated extracellular signal-regulated kinases (pERK) reveals low expression in the middle of the suprachiasmatic nucleus (SCN) during the subjective day (CT10) and high expression at night (CT16). This circadian cycling is abolished when the eyes are removed, while the cycling of pERK persists in the cells around the edges of the SCN. Dashed circle highlights a group of SCN cells which lose rhythmicity in the absence of ocular input. (**a**) Modified from [38]. (**b**) Reproduced with permission from [57]

Similarly, SCN neurons are not all alike in their daily rhythms. Early studies noted that SCN neurons differ in their periods and amplitudes in vitro [47–49]. Within the SCN tissue, most neurons reach their peak in *Period* gene transcription and firing rate near midday, but others peak closer to dawn or dusk [50–53]. Some SCN neuron types do not express rhythms in *Period* gene expression or membrane potential [54–56]. Some SCN neurons become arrhythmic following enucleation indicating that their rhythmicity is driven by retinal inputs [57]. Loss of VIP signaling results in as few as 25% of SCN neurons showing rhythms in firing rate or *Period* gene expression [7, 51, 58, 59]. Thus, these in vivo and in vitro results have led to the conclusion that SCN neurons are not all alike in their intrinsic capacity to generate rhythms.

Importantly, there is evidence for a hierarchy of importance among rhythmic neurons. For example, the calbindin (CalB) neurons in the caudal portion of the hamster SCN appear essential for circadian rhythms in behavior based on lesioning studies [10, 60] and loss of calbindin in mice disrupts circadian rhythms in mice [61]. Loss of the PDF-expressing neurons in flies produces a gradual loss of circadian rhythmicity [5]. These results have helped guiding the efforts to model the network of pacemaking neurons.

Models highlight the range of possible configurations in the circadian network. One set of models assumes, based on data in flies and mammals, that sustained rhythmicity requires intercellular interactions (synaptic and/or nonsynaptic) [62, 63]. Other models extend that idea to suggest that the specialized neurons are self-sufficient circadian pacemakers which drive or entrain oscillations in all the other pacemaker cells [58, 64]. A variation on that theme explicitly divides SCN neurons into gates and oscillators and organizes the network so that outputs from intrinsically oscillatory neurons (primarily in the dorsal SCN) feedback to modulate the function of otherwise arrhythmic neurons (primarily in the ventral SCN) [65, 66]. Thus, the field is currently testing diverse models of the circadian network where the cellular nodes are distinguished either as self-sustained, damped or as driven circadian oscillators.

8.4 Circadian Synchronizing Pathways: Network Connections

Synchrony among neurons has attracted attention from scientists interested in the underlying conductances, synapses, gap junctions, transmitters, and modulators as well as those seeking to understand the consequences of synchrony on information coding, coordinated motor outputs, and even consciousness [67–69]. Principles from oscillations at time scales of milliseconds to days have highlighted the relative importance on membrane potential oscillations of inhibitory and excitatory communication, extremely fast effects of electrical coupling (within 1 ms) and slower effects of neurotransmitters (milliseconds) and even slower effects of neuromodulators (up to minutes) [70–72]. In all systems, ongoing efforts focus on establishing whether connections are numerous or rare, reciprocal or directed, and strong or weak.

In the circadian systems of mammals and flies, anatomical data begin to be complimented by physiological data (Fig. 8.3).

The phenotypes of flies mutant for *pdf* and for the *pdf receptor* are highly congruent [5, 41, 73–75]. That suggests the PDF-R is the major receptor underlying PDF

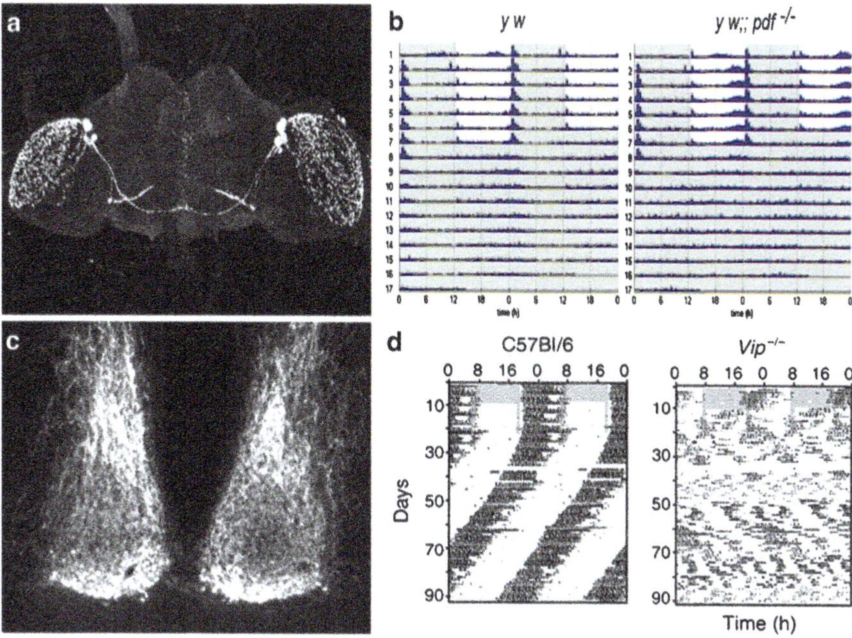

Fig. 8.3 Similar neuropeptide signaling among circadian neurons in diverse animals. (**a**) Image of a fly brain stained for PDF. Approximately four large and four small ventral lateral neurons on each side of the brain express PDF in their soma and projections. The large neurons project tangentially within the ipsilateral and contralateral optic lobes; the small neurons project a more discrete terminal field within the dorsal protocerebrum. (**b**) Average locomotion pattern of WT flies (yw; $n=16$) and flies lacking PDF (yw;;pdf–/–; $n=15$) in a 12 h:12 h light:dark cycle for 1 week and in constant darkness for 10 days. These actograms show the daily peaks of activity anticipating dawn and dusk in WT flies and loss of anticipation around dawn and advanced activity around dusk in the pdf-mutant flies. In constant darkness (*gray shading*), the mutant flies display fast or arrhythmic locomotor patterns, while the WT flies show persistent near-24 h rhythms. (**c**) Image of a mouse SCN stained for VIP. Approximately 3,000 neurons in the bilateral SCN, predominantly near the ventral margin on top of the optic chiasm, express VIP in their soma and projections. These neurons project to each other, throughout the ipsilateral SCN, to the contralateral SCN, and to areas dorsal to the SCN (unpublished data from Dr. Christian Beaulé). Locomotion pattern of a WT mouse (C57Bl/6) and a mouse lacking VIP (Vip–/–) in a 12 h:12 h light:dark cycle for 10 days, a 1 h light:11 h dark cycle for 10 days and in constant darkness (*absence of gray shading*) for 70 days. These actograms show the typical daily activity predominantly during the dark in the WT mouse and advanced activity around dusk in the Vip-mutant mouse. In constant darkness, the mutant mouse becomes arrhythmic or shows multiple circadian periods while the WT mouse show persistent near-24 h rhythms with activity predominantly during the subjective night. (**d**) Reproduced with permission from [110]

support of circadian behavioral rhythms. The behavioral effects upon loss of PDF signaling are a syndrome that includes, in a light–dark cycle, a loss of the morning anticipatory peak and an advance by several hours of the major peak which occurs in the late day and an inability to adjust to changes in photoperiod [45]. These flies are weakly rhythmic or arrhythmic in total darkness. The distribution of the PDF-R remains enigmatic [41, 74, 75], but cAMP activation and sleep studies in vivo suggest it is widely expressed among pacemaker neuron groups [76–79].

In the mammalian SCN, neuropeptide signaling has also been implicated in synchronizing circadian cells. Loss of VIP or its receptor, VPAC2R, results in two striking changes in the SCN: a reduction in the percentage of rhythmic neurons and a loss of synchrony among the circadian neurons. Mice deficient for VIP signaling also show desynchronized daily patterns of locomotor activity [7, 50, 51, 58, 59]. How VIP mediates synchrony is not known. One possibility was that VIP stimulates the release of GABA and that GABA acts as the circadian coupling factor. Indeed, blocking GABA signaling slows the rate at which the dorsal SCN will adjust to large shifts in the light cycle [80]. However, synchronous daily rhythms in gene expression and firing rate do not require GABA and daily application of a VIP agonist can restore synchrony to VIP-deficient SCN in the presence of GABA antagonists [81]. Gap junctions present another potential and under-explored pathway for entrainment among the circadian oscillators. The basal retinal neurons of the snail eye use gap junctions so that they fire compound action potentials simultaneously in a circadian pattern [82]. Global loss of connexin 36, a neuron-specific component of gap junctions, abolishes simultaneous firing in the mouse SCN and weakens daily rhythms in running wheel activity [83]. It remains to be seen how these various transmitter systems ultimately impinge on the progression of circadian timing in individual cells, but functional similarities of connections in the SCN and in the fly brain suggests some rules for the road to synchrony.

8.5 Circadian Organization: In and Out of the Network

8.5.1 Inputs

For a clock to be useful, it must reset to local time. The network of circadian oscillators can entrain to daily cues including light, food and temperature (see also Chap. 4). In flies, it appears that many of the *Period*-expressing neurons entrain *directly* to ambient light cues through the Cryptochrome-Timeless protein complex [84–86]. Additional light-input pathways to clock cells derive from various light-sensing organs, including the retina and the ocelli [87]. Temperature cycles, interestingly, appear to entrain the E and the M oscillators differentially [88], although it is unclear if this effect is direct [89]. There is also evidence for temperature-mediated, cell-autonomous changes in pacemaker function through an untranslated sequence upstream of the *Period* gene [90]. This work has highlighted how natural selection, in this case operating at the level of splicing signals, can play an important role in environmental adaptation by comparing the circadian patterns of flies from different climates.

Still, it appears that not all cells are equally sensitive to these environmental cues. Colin Pittendrigh, a founder of modern chronobiology, once noted that a temperature step leads to a transient, but not steady state change in the timing of eclosion [91]. This led him to the remarkable prediction that there is a temperature-sensitive circadian oscillator which interacts with a stronger, light-sensitive pacemaker. This dual-oscillator model has not been proven in its anatomical or physiological specifics, but has been instrumental in guiding the field. The idea remains that environmental inputs act differentially on various circadian oscillators (e.g. shifting some more than others) which then interact to shape a coordinate rhythmic output. For example, under constant light, flies become arrhythmic, but *cryptochrome* mutant flies remain strongly rhythmic [92]. Often the rhythmicity decomposes into at least two separable rhythms – a fast one that is normally decelerated by light and a slower one normally accelerated by light with M cell cycling correlated with the fast rhythm of behavior while E pacemaker cycling correlated with the slower one [93].

In mammals, light acts through the retinal melanopsin cells to entrain *indirectly* circadian rhythms in the brain and body. The elegant work of several labs has shown that deletion of the melanopsin neurons eliminates the ability of light to synchronize behavioral rhythms (Chap. 4). These neurons project to the SCN and a few other brain areas. It seems likely, based on tract tracing and light-induced gene expression, that these inputs act predominantly in the ventral SCN which then signals to the dorsal SCN to shift the circadian rhythms to local time. Interestingly, although environmental temperatures typically have little effect on circadian timing in homeothermic mammals, the SCN (and peripheral oscillators) will entrain to daily temperature cycles in vitro that mimic the temperature profiles in the brain and body [94, 95]. This raises the possibility that brain temperature is both an output from and an input to the SCN. The availability of food for mammals can be a potent synchronizing signal [96] which contrasts with its weak effects on flies [97]. The story is complicated further because restricted feeding entrains peripheral tissues in mammals, but has little to no effect on SCN timing [98–101]. Thus, the network of circadian oscillators in flies and mammals entrains to salient, daily environmental signals, but each oscillator can differ in its sensitivity and responsivity to these cues.

8.5.2 Outputs

For years, the study of clock control of physiology and behavior (outputs) was treated largely as a stamp collecting activity. "Everything is circadian" seemed to be the message. With the discovery of multiple and heterogeneous clocks in the circadian network, there has been a renaissance of interest and studies on circadian outputs (see Chaps. 5–6, 9–10). In the context of this chapter on circadian neural networks, we focus on the recent discoveries showing (1) subgroups of rhythmic neurons that control features of the same circadian behavior and (2) the distinct cell types which control divergent circadian pathways. In *Drosophila*, there is as yet very little data yet to indicate the cellular pathways that connect the pace making centers with the premotor centers that drive motorneurons activities. Certain observations suggest the neuroendocrine Pars Intercerebralis [41, 102] and the Central Complex [77] may help linking the pacemakers to such output pathways, but the evidence to date remains preliminary.

The VIP neurons of the SCN provide an example of a distinct population which appears to control different outputs. SCN VIP projections to the preoptic area have been implicated in the control of daily release of gonadotropin releasing hormone, prolactin, and glucocorticoids [103–105]. This simple example illustrates how the timing of a daily event can be determined by interactions presumably downstream of the master pace making cells. Importantly, loss of VIP likely impairs many but not all circadian rhythms in the body as has been shown for fear memory [106].

8.5.3 Feedback and Zeitnehmers

It is clear that the tradition of dividing the circadian system into input, oscillator and output components requires revision. An important addition has been to recognize that some outputs modulate input pathways. These time-takers, or Zeitnehmers, gate sensitivity of the pacemaker to entraining signals. Examples exist at the intracellular and intercellular levels of organization. In the plant, Arabidopsis thaliana, the protein product of the Elf3 gene gates light input to the circadian oscillator, but is not required for oscillations [107]. In retinal neurons of the marine snail, *Bulla gouldiana*, membrane depolarization is both on the output and the input of the circadian pacemaker [82]. In SCN neurons, phosphorylation of CREB appears to be both a critical step in photic entrainment and a clock regulated event [108]. VIP is released in a circadian pattern and entrains circadian rhythms in firing rate [109, 110]. Loss of VIP abolishes circadian rhythms in photic sensitivity [111]. Yet, neurons can sustain rhythms in the absence of VIP placing VIP outside the pacemaker, on the output and on the input. These feedback loops onto the core timing loop are likely to impart new functions and stability to the circadian system.

8.6 Areas of Opportunity

Which cells are pacemakers for which circadian behaviors? It is a striking conclusion to say that the components that are necessary to generate 24-h rhythms are entirely intracellular – that they represent a cell-autonomous oscillatory mechanism. The lack of direct evidence, however, should compel further investigations into the intrinsic properties of circadian cells. It is satisfying that the cell-autonomous circadian cycling is found in unicellular organisms and models of transcription-translation negative feedback loops, but whether these fully reflect circadian timing in any or all neurons, for example, is unclear. Although many cells express circadian rhythms in gene expression, many have yet to be investigated. *Per*-expressing neurons in the mammalian spinal cord and enteric nervous system, and the fly gut have not been examined for their intrinsic circadian rhythmicity. Nonneural cells (e.g. glia in *Drosophila*, fibroblasts and glia in rodents) also express clock genes rhythmically and likely participate in the circadian network [112–114]. Finally, we do not know which cells are responsible for most circadian behaviors like waking, falling asleep or daily secretion of more than 100 neuropeptides.

What is the role of network interactions in rhythm generation? Rhythmicity in many cells is lost in mice deficient for VIP or its receptor, VPAC2R, and flies lacking PDF or its receptor, PDF-R. These results have led to the hypothesis that many cells depend on network interactions and, perhaps, membrane currents to amplify or sustain their circadian cycling. Given the dynamic nature of neuronal connections and their adaptation to experience, it will be critical to understand how these connections change and how circadian circuits develop, adapt and anticipate in the context of a changing environment. For now, it is clear that model systems have revealed common design principles at the molecular, cellular and intercellular levels. They have also illustrated that there is no single wiring diagram for a given circadian circuit.

References

1. Tei H, Okamura H, Shigeyoshi Y, Fukuhara C, Ozawa R, Hirose M et al (1997) Circadian oscillation of a mammalian homologue of the Drosophila period gene. Nature 389:512–516
2. Sun ZS, Albrecht U, Zhuchenko O, Bailey J, Eichele G, Lee CC (1997) RIGUI, a putative mammalian ortholog of the Drosophila period gene. Cell 90:1003–1011
3. Yu W, Hardin PE (2006) Circadian oscillators of Drosophila and mammals. J Cell Sci 119:4793–4795
4. Van Gelder RN, Herzog ED, Schwartz WJ, Taghert PH (2003) Circadian rhythms: in the loop at last. Science 300:1534–1535
5. Renn SC, Park JH, Rosbash M, Hall JC, Taghert PH (1999) A pdf neuropeptide gene mutation and ablation of PDF neurons each cause severe abnormalities of behavioral circadian rhythms in Drosophila. Cell 99:791–802
6. Taghert PH, Shafer OT (2006) Mechanisms of clock output in the Drosophila circadian pacemaker system. J Biol Rhythms 21:445–457
7. Vosko AM, Schroeder A, Loh DH, Colwell CS (2007) Vasoactive intestinal peptide and the mammalian circadian system. Gen Comp Endocrinol 152:165–175
8. Li JD, Hu WP, Boehmer L, Cheng MY, Lee AG, Jilek A et al (2006) Attenuated circadian rhythms in mice lacking the prokineticin 2 gene. J Neurosci 26:11615–11623
9. Antle MC, Silver R (2005) Orchestrating time: arrangements of the brain circadian clock. Trends Neurosci 28:145–151
10. LeSauter J, Silver R (1999) Localization of a suprachiasmatic nucleus subregion regulating locomotor rhythmicity. J Neurosci 19:5574–5585
11. Moore RY, Silver R (1998) Suprachiasmatic nucleus organization. Chronobiol Int 15:475–487
12. Nitabach MN, Taghert PH (2008) Organization of the Drosophila circadian control circuit. Curr Biol 18:R84–R93

13. Michel S, Geusz ME, Zaritsky JJ, Block GD (1993) Circadian rhythm in membrane conductance expressed in isolated neurons. Science 259:239–241
14. Shafer OT, Helfrich-Forster C, Renn SC, Taghert PH (2006) Reevaluation of *Drosophila melanogaster*'s neuronal circadian pacemakers reveals new neuronal classes. J Comp Neurol 498:180–193
15. Helfrich-Forster C (2005) Organization of endogenous clocks in insects. Biochem Soc Trans 33:957–961
16. Zhao J, Kilman VL, Keegan KP, Peng Y, Emery P, Rosbash M et al (2003) Drosophila clock can generate ectopic circadian clocks. Cell 113:755–766
17. Stoleru D, Peng Y, Agosto J, Rosbash M (2004) Coupled oscillators control morning and evening locomotor behaviour of Drosophila. Nature 431:862–868
18. Murad A, Emery-Le M, Emery P (2007) A subset of dorsal neurons modulates circadian behavior and light responses in Drosophila. Neuron 53:689–701
19. Stoleru D, Nawathean P, Fernandez ML, Menet JS, Ceriani MF, Rosbash M (2007) The Drosophila circadian network is a seasonal timer. Cell 129:207–219
20. Tanoue S, Krishnan P, Chatterjee A, Hardin PE (2008) G protein-coupled receptor kinase 2 is required for rhythmic olfactory responses in Drosophila. Curr Biol 18:803–807
21. Tanoue S, Krishnan P, Krishnan B, Dryer SE, Hardin PE (2004) Circadian clocks in antennal neurons are necessary and sufficient for olfaction rhythms in Drosophila. Curr Biol 14:638–649
22. Plautz JD, Kaneko M, Hall JC, Kay SA (1997) Independent photoreceptive circadian clocks throughout Drosophila. Science 278:1632–1635
23. Kowalska E, Brown SA (2007) Peripheral clocks: keeping up with the master clock. Cold Spring Harb Symp Quant Biol 72:301–305
24. Reppert SM, Weaver DR (2002) Coordination of circadian timing in mammals. Nature 418:935–941
25. Hastings MH, Maywood ES, Reddy AB (2008) Two decades of circadian time. J Neuroendocrinol 20:812–819
26. Stephan FK, Zucker I (1972) Circadian rhythms in drinking behavior and locomotor activity of rats are eliminated by hypothalamic lesions. Proc Natl Acad Sci U S A 69:1583–1586
27. Moore RY, Eichler VB (1972) Loss of a circadian adrenal corticosterone rhythm following suprachiamatic lesions in rat. Brain Res 42:201–206
28. Guo H, Brewer JM, Lehman MN, Bittman EL (2006) Suprachiasmatic regulation of circadian rhythms of gene expression in hamster peripheral organs: effects of transplanting the pacemaker. J Neurosci 26:6406–6412
29. Meyer-Bernstein EL, Jetton AE, Matsumoto SI, Markuns JF, Lehman MN, Bittman EL (1999) Effects of suprachiasmatic transplants on circadian rhythms of neuroendocrine function in golden hamsters. Endocrinology 140:207–218
30. LeSauter J, Silver R (1998) Output signals of the SCN. Chronobiol Int 15:535–550
31. Ralph MR, Foster RG, Davis FC, Menaker M (1990) Transplanted suprachiasmatic nucleus determines circadian period. Science 247:975–978
32. Welsh DK, Logothetis DE, Meister M, Reppert SM (1995) Individual neurons dissociated from rat suprachiasmatic nucleus express independently phased circadian firing rhythms. Neuron 14:697–706
33. Ikeda M, Sugiyama T, Wallace CS, Gompf HS, Yoshioka T, Miyawaki A et al (2003) Circadian dynamics of cytosolic and nuclear Ca(2+) in single suprachiasmatic nucleus neurons. Neuron 38:253–263
34. Yamaguchi S, Isejima H, Matsuo T, Okura R, Yagita K, Kobayashi M et al (2003) Synchronization of cellular clocks in the suprachiasmatic nucleus. Science 302:1408–1412
35. Abe M, Herzog ED, Yamazaki S, Straume M, Tei H, Sakaki Y et al (2002) Circadian rhythms in isolated brain regions. J Neurosci 22:350–356
36. Chaudhury D, Wang LM, Colwell CS (2005) Circadian regulation of hippocampal long-term potentiation. J Biol Rhythms 20:225–236

37. Yang Z, Emerson M, Su HS, Sehgal A (1998) Response of the timeless protein to light correlates with behavioral entrainment and suggests a nonvisual pathway for circadian photoreception. Neuron 21:215–223
38. Shafer OT, Rosbash M, Truman JW (2002) Sequential nuclear accumulation of the clock proteins period and timeless in the pacemaker neurons of *Drosophila melanogaster*. J Neurosci 22:5946–5954
39. Peng Y, Stoleru D, Levine JD, Hall JC, Rosbash M (2003) Drosophila free-running rhythms require intercellular communication. PLoS Biol 1:13
40. Lin Y, Stormo GD, Taghert PH (2004) The neuropeptide pigment-dispersing factor coordinates pacemaker interactions in the Drosophila circadian system. J Neurosci 24:7951–7957
41. Lear BC, Merrill CE, Lin JM, Schroeder A, Zhang L, Allada R (2005) A G protein-coupled receptor, groom-of-PDF, is required for PDF neuron action in circadian behavior. Neuron 48:221–227
42. Wu Y, Cao G, Pavlicek B, Luo X, Nitabach MN (2008) Phase coupling of a circadian neuropeptide with rest/activity rhythms detected using a membrane-tethered spider toxin. PLoS Biol 6:e273
43. Nitabach MN, Blau J, Holmes TC (2002) Electrical silencing of Drosophila pacemaker neurons stops the free-running circadian clock. Cell 109:485–495
44. Cao G, Nitabach MN (2008) Circadian control of membrane excitability in *Drosophila melanogaster* lateral ventral clock neurons. J Neurosci 28:6493–6501
45. Yoshii T, Wulbeck C, Sehadova H, Veleri S, Bichler D, Stanewsky R et al (2009) The neuropeptide pigment-dispersing factor adjusts period and phase of Drosophila's clock. J Neurosci 29:2597–2610
46. Lin Y, Han M, Shimada B, Wang L, Gibler TM, Amarakone A et al (2002) Influence of the period-dependent circadian clock on diurnal, circadian, and aperiodic gene expression in *Drosophila melanogaster*. Proc Natl Acad Sci U S A 99:9562–9567
47. Liu C, Weaver DR, Strogatz SH, Reppert SM (1997) Cellular construction of a circadian clock: period determination in the suprachiasmatic nuclei. Cell 91:855–860
48. Herzog ED, Takahashi JS, Block GD (1998) Clock controls circadian period in isolated suprachiasmatic nucleus neurons. Nat Neurosci 1:708–713
49. Honma S, Shirakawa T, Katsuno Y, Namihira M, Honma K-I (1998) Circadian periods of single suprachiasmatic neurons in rats. Neurosci Lett 250:157–160
50. Ciarleglio CM, Gamble KL, Axley JC, Strauss BR, Cohen JY, Colwell CS et al (2009) Population encoding by circadian clock neurons organizes circadian behavior. J Neurosci 29:1670–1676
51. Maywood ES, Reddy AB, Wong GK, O'Neill JS, O'Brien JA, McMahon DG et al (2006) Synchronization and maintenance of timekeeping in suprachiasmatic circadian clock cells by neuropeptidergic signaling. Curr Biol 16:599–605
52. Inagaki N, Honma S, Ono D, Tanahashi Y, Honma KI (2007) Separate oscillating cell groups in mouse suprachiasmatic nucleus couple photoperiodically to the onset and end of daily activity. Proc Natl Acad Sci U S A 104:7664–7669
53. Jagota A, De la Iglesia HO, Schwartz WJ (2000) Morning and evening circadian oscillations in the suprachiasmatic nucleus in vitro. Nat Neurosci 3:372–376
54. Hamada T, LeSauter J, Venuti JM, Silver R (2001) Expression of period genes: rhythmic and nonrhythmic compartments of the suprachiasmatic nucleus pacemaker. J Neurosci 21:7742–7750
55. Yan L, Hochstetler KJ, Silver R, Bult-Ito A (2003) Phase shifts and Per gene expression in mouse suprachiasmatic nucleus. Neuroreport 14:1247–1251
56. Jobst EE, Allen CN (2002) Calbindin neurons in the hamster suprachiasmatic nucleus do not exhibit a circadian variation in spontaneous firing rate. Eur J NeuroSci 16:2469–2474
57. Lee HS, Nelms JL, Nguyen M, Silver R, Lehman MN (2003) The eye is necessary for a circadian rhythm in the suprachiasmatic nucleus. Nat Neurosci 6:111–112
58. Aton SJ, Herzog ED (2005) Come together, right...now: synchronization of rhythms in a mammalian circadian clock. Neuron 48:531–534

59. Brown TM, Piggins HD (2007) Electrophysiology of the suprachiasmatic circadian clock. Prog Neurobiol 82:229–255
60. Kriegsfeld LJ, LeSauter J, Silver R (2004) Targeted microlesions reveal novel organization of the hamster suprachiasmatic nucleus. J Neurosci 24:2449–2457
61. Kriegsfeld LJ, Mei DF, Yan L, Witkovsky P, LeSauter J, Hamada T et al (2008) Targeted mutation of the calbindin D28K gene disrupts circadian rhythmicity and entrainment. Eur J NeuroSci 27:2907–2921
62. Roenneberg T, Chua EJ, Bernardo R, Mendoza E (2008) Modelling biological rhythms. Curr Biol 18:R826–R835
63. Bernard S, Gonze D, Cajavec B, Herzel H, Kramer A (2007) Synchronization-induced rhythmicity of circadian oscillators in the suprachiasmatic nucleus. PLoS Comput Biol 3:e68
64. To TL, Henson MA, Herzog ED, Doyle FJ III (2007) A molecular model for intercellular synchronization in the mammalian circadian clock. Biophys J 92:3792–3803
65. Antle MC, Foley DK, Foley NC, Silver R (2003) Gates and oscillators: a network model of the brain clock. J Biol Rhythms 18:339–350
66. Antle MC, Foley NC, Foley DK, Silver R (2007) Gates and oscillators II: zeitgebers and the network model of the brain clock. J Biol Rhythms 22:14–25
67. Herzog ED (2007) Neurons and networks in daily rhythms. Nat Rev Neurosci 8:790–802
68. Schmidt C, Collette F, Cajochen C, Peigneux P (2007) A time to think: circadian rhythms in human cognition. Cogn Neuropsychol 24:755–789
69. Kuhlman SJ, McMahon DG (2006) Encoding the ins and outs of circadian pacemaking. J Biol Rhythms 21:470–481
70. Prinz AA, Thirumalai V, Marder E (2003) The functional consequences of changes in the strength and duration of synaptic inputs to oscillatory neurons. J Neurosci 23:943–954
71. Traub RD, Kopell N, Bibbig A, Buhl EH, LeBeau FE, Whittington MA (2001) Gap junctions between interneuron dendrites can enhance synchrony of gamma oscillations in distributed networks. J Neurosci 21:9478–9486
72. Goldman MS, Golowasch J, Marder E, Abbott LF (2001) Global structure, robustness, and modulation of neuronal models. J Neurosci 21:5229–5238
73. Blanchardon E, Grima B, Klarsfeld A, Chelot E, Hardin PE, Preat T et al (2001) Defining the role of Drosophila lateral neurons in the control of circadian rhythms in motor activity and eclosion by targeted genetic ablation and PERIOD protein overexpression. Eur J NeuroSci 13:871–888
74. Hyun S, Lee Y, Hong ST, Bang S, Paik D, Kang J et al (2005) Drosophila GPCR Han is a receptor for the circadian clock neuropeptide PDF. Neuron 48:267–278
75. Mertens I, Vandingenen A, Johnson EC, Shafer OT, Li W, Trigg JS et al (2005) PDF receptor signaling in Drosophila contributes to both circadian and geotactic behaviors. Neuron 48:213–219
76. Shafer OT, Kim DJ, Dunbar-Yaffe R, Nikolaev VO, Lohse MJ, Taghert PH (2008) Widespread receptivity to neuropeptide PDF throughout the neuronal circadian clock network of Drosophila revealed by real-time cyclic AMP imaging. Neuron 58:223–237
77. Parisky KM, Agosto J, Pulver SR, Shang Y, Kuklin E, Hodge JJ et al (2008) PDF cells are a GABA-responsive wake-promoting component of the Drosophila sleep circuit. Neuron 60:672–682
78. Shang Y, Griffith LC, Rosbash M (2008) Feature Article: Light-arousal and circadian photoreception circuits intersect at the large PDF cells of the Drosophila brain. Proc Natl Acad Sci U S A 105:19587–19594
79. Chung BY, Kilman VL, Keath JR, Pitman JL, Allada R (2009) The GABA(A) receptor RDL acts in peptidergic PDF neurons to promote sleep in Drosophila. Curr Biol 19:386–390
80. Albus H, Vansteensel MJ, Michel S, Block GD, Meijer JH (2005) A GABAergic mechanism is necessary for coupling dissociable ventral and dorsal regional oscillators within the circadian clock. Curr Biol 15:886–893
81. Aton SJ, Huettner JE, Straume M, Herzog ED (2006) GABA and Gi/o differentially control circadian rhythms and synchrony in clock neurons. Proc Natl Acad Sci U S A 103:19188–19193

82. Block GD, Geusz M, Khalsa SBS, Michel S, Whitmore D (1996) Circadian rhythm generation, expression and entrainment in a molluscan model system. Prog Brain Res 111:93–102
83. Long MA, Jutras MJ, Connors BW, Burwell RD (2005) Electrical synapses coordinate activity in the suprachiasmatic nucleus. Nat Neurosci 8:61–66
84. Ashmore LJ, Sehgal A (2003) A fly's eye view of circadian entrainment. J Biol Rhythms 18:206–216
85. Benito J, Houl JH, Roman GW, Hardin PE (2008) The blue-light photoreceptor CRYPTOCHROME is expressed in a subset of circadian oscillator neurons in the Drosophila CNS. J Biol Rhythms 23:296–307
86. Yoshii T, Todo T, Wulbeck C, Stanewsky R, Helfrich-Forster C (2008) Cryptochrome is present in the compound eyes and a subset of Drosophila's clock neurons. J Comp Neurol 508:952–966
87. Rieger D, Stanewsky R, Helfrich-Forster C (2003) Cryptochrome, compound eyes, Hofbauer-Buchner eyelets, and ocelli play different roles in the entrainment and masking pathway of the locomotor activity rhythm in the fruit fly *Drosophila melanogaster*. J Biol Rhythms 18:377–391
88. Miyasako Y, Umezaki Y, Tomioka K (2007) Separate sets of cerebral clock neurons are responsible for light and temperature entrainment of Drosophila circadian locomotor rhythms. J Biol Rhythms 22:115–126
89. Glaser FT, Stanewsky R (2007) Synchronization of the Drosophila circadian clock by temperature cycles. Cold Spring Harb Symp Quant Biol 72:233–242
90. Low KH, Lim C, Ko HW, Edery I (2008) Natural variation in the splice site strength of a clock gene and species-specific thermal adaptation. Neuron 60:1054–1067
91. Pittendrigh CS, Bruce VG, Kaus P (1958) On the significance of transients in daily rhythms. Proc Natl Acad Sci U S A 44:965–973
92. Yoshii T, Funada Y, Ibuki-Ishibashi T, Matsumoto A, Tanimura T, Tomioka K (2004) Drosophila cry(b) mutation reveals two circadian clocks that drive locomotor rhythm and have different responsiveness to light. J Insect Physiol 50:479–488
93. Rieger D, Shafer OT, Tomioka K, Helfrich-Forster C (2006) Functional analysis of circadian pacemaker neurons in *Drosophila melanogaster*. J Neurosci 26:2531–2543
94. Herzog ED, Huckfeldt RM (2003) Circadian entrainment to temperature, but not light, in the isolated suprachiasmatic nucleus. J Neurophysiol 90:763–770
95. Brown S, Zumbrunn G, Fleury-Olela F, Preitner N, Schibler U (2002) Rhythms of mammalian body temperature can sustain peripheral circadian clocks. Curr Biol 12:1574
96. Stephan FK (2002) The "other" circadian system: food as a Zeitgeber. J Biol Rhythms 17:284–292
97. Oishi K, Shiota M, Sakamoto K, Kasamatsu M, Ishida N (2004) Feeding is not a more potent Zeitgeber than the light-dark cycle in Drosophila. Neuroreport 15:739–743
98. Schibler U (2009) The 2008 pittendrigh/aschoff lecture: peripheral phase coordination in the Mammalian circadian timing system. J Biol Rhythms 24:3–15
99. Mistlberger RE (2006) Circadian rhythms: perturbing a food-entrained clock. Curr Biol 16:R968–R969
100. Storch KF, Weitz CJ (2009) Daily rhythms of food-anticipatory behavioral activity do not require the known circadian clock. Proc Natl Acad Sci U S A 106:6808–6813
101. Waddington LE, Harbour VL, Barry-Shaw J, Renteria DL, Robinson B, Stewart J et al (2006) Restricted access to food, but not sucrose, saccharine, or salt, synchronizes the expression of Period2 protein in the limbic forebrain. Neuroscience 144:402–411
102. Williams JA, Sehgal A (2001) Molecular components of the circadian system in Drosophila. Ann Rev Physiol 63:729–755
103. Gerhold LM, Wise PM (2006) Vasoactive intestinal polypeptide regulates dynamic changes in astrocyte morphometry: impact on gonadotropin releasing hormone neurons. Endocrinology 147:2197–21202
104. Kennett JE, Poletini MO, Freeman ME (2008) Vasoactive intestinal polypeptide modulates the estradiol-induced prolactin surge by entraining oxytocin neuronal activity. Brain Res 1196:65–73

105. Loh DH, Abad C, Colwell CS, Waschek JA (2008) Vasoactive intestinal peptide is critical for circadian regulation of glucocorticoids. Neuroendocrinology 88:246–255
106. Chaudhury D, Loh DH, Dragich JM, Hagopian A, Colwell CS (2008) Select cognitive deficits in vasoactive intestinal peptide deficient mice. BMC Neurosci 9:63
107. McWatters HG, Bastow RM, Hall A, Millar AJ (2000) The ELF3 zeitnehmer regulates light signalling to the circadian clock. Nature 408:716–720
108. Akashi M, Hayasaka N, Yamazaki S, Node K (2008) Mitogen-activated protein kinase is a functional component of the autonomous circadian system in the suprachiasmatic nucleus. J Neurosci 28:4619–4623
109. Shinohara K, Honma S, Katsuno Y, Abe H, Honma K-I (1995) Two distinct oscillators in the rat suprachiasmatic nucleus in vitro. Proc Natl Acad Sci U S A 92:7396–7400
110. Aton SJ, Colwell CS, Harmar AJ, Waschek J, Herzog ED (2005) Vasoactive intestinal polypeptide mediates circadian rhythmicity and synchrony in mammalian clock neurons. Nat Neurosci 8:476–483
111. Hughes AT, Fahey B, Cutler DJ, Coogan AN, Piggins HD (2004) Aberrant gating of photic input to the suprachiasmatic circadian pacemaker of mice lacking the VPAC2 receptor. J Neurosci 24:3522–3526
112. Ewer J, Frisch B, Hamblen-Coyle MJ, Rosbash M, Hall JC (1992) Expression of the period clock gene within different cell types in the brain of Drosophila adults and mosaic analysis of these cells' influence on circadian behavioral rhythms. J Neurosci 12:3321–3349
113. Suh J, Jackson FR (2007) Drosophila ebony activity is required in glia for the circadian regulation of locomotor activity. Neuron 55:435–447
114. Prolo LM, Takahashi JS, Herzog ED (2005) Circadian rhythm generation and entrainment in astrocytes. J Neurosci 25:404–408
115. Grima B, Chélot E, Xia R, Rouyer F (2004) Morning and evening peaks of activity rely on different clock neurons of the Drosophila brain. Nature 431:869–873
116. Yang Z, Sehgal A (2001) Role of molecular oscillations in generating behavioral rhythms in Drosophila. Neuron 29:453–467

Chapter 9
The Circadian Clock and the Homeostatic Hourglass: Two Timepieces Controlling Sleep and Wakefulness

Sarah Laxhmi Chellappa and Christian Cajochen

9.1 Introduction

Two facets of time play a pivotal role in the prediction of sleep propensity in humans: *internal* biological time and *external* time that is elapsed time spent awake and asleep [1, 2]. Briefly, *internal time* is determined by the endogenous circadian pacemaker, and needs to undergo a daily synchronization with the *external* time. On the other hand, "elapsed time" awake and asleep mirrors the homeostatic process and accumulates exponentially during wakefulness and is subsequently discharged during sleep in a faster exponential fashion.

Biological rhythms (*internal time*) comprise repetitive biological events with three main features: period, which corresponds to the length of a given rhythm; phase, which consists in the timing of a given rhythm with respect to a stimulus; and amplitude, which is the measure of the amount of a rhythmic event [3]. Circadian rhythms in humans have a period of approximately 24 h that is in tune with the 24-h solar light–dark cycle, the most important recurring stimulus in our environment [4]. Moreover, these rhythms have been recently shown to depend on genetically controlled rhythmic molecular events that rule several dimensions of cellular, system, and behavioral functions [5–7].

One fundamental behavioral circadian rhythm is the sleep–wake cycle. While wake during the diurnal period enables optimal use of vision to direct behavior, sleep during the night provides the best time for critical restorative behavior. Thus, the physiological rationale for the circadian regulation is to provide an optimal

S.L. Chellappa (✉)
The CAPES Foundation / Ministry of Education of Brazil, Brasilia-DF, Brazil;
Centre for Chronobiology, Psychiatric University Clinics, Wilhelm Kleinstrasse 27, CH-4025 Basel, Switzerland
e-mail: sarah.chellappa@upkbs.ch

C. Cajochen
Centre for Chronobiology, Psychiatric University Clinics, Wilhelm Kleinstrasse 27, CH-4025 Basel, Switzerland
e-mail: Christian.Cajochen@upkbs.ch

U. Albrecht (ed.), *The Circadian Clock*, Protein Reviews 12,
DOI 10.1007/978-1-4419-1262-6_9, © Springer Science+Business Media, LLC 2010

temporal organization for the sleep–wake cycle on several domains, such as the timing and architecture of sleep [8].

Besides this internal circadian control of the sleep–wake rhythm, the amount of time spent awake and asleep per 24 h is under a homeostatic control such that sleep propensity increases with elapsed time awake and dissipates with elapsed time asleep during the following sleep episode [9]. Independent of internal time (i.e., circadian phase), the maximal capacity to stay awake in an adult human is around 16 h, and the maximal capacity to maintain a sleep efficiency of 90% is around 8 h. Thus, episodes of more than 16 h of wakefulness lead to more consolidated and longer sleep episodes and vice versa (i.e., homeostasis). It is assumed that sleep is required to prevent the waking brain of "synaptic overload" or cellular stress. Indeed, a recent hypothesis argues that the homeostatic sleep process plays an important role in the regulation of cortical synaptic plasticity [10, 11]. However, while the physiological framework for the circadian timing system is rather well-known, the underlying basis for the sleep homeostatic function remains uncertain and is likely to be fairly complex.

Circadian and homeostatic aspects of sleep regulation are not independent of each other. Quantitative analyzes of EEG activity have consistently shown that these two systems interact, not simply adding up [1, 12, 13]. This non-additive interaction suggests that minor changes in the circadian phase can have dramatic repercussions on sleep measures, if sleep pressure is high, and vice-versa.

In the following sections, we will address: (1) the circadian clock, its neuroanatomical framework, and how it regulates sleep; (2) the sleep homeostat and possible mechanisms that underlie this process; (3) the interaction of circadian and homeostatic control of sleep; (4) examples of this interaction, such as morning and evening types, short and long sleepers and healthy aging; (5) clinical conditions in which these processes go out of sync.

9.2 Circadian Sleep Process

9.2.1 How Does the Circadian Clock Work?

The circadian system modulates a wide array of human physiology and behavior patterns (for a review see [14]). Briefly, the master pacemaker driving circadian rhythms, the suprachiasmatic nuclei (SCN), acts as the central neural pacemaker for the generation and/or synchronization of circadian rhythms [15, 16]. These circadian rhythms are self-sustained and persist in the absence of environmental time cues with a remarkable precision and represent a cyclic process that can be described by the period, phase, and amplitude of the oscillation, together with its resetting sensitivity to various circadian synchronizers [4]. Under normal conditions, circadian rhythms are entrained to the 24-h day, thus enabling that behavioral, physiologic, and genetic rhythms are aptly timed with the daily changes in the environment. In order to obtain the circadian entrainment, the SCN, is synchronized to the external light–dark cycle through retinal light input (light being the main synchronizer or "zeitgeber") [17]. A specialized nonvisual retinohypothalamic tract provides a direct neuronal connection to the SCN from novel photoreceptors in the retinal ganglion cells that measure luminance [18, 19]. Thus, a daily resetting of the circadian pacemaker is necessary in such a manner that the imposed period of the environmental synchronizer, like the 24-h day, corresponds to the intrinsic period of the circadian oscillator, therefore ensuing a stable entrainment.

9.2.1.1 What Are the Neuroanatomical Underpinnings of the Circadian Pacemaker?

Figure 9.1 schematically illustrates the structural inputs and some of the behavioral and neuroendocrine outputs of the circadian timing system [20]. The key input structure for entrainment of the mammalian circadian pacemaker is the eye. This is clearly demonstrated when the optic nerve is damaged, which results in loss of

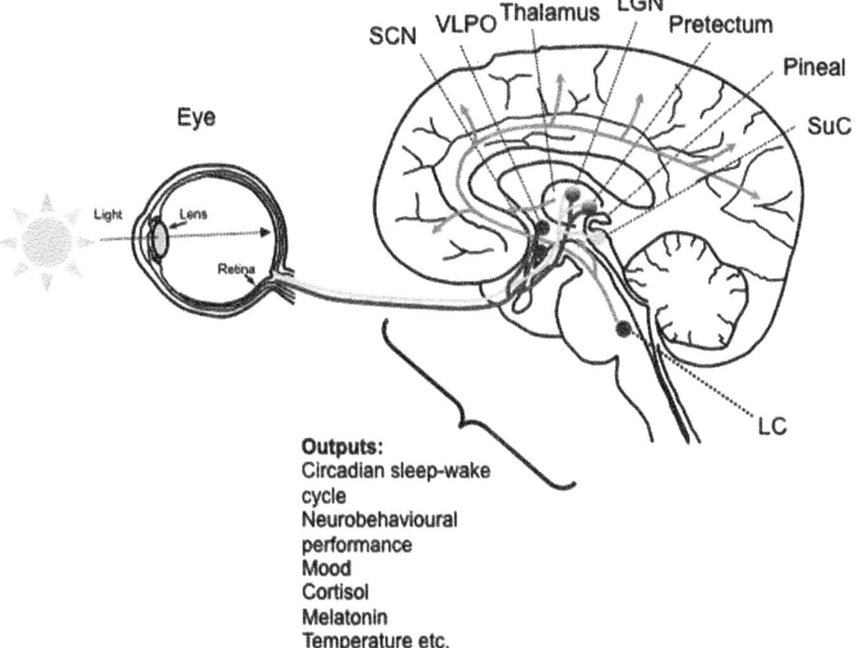

Fig. 9.1 *Structural inputs and behavioral and neuroendocrine outputs of the circadian timing system.* Light activates melanopsin-containing intrinsically photosensitive retinal ganglion cells and rod and cone classical ganglion cells. Melanopsin-containing ganglion cells (*dark blue*) project to several nonvisual areas of the brain, including the suprachiasmatic nuclei (SCN), which project multisynaptically to the pineal gland and to other areas that share input from the visual photoreceptor system (*dark yellow*), like the lateral geniculate nucleus (LGN), pretectum and superior colliculus (SuC). Light stimulates the ascending arousal system and eventually the cortex to optimize alertness and cognition. Moreover, light information also reaches sleep-promoting neurons of the ventrolateral preoptic nucleus (VLPO) and the noradrenergic locus coeruleus (LC) system, which is involved in the circadian regulation of arousal (modified from [165] Cajochen 2007)

entrainment, as observed in free-running rhythms in enucleated blind individuals [21]. Thus, the entrainment process involves both photoreception and central visual projection from the retina to the SCN. Initially, it was postulated that conventional photoreceptor-mediated function was essential for entrainment. However, in transgenic mice, a knockout of genes for both rods and cones does not affect entrainment [22]. It is far more probable that this process is mediated by a set of ganglion cells that contain photopigments, particularly melanopsin [19, 23]. The entrainment pathway consists in a direct projection from the retina to the SCN by the retinohypothalamic tract [24]. Furthermore, there is an important secondary visual pathway that projects to the SCN from the intergeniculate leaflets, a ventral thalamic component of the lateral geniculate complex. This pathway, the geniculo-hypothalamic tract, contains GABA [3] and appears to modulate the effect of the retinal input on the SCN. Furthermore, the intergeniculate leaflets obtain additional information from structures like the dorsal raphé nuclei [25]. This indicates that this path allows the tight regulation of photic and nonphotic (serotonergic pathway from the raphé nuclei) entrainment of the circadian clock.

The main circadian pacemaker is the SCN for four fundamental reasons: the SCN is where the entrainment pathway ends [3]; SCN lesions can disrupt the temporal pattern of the sleep–wake cycle and core body temperature [26]; if the SCN is isolated, it may not interfere with the SCN rhythmicity, although it certainly abolishes the sleep–wake cycle [27]; isolated SCN neurons preserve the circadian control of their firing rate [15]. The SCN is comprised of individual neuronal oscillators that form a group of pacemakers and consists of two major anatomical subdivisions, shell and core [17]. The core localizes above the optic chiasm, comprises vasoactive intestinal polypeptide-producing neurons, and receives RHT inputs, whereas the shell surrounds the core, comprises vasopressin-producing neurons, and receives hypothalamus, brainstem, and basal forebrain inputs. It is the conjunction of these subdivisions that generate the overt circadian expression of the SCN.

The SCN innervates several brain areas mostly located within the thalamus and the hypothalamus. The SCN has indirect projections via the dorsomedial hypothalamus (DMH) to the ventrolateral preoptic nucleus of the hypothalamus (VLPO) and to arousal-promoting cell groups [28]. Recently, it has been described that the VLPO, together with the wake-maintaining posterior lateral hypothalamus, can generate a "flip-flop" switch for sleep–wake control [29]. According to this model, monoaminergic nuclei, such as the histaminergic tuberomammilary neurons (TMN), locus coeruleus (LC) and the serotonergic dorsal and median raphé nuclei (DR) promote wakefulness by direct excitatory effects on the cortex and by inhibition of sleep promoting neurons of the VLPO. With an increasing reduction of the circadian drive for arousal in the later part of the waking period, there is a substantial increase in the neuron firing rate of VLPO, which is comprised of GABA neurons that project to wake-promoting areas. During sleep, the VLPO inhibits the monoaminergic-mediated arousal regions through GABAergic and galaninergic projections. This leads to a progressive synchronization in the thalamo-cortical network through a synchronous discharge of the thalamic reticular nucleus [29, 30]. As a result, this strongly enhances the generation of sleep spindles and deeper stages of NREM sleep [31]. Intermediate states between sleep and wakefulness are, thus, avoided through the reciprocal inhibition of VLPO neurons and monoaminergic cell groups, which reinforce their own firing rates in a parallel manner.

The overt importance of the "flip-flop" system can be provided by the ablation of VLPO, which results in prolonged episodes of wakefulness [29, 30]). Similarly, orexin/hypocretin, hypothalamic peptides that play an important role in maintaining wakefulness, is under the direct control of the SCN [32, 33] and has peak levels of activity at the same time of the circadian alertness signal [34, 35]. Reduced levels of these peptides can lead to the very well-known sleep attacks that occur in narcolepsy [36]. The SCN must also influence the "flip-flop" switch to produce the extended sleep–wake phases [37]. The circadian influence may be strengthened by the presence of SCN outputs by either simultaneously pressing down on one side of the "flip-flop" switch, while pushing up on the other, or sequentially pressing down or pushing up on one side, and then switching to the other side, thus regulating when sleep or wakefulness should occur.

The clock output is not only assured by neuronal projections but also involves the pituitary gland and the autonomic nervous system [28]. A classical example of the association between the SCN and the physiological responses that undergo its regulation is the secretion of melatonin by the pineal gland [38]. Melatonin is synthesized in a circadian fashion, in which maximum levels are secreted at night, with the onset of production in the later part of the day, while the lowest levels occur during the day. The increase in melatonin secretion leads to an inhibition of the firing rate of the SCN neurons, with a subsequent decline of the circadian force for arousal, thus enhancing sleep [28].

9.2.1.2 Hands of the Circadian Clock in Humans

On a behavioral level, clock outputs can be assessed by measuring rhythms, such as the circadian rhythm of core body temperature, plasma melatonin or cortisol concentration, etc. These variables, when assessed under conditions in which the confounding effects of variations in behavior and environment are controlled for, represent the "classical" markers of circadian phase, period and amplitude, which comprise parameters of "internal time" [39, 40].

9.2.1.3 Does the SCN Play an *Exclusive* Role as a Circadian Pacemaker?

There is a current change of opinion about the classical view of a unique circadian pacemaker orchestrating all physiological and behavioral rhythms [6]. In its place, the novel conception is that the SCN synchronizes numerous peripheral oscillators in order to achieve temporally coordinated physiology, through neural and endocrine control mediated by continuous synchronization of the SCN. In fact, neuronal signals emitted from the SCN are essential for the long-term maintenance of circadian gene expression in the periphery [41]. Thus, peripheral clocks appear to adjust their phase and period length according to the rhythm imposed by the SCN [6]. Furthermore, these peripheral oscillators may be connected among themselves and feedback to the SCN by means of auto-regulatory feedback loops [15, 42]. How these peripheral oscillators impact on human circadian sleep–wake rhythms is not yet known. Taken together, this implies a web of pacemakers that regulates the circadian rhythms.

9.2.1.4 How Does the Circadian Timing System Modulate Sleep?

The sleep/wake cycle is perhaps the most obvious expression of the circadian rhythms in humans and the timing of this sleep/wake cycle is assumed to reflect the output of the circadian pacemaker [1]. A hallmark of this circadian regulation is the peak of REM sleep during the early hours of the morning. Within a causal framework, this peak of REM sleep may represent a circadian sleep-promoting signal to ensure the normal sleep duration. The circadian activation of REM sleep may occur by indirect projections from the SCN to the mesopontine tegmental nuclei, directly implicated with REM sleep generation [43].

Interestingly, even though the SCN plays a major role in sleep regulation, it has rather limited monosynaptic outputs to sleep-regulatory centers, like VLPO and the lateral hypothalamus [44]. Therefore, the circadian sleep regulation might be mediated by multisynaptic projections from the SCN to sleep–wake centers, such as the subparaventricular zone and the dorsomedial hypothalamic nucleus [45]. The latter, for instance, sends an intense GABAergic projection to the VLPO, thus ensuring a putative mechanism for the circadian sleep regulation [46].

However, sleep propensity depends not only on the circadian rhythmicity, but also on sleep satiety or sleep pressure, as indexed by the level of homeostatic sleep drive [47, 48]. Acute sleep loss, sleep interruptions, sleep disorders, and chronic under-sleeping dramatically increase the homeostatic sleep pressure, which, in turn, increases sleep propensity and impairs neurobehavioral performance [49, 50]. For instance, when sleep propensity is elevated, attention failures increase, irrespective of whether this is a result of acute sleep deprivation, chronic sleep loss or indeed a misalignment of circadian phase [51].

This leads to the following question: what is the role of the sleep/wake homeostat?

9.3 Sleep/Wake Homeostatic Process

9.3.1 Overview

Sleep homeostasis implies the enhancement of sleep propensity when sleep is curtailed, and its reduction when there is an excess of sleep [1, 2]. This is clearly demonstrated in sleep deprivation protocols utilized to challenge homeostatic sleep mechanisms [52–55]. Accordingly, as elapsed time awake during sleep deprivation is extended, the increase in sleep pressure augments low-frequency EEG activity in the range of 0.75–8 Hz during NREM sleep and similarly during REM sleep. This increase can be reversed by short nap episodes, by which the high sleep pressure is therefore attenuated [56, 57]. When considering the topographical distribution of EEG activity, the homeostatic increase in low-EEG components appears to be more prominent in frontal brain areas during sleep [54, 58, 59]. Likewise, EEG activity in the sleep spindle range during NREM sleep (12–15 Hz) is modified with increased sleep pressure, in such a manner that high spindle frequency activity (>13.5 Hz) decreases, while low spindle activity (<13.5 Hz) increases [60]. Taken together, quantitative EEG analyzes in sleep deprivation protocols show that the low EEG components during both NREM and REM sleep, together with sleep-spindle activity in NREM sleep, are particularly sensitive to changes in the duration of prior wakefulness and sleep. This, in turn, implies that these specific frequency activities can act as correlates of the homeostatic sleep process. Another point to be addressed is that the decay rate of the homeostatic pressure during sleep appears to be dependent on the sleep stage. In other words, a faster decay rate happens during slow-wave sleep rather than during REM sleep [61].

While the neuroanatomical and molecular substrates for the circadian sleep regulation are rather well-known, there remains controversy concerning the potential neuronal structures responsible for sleep homeostasis [40]. Thus in Sect. 9.3.2, possible physiological underpinnings for the sleep homeostat are summarized.

9.3.2 *How Does Sleep Homeostasis Occur?*

Within a molecular framework, the adenosinergic system can be implicated with the sleep homeostat [62–64]. During wakefulness, increased metabolic and neural activity leads to higher extra-cellular adenosine concentrations, whereas, during sleep, there is a substantial decline in adenosine concentrations. This suggests that adenosine may be related to sleep regulation by inhibition of neuronal activity. Similarly, in humans, a genetic variant of adenosine deaminase, which is associated with the reduced metabolism of adenosine to inosine, particularly enhances slow-wave sleep and slow-wave activity during sleep [62]. Taken together, the adenosinergic system can contribute to the inter-individual variability in sleep homeostasis regulation.

The neuroanatomical underpinnings for the sleep homeostatic process are still fairly unknown. Converging lines of evidence support that local adenosine levels rise in certain cortical areas during waking and decline during sleep [65, 66]. Given that these changes are more likely to be predominant in the basal forebrain in detriment to the other cortical regions [67], local release of adenosine in this structure has been proposed as a signal for the homeostatic regulation of NREM sleep (for a review see [68]. Alternatively, electrophysiological data indicate that adenosine may disinhibits and/or actively induce sleep-promoting neurons in the VLPO area of the hypothalamus [69, 70]. Furthermore, adenosine may contribute to global cortical inhibition, due to reduced activating input from ascending cholinergic and monoaminergic pathways and as a result of long-lasting hyperpolarizing potentials during NREM sleep [71], which consists in a type of inhibition due to a reduced activating input from the ascending cholinergic and monoaminergic pathways.

Intriguing evidence has shown that genetic processes influence the liability to sleep loss at transcriptional level, as observed in the cortical levels of Homer1a expression that strongly reflect the response to sleep loss [72]. Accordingly, the time-course gene expression analysis indicated that while more than 2,000 cortical transcripts undergo circadian control, less than 400 continue rhythmic during the course of sleep deprivation. This indicates that most diurnal gene transcription changes are sturdily dependent on the sleep/wake homeostat. Moreover, after sleep deprivation, Homer1a and three other genes – Ptgs2, Jph3, and Nptx2 – are over-expressed, all of which are known to play a pivotal role in recovery from glutamate-induced neuronal hyperactivity. The activation of Homer1a points towards a functional role for sleep in intracellular calcium homeostasis, which, in turn, provides a protective and recovering effect for the neuronal activation during wakefulness.

9.3.2.1 The Functional Role of the Sleep/Wake Homeostat

Recently, an interesting hypothesis – the synaptic homeostasis hypothesis – argues that sleep plays a key role in the regulation of cortical synaptic plasticity [10, 11]. A schematic diagram of this hypothesis is represented in Fig. 9.2. The novel aspect concerning this hypothesis is that, besides the proposed association between the homeostatic process and synaptic strength, it tries to unravel the specific mechanisms for this relationship. Accordingly, this hypothesis claims that wakefulness encompasses synaptic potentiation in several cortical circuits, that synaptic potentiation is intertwined to the homeostatic regulation of slow-wave activity, which, in turn, is associated with synaptic downscaling. As a result, this downscaling can relate to the beneficial effects of sleep on neural function.

Process S [9] describes the process of synaptic homeostasis, mainly by reflecting how the amount of synaptic strength in the cortex modifies due to wakefulness and sleep [10, 11]. Under normal sleep–wake conditions, synaptic strength increases during wakefulness and peaks prior to sleep, while during the sleep episode it substantially decreases until it reaches a baseline level when sleep ends. During wakefulness, the neuromodulatory setting, such as the high levels of noradrenaline, favors information storage, namely by long-term potentiation of synaptic strength. This potentiation starts when the firing of a presynaptic neuron is followed by the depolarization or firing of a postsynaptic neuron. As a result of the net increase in synaptic strength, waking plasticity progressively saturates several domains, such as learning.

On the other hand, during sleep, slow oscillations are triggered in membrane potential, which comprise the depolarized and hyperpolarized phases that enable synaptic activity to not be followed by synaptic potentiation. Furthermore, stronger cortico-cortical connections increase the degree of synchronization among populations of neurons, which reflect in slow waves of larger amplitude, particularly during the initial sleep stage [73–76]. Similarly, this may also explain why the increase in

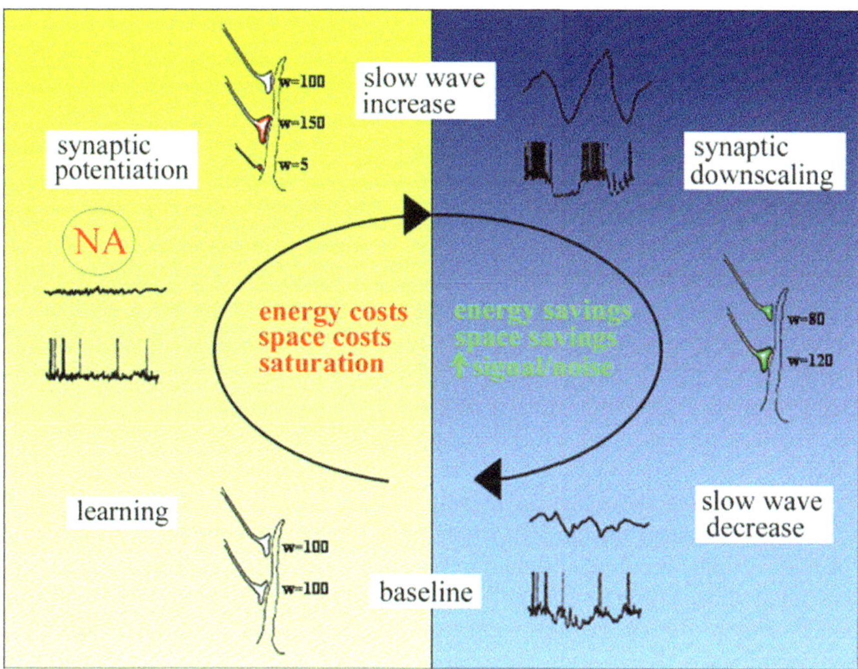

Fig. 9.2 *The synaptic homeostasis hypothesis* – During wakefulness (*yellow background*), the neuromodulatory milieu favors the storage of information, through long-term potentiation of synaptic strength (*in red*). Due to the net increase in synaptic strength, waking plasticity has a cost in terms of energy requirements, space requirements, and progressively saturates learning capacity. During sleep (*blue background*), changes in the neuromodulatory milieu trigger the occurrence of slow oscillations in membrane potential, comprising depolarized and hyperpolarized phases, which affect every neuron in the cortex, and which are reflected in the EEG as SWA (With permission from [10])

EEG spectral power density after wakefulness extends to other frequency bands besides the slow-wave or delta band [54, 77–79], although it remains to be explained why the time course of the increase varies for different frequency bands [80]. The repeated sequences of depolarization – hyperpolarization can result in the downscaling of the synapses impinging on each neuron, which implies decrease in synaptic strength. This reduced synaptic strength leads to a decrease in the amplitude and synchronization of the slow oscillations in the membrane potential [10, 11]. As a consequence, slow-wave activity progressively decreases during the rest of the sleep episode. With the attenuation of slow-waves, downscaling is gradually reduced till synaptic strength reaches a suitable baseline level. By returning synaptic plasticity to an adequate baseline level, sleep enforces synaptic homeostasis. The main advantage is that the neural circuits keep track of previous experiences, although they remain efficient at a recalibrated level of synaptic strength, which, in turn, enables the restarting of this entire process on the following waking episode.

There is consistent evidence for this intriguing hypothesis. By increasing the total duration of wakefulness, there is a concomitant increase in the long-term potentiation related genes, followed by an increase in slow-wave activity [81–83]. Accordingly, if wakefulness is not accompanied by long-term potentiation changes in synaptic strength, the homeostatic increase in slow-wave sleep can be eliminated. Thus, it is likely that it is not wakefulness *per se*, but rather the induction of long-term potentiation molecules associated with wakefulness that is responsible for the homeostatic increase in slow-wave activity. Another evidence builds-up from the fact that, if synaptic potentiation is stronger in specific brain areas, slow-wave activity during subsequent sleep increases substantially in that given area, which suggests that sleep may be locally regulated [84–87]. This may imply that the well-known topographic differences in slow-wave homeostasis, with a predominance for frontal regions to exhibit a stronger response to sleep deprivation [54, 58, 59], can be associated to topographical differences in the susceptibility to plastic changes [88].

There are several reasons as to why sleep might be necessary for synaptic homeostasis. Perhaps the most important is that the neuronal network has to assess the total synaptic input independent of behavioral needs, in order to determine the amount of downscaling for synaptic homeostasis [10, 11]. Contrary to wakefulness, neural activity during sleep is spontaneous and happens almost detached from behavioral needs. As a result, the neuronal network can downscale appropriately. Another point is that downscaling is basically promoted by repetitive depolarization – hyperpolarization sequences, which are compatible with sleep, but would interfere with behavioral needs if it had to happen during wakefulness. Taken together, it might be that sleep plays a pivotal role for plasticity, through the homeostatic regulation of the synaptic plasticity of neuronal networks.

As illustrated in these sections, both the circadian clock and the sleep homeostat are essential for sleep regulation. However, are these aspects of sleep regulation really independent of each other? A large body of evidence argues against this concept. In Sect. 9.4, the interplay of these two systems will be described and how both these systems contribute to the regulatory mechanisms that underlie sleep.

9.4 Circadian and Homeostatic Processes: Is It Really a Tug of War?

9.4.1 The Two-Process Model of Sleep Regulation

As illustrated in the previous sections, despite the usual controversy regarding whether it is the circadian or the homeostatic process that really underpins sleep, there is mounting evidence in support of the interaction of these two systems for sleep regulation [1, 89]. In fact, the homeostatic sleep–wake regulation cannot exclusively account for changes in sleep propensity, which leads to the assumption

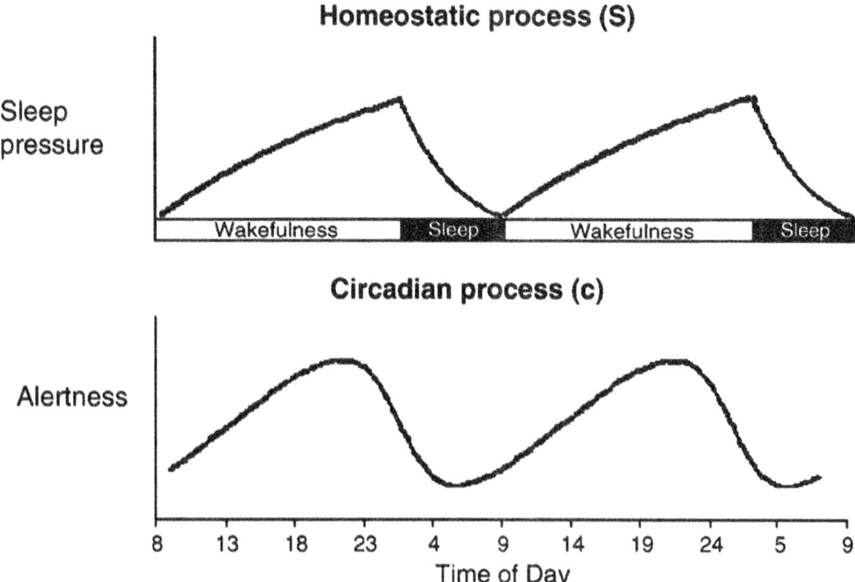

Fig. 9.3 *Interaction of process S and process C* – The homeostatic sleep pressure builds up during wakefulness and dissipates exponentially during sleep. The circadian process is related to the time of day, irrespective of previous sleep duration, and opposes the homeostatic process (modified from [9])

that, in addition to this system, the circadian system is equally involved in sleep regulation [90]. Indeed it is the combination of these two oscillatory processes with very different properties that best explained the timing of human sleep/wake behavior in humans living in the absence of time cues [91].

Given this important interface, the circadian and sleep homeostatic processes have been conceptualized in the two process model of sleep regulation to better understand the timing and architecture of sleep [9, 92]. According to this model, as illustrated in Fig. 9.3, the homeostatic sleep drive accumulates with each waking hour and is dissipated by sleep itself on an exponential manner. This process has properties very different from those of the circadian oscillator, which opposes the increasing homeostatic drive for sleep that builds near the end of the habitual wake day [93]. A similar process may happen during the end of the sleep episode, when sleep pressure has dramatically decreased. In order to counteract a possible arousal during these early morning hours, the circadian oscillator probably ticks in through a sleep-promoting signal, which opposes this decrease in the homeostatic sleep pressure, thus ensuring a longer sleep.

Figure 9.4 summarizes how sleep–wake cycles are regulated by circadian and homeostatic factors. Examples of this interaction can be drawn from daily variations in measures, such as alertness and neurobehavioral performance that reflect the

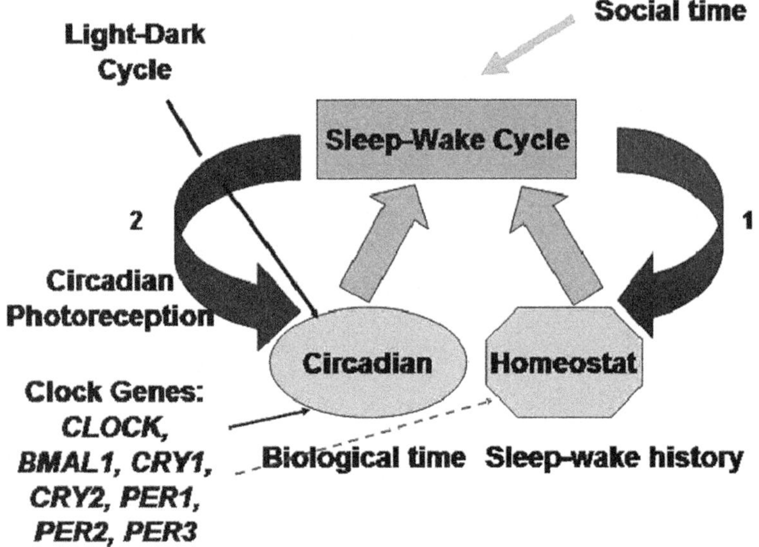

Fig. 9.4 *Sleep–wake cycles are regulated by circadian and homeostatic factors* – The circadian pacemaker and the sleep homeostat modulate of the timing and structure of the human sleep–wake. The oscillation of the sleep homeostat is strongly determined by the sleep–wake cycle (*arrow 1*). Light input to the circadian clock is mediated by circadian photoreception and the sleep–wake cycle is a major determinant of this light input (*arrow 2*). Furthermore, social time and clock genes can underlie this process (modified from [132])

output of the circadian pacemaker in humans [47, 94, 95]. Drowsiness and attention are, respectively, increased and impaired immediately after core body temperature nadir, postulated to be near regular wake-time, the opposite being true when these measures are performed before the time at which subjects would normally sleep, as illustrated in Fig. 9.5 [94]. Paradoxically, the circadian drive for wakefulness peaks before habitual bedtime in humans entrained to the 24-h day. This contradictory phase relationship between the timing of the circadian sleep propensity rhythm and the timing of sleep and wakefulness during entrainment to the 24-h day is postulated to assist the consolidation of sleep and wakefulness in humans [96]. The circadian pacemaker opposes this increasing drive for sleep in the early evening by increasing the drive for wakefulness. Similarly, some hours prior to bedtime, the sleep promoting hormone melatonin is released in higher levels and, afterwards, melatonin receptors in the SCN suppress the firing of the SCN neurons [97]. It is possible that melatonin may serve to reduce the wake-promoting signal from the SCN, thereby promoting sleep immediately after the peak of the circadian drive for wakefulness [98]. In turn, the SCN actively promotes sleep prior to the habitual wake time, when homeostatic sleep pressure is at the lowest levels. This can be illustrated by the fact that the peak in the circadian rhythm of REM sleep happens prior to [99].

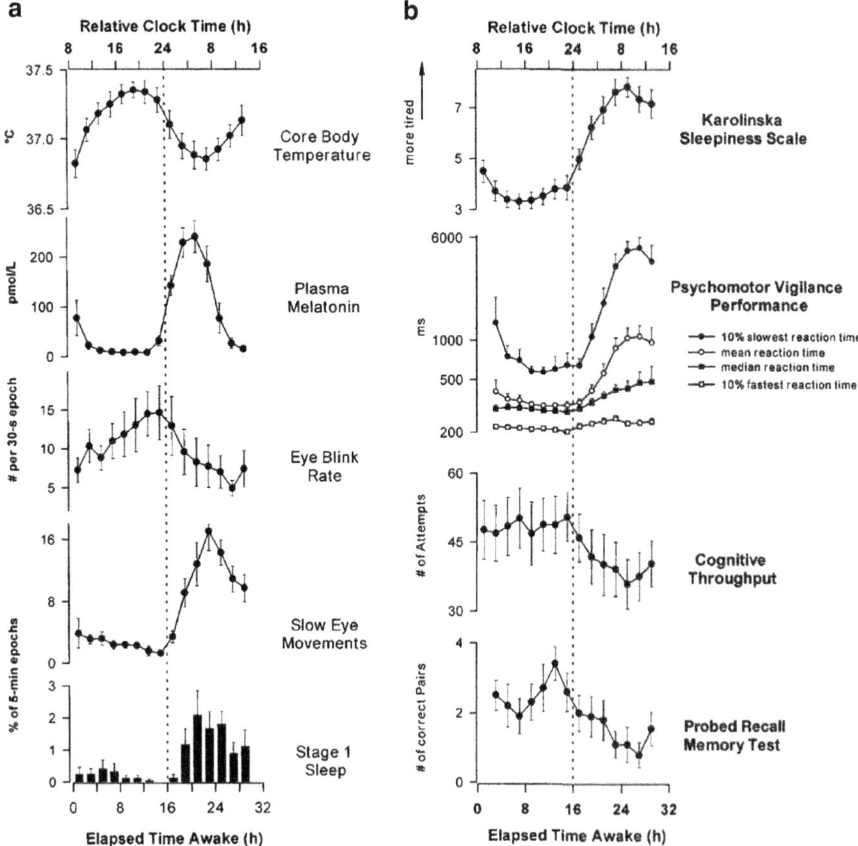

Fig. 9.5 *Interaction of the circadian oscillator and the sleep homeostat* – Left panel (**a**) depicts the time courses of core body temperature, endogenous plasma melatonin, mean eye blink rate per 30-s epoch during Karolinska drowsiness test, incidence of slow eye movements (SEMs, percentage of 30-s epochs containing at least 1 SEM/5-min interval), and incidence of stage 1 sleep (percentage of 30-s epochs containing at least 15 s of stage 1 sleep per 5-min interval) are shown, averaged across ten subjects ± SE. Right panel (**b**) illustrates the time course of subjective sleepiness as assessed on Karolinska sleepiness scale (KSS; highest possible score = 9, lowest possible score = 1), psychomotor vigilance performance [mean, median, 10% slowest and fastest reaction times in ms (logarithmic scale)], cognitive performance (numbers of attempts in 4-min two-digit addition task), and memory performance (number of correct word pairs in probed recall memory task) are shown averaged across ten subjects ± SE. All data were binned in 2-h intervals and expressed with respect to elapsed time since scheduled wake-time. Vertical reference line indicates transition of subjects' habitual wake- and bedtime (with permission from [94])

9.4.2 Are These Two Systems Independent or Complementary?

How the circadian and the homeostatic process interact is still a matter of intense debate. For instance, it is not yet clear where in the central nervous system this possible interaction occurs and whether the SCN directly or indirectly interacts with the brain centers responsible for the sleep homeostatic process, such as the basal forebrain and the VLPO. Prior studies focusing on SCN-lesioned animals provide evidence that the homeostatic process is still operative and is not drastically changed [93, 100]. This argues in favor of the two processes being probably independent from each other. On the other hand, it might be that the circadian and the homeostatic processes interact more downstream in the cascade or that the output variables of measurement do not reflect an accurate interaction [13, 101].

In rats, the combination of SCN neuronal activity with EEG recordings revealed that changes in vigilance states are paralleled by strong variations in SCN activity [102]. During REM sleep, SCN activity was elevated, whereas in NREM sleep it was substantially decreased. In concomitance, two types of sleep deprivation – of slow-wave sleep in NREM and of total REM sleep – were carried out to prove that variations in SCN activity are related to changes in vigilance state [102]. Accordingly, SCN activity appeared to be differentially affected by the alternations of sleep states and these sleep-dependent changes were superimposed on the circadian modulation in SCN activity. Given that the SCN neuronal activity was tightly correlated to fluctuations in slow-wave activity, it is likely that the SCN receives continuous input about changes in sleep homeostasis.

To conclusively illustrate in humans how these processes interact and in order to quantify their strength in the control of sleep and wakefulness, protocols must be applied to disentangle these two processes [2]. Why? The sleep–wake cycle can be usually observed directly. The challenge, though, has been to separate the relative contribution of the sleep homeostat, the circadian clock and the light input to sleep–wake characteristics. Such an assessment mostly requires controlled laboratory conditions, since in the daily world the timing of this sleep–wake cycle is strongly modulated by our social demands and artificial lighting environments.

9.4.2.1 What Can Happen When One Disentangles These Two Systems?

Devoid of the imposed light–dark and social cycles, sleep remains consolidated, although it desynchronizes from the 24-h solar day. In concomitance, one observes dramatic changes in the internal phase relationship between core body temperature and the sleep–wake cycle [103]. For instance, sleep can initiate near the core body temperature nadir, instead of 6 h before. This drastic change in the phase relationship suggests that some distinguished oscillators rule these rhythms. A thrilling example is the spontaneous internal desynchrony [104, 105]. Under this free-running condition, subjects live in an environment free of time cues and self-select the light–dark cycle to which they are exposed. The sleep–wake cycle then oscillates with a longer or shorter period when compared with the other physiological measures, although the period of core body temperature rhythm remains rather stable. Accordingly, the longest spontaneous sleep duration habitually occurs when sleep starts near to the wake-maintenance zone and at the peak of the core body temperature [106–108]. Similarly, the crest of the REM sleep propensity rhythm is situated close to core body temperature nadir [106–108]. As a consequence, sleep is initiated on a broader range of circadian phases. Under such conditions, waking activities, such as food intake and hourly time estimation, are associated with the sleep–wake cycle period and perhaps are not ruled by the circadian pacemaker.

In the forced desynchrony protocol (FD), participants live on artificially either very long or very short days [39]. This imposed desynchrony between the sleep–wake schedule and the output of the circadian pacemaker occurs only under conditions in which the sleep–wake schedule is outside the range of human entrainment. As a consequence, the circadian system cannot be entrained to the new imposed sleep–wake cycle. Accordingly, scheduled sleep and wake episodes happen at nearly all circadian phases, and when light intensity during scheduled waking episodes is kept low, the pacemaker can free-run with a stable period in the range of 23.9–24.5 h [4]. Since subjects are scheduled to stay in bed in darkness, the variation in the amount of wakefulness prior to each sleep episode is significantly minimized. This enables to average data either over successive circadian cycles or over successive sleep or wake episodes, therefore separating the circadian profile of a given variable by removing the confounding sleep–wake-dependent contribution or vice versa.

A wide array of interactions between the sleep–wake cycle and circadian processes in the regulation of sleep, the EEG during sleep and wakefulness, alertness, motivation, and neurobehavioral and physiological variables can be unraveled though this protocol [12, 90, 109–111]. For example, forced desynchrony studies have revealed that sleep efficiency is maximum when subjects sleep at the circadian phase during which endogenous melatonin is released [112]. In contrast, wakefulness during the sleep episode is highest when sleep is scheduled to happen at a circadian phase during which endogenous melatonin secretion is absent [112]. This led to the hypothesis that melatonin administration may improve sleep efficiency at such times, which in fact happened when melatonin was administered 30 minutes prior to scheduled sleep, leading subjects to experience 30 minutes of more sleep when they slept at a circadian phase without the release of endogenous melatonin [113]. Contrariwise, there was no effect of melatonin administration on sleep efficiency and sleep duration when melatonin was administered at the circadian phase of endogenous melatonin production. Thus, melatonin efficacy appears to be dependent on the circadian phase and serves as an interesting example as to how strongly the circadian and homeostatic mechanisms underpinning sleep are intertwined.

Taken together, quantitative analyzes of the interaction between these two systems provide strong evidence that these two factors interact, and not merely add up [1, 2], as illustrated in Fig. 9.6 ([166]). This implies that the amplitude of the circadian modulation of sleep depends on homeostatic sleep pressure [110]. As an illustration, when sleep pressure is low the circadian variation in performance is equally small. Similarly, as will be depicted in Sect. 9.4.3, changes in circadian amplitude together with variations in sleep pressure may happen for sleep spindle activity, REM density, sleep consolidation and so forth. This non-additive nature implies that relatively slight variations in the circadian phase can have major repercussions on a given variable, if

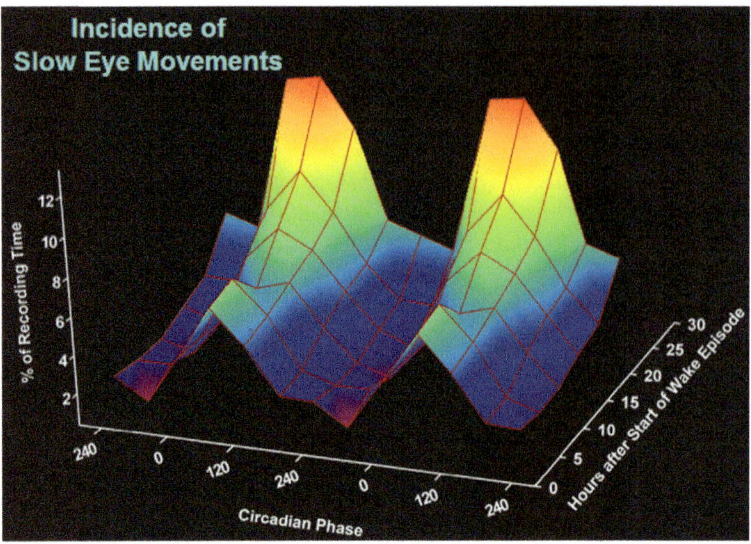

Fig. 9.6 *Non-additive nature of the circadian oscillator and the sleep homeostat* – Incidence of slow eye movements expressed in terms of percentage (%) of recording time (*y*-axis) and hours after start of sleep episode (*z*-axis) in relation to the circadian phase (*x*-axis, 0 indicating core body temperature nadir) (with permission from [166])

sleep pressure is high, the contrary also being true. In Sect. 9.4.3, we will summarize the possible homeostatic and circadian influences on the sleep EEG.

9.4.3 What is Homeostatic and/or Circadian in the Sleep EEG Activity?

The circadian and the homeostatic processes play a multitude of roles during sleep. An indication about the strength of these processes on EEG activity builds up from previous FD studies, in which the period of the sleep–wake cycle was either 28 or 42.85 h [4, 110]. Accordingly, the circadian rhythm of endogenous melatonin oscillated within a 24-h basis and with a similar period of the circadian rhythm of core body temperature, thus leading to a desynchronization of these rhythms [2, 12]. EEG power spectra during wakefulness, REM sleep and NREM indicated high alpha power during wakefulness, predominance of low frequencies and in the spindle range during NREM sleep, and lower values in the same frequency bins during REM sleep. This suggests that the relative contribution of the circadian and homeostatic processes exhibited frequency-specific modulation. During all stages of vigilance, low EEG components were predominantly modulated by the homeostatic factor. On the other hand, the circadian modulation of the EEG differed across these states, such that the maximum circadian variance was shown for REM sleep in the alpha activity range and for NREM sleep in the low spindle frequency activity (Fig. 9.7) [112]. An interesting aspect in the EEG power spectra regards sleep spindles, which are primarily generated and modulated by a thalamocortical network, which comprises the interplay between reticular thalamic, cortical pyramidal, and thalamocortical cells [115]. Initially, the progressive hyperpolarization of thalamocortical cells after sleep onset results in the appearance of spindle oscillations, which are then replaced by slow-wave oscillations, when deepening of sleep proceeds, and thalamocortical neurons achieve a voltage range at which slow-wave oscillations are triggered [116]. Thus, sleep spindles are deemed to play a key role in neuronal plasticity and sleep maintenance, basically by inhibiting the sensory information that reaches the cerebral cortex [117].

When the sleep episode coincides with the circadian phase of endogenous melatonin secretion and when it is highly consolidated, mild reductions in the frequency range of slow-waves and theta activity are observed in the NREM sleep, while profound variations occur in the spindle frequency activity [48, 90]. Low frequency sleep spindle activity (12.25–13 Hz) exhibits an outstanding circadian modulation, with maximum levels during the circadian phase of melatonin secretion [112]. This has been interpreted as evidence for the circadian modulation of the frequency of sleep spindles.

Taken together, spectral hallmarks of EEG activity during sleep exhibit a frequency-specific homeostatic and circadian modulation. These two independent oscillatory processes correspond to an essential component of cortical activation during sleep, which is likely to be related to the processing of external sensory stimuli and behavioral responses [2].

Given the fundamental role played by these systems on the EEG activity, questions emerge: What about genes? Can the molecular machinery be implicated with the circadian and homeostatic modulation? If so, how and to what extent is its role? In Sect. 9.4.4, we will address some of the exciting ideas that are going on in this field.

9.4.4 What is the Role of Genes in the Circadian and Homeostatic SLEEP Modulation?

The molecular machinery that underlies circadian and homeostatic sleep regulation still has many gaps of uncertainty, although many of its features are being unraveled. Human circadian oscillators are considered to rely on the transcription/translation feedback loops in clock gene expression [15, 118, 119]. The molecular interplay of clock genes consists largely of clock RNA and proteins that oscillate in a circadian style, through positive and negative feedback loops of transcription and translation. The encoded protein of two core clock genes, *clock* and *bmal1*, binds as a transcription factor to the DNA. The *Clock-Bmal1*-complex translocates to the nucleus whereby it induces transcription of the other core clock genes like *per* 1, 2 and 3 and *cry* 1 and 2. m*cry*1 and 2, in turn, inhibits the transcriptional activity of *Clock* and/or *Bmal1*, thereby inhibiting their own transcription, whereas m*per*2 is involved in transcription of *Bmal1*. *Per*3 forms heterodimers with *Per*1/2 and *Cry*1/2, respectively, and enters into the nucleus, resulting in the inhibition of *Clock-Bmal1*-mediated

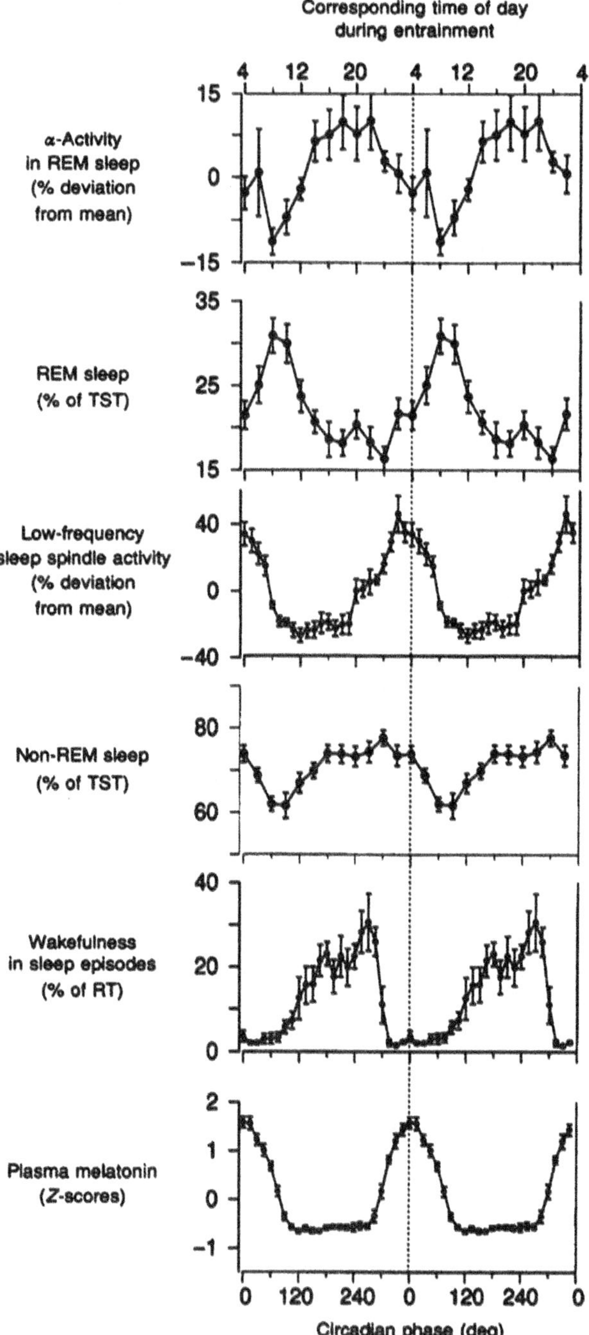

Fig. 9.7 *Circadian EEG components* – Phase relationships between the circadian modulation of Alfa activity in REM sleep (*first panel*), REM sleep (*second panel*), low-frequency spindle activity in NREM sleep (*third panel*), and plasma melatonin (*last panel*). All data are plotted against circadian phase of the plasma melatonin rhythm (O deg corresponds to the fitted maximum, bottom x-axis). Top x-axis indicates the average clock time of the circadian melatonin rhythm during the

transcription [120]. Furthermore, the major and crucial loop involves the autorepression of cryptochrome (Cry1, Cry2) and period (*Per1, Per2*) genes and this rhythm-generating circuitry is functional in most cell types [15, 118, 119].

Besides the circadian system, clock genes also affect sleep homeostasis. For instance, the mutation in mice clock gene can lead to alterations of sleep homeostasis [121, 122]. Indeed, when compared with the heterozygous or wild type, homozygous mice showed less and shorter NREM sleep episodes and significantly lower delta frequency during light–dark cycles. Similarly, mice tested on their incapacity to respond to sleep pressure by being sleep deprived for 6 h revealed that the homozygotes slept less than the heterozygotes or the wild type. Analysis of recovery night supported the concept that homozygous mice responded to sleep deficit by increasing sleep time and total amount of sleep [121].

The functional role of m*Per1* and m*Per2* genes in the homeostatic regulation of sleep has been unraveled by the comparison of mice deficient in mPER1 – m*Per1* mutants – or in mPER2 – m*Per2* mutants – with wild-type controls after sleep deprivation [123]. Under this condition, all mice exhibited an increase of slow-wave activity in NREM sleep, which is known to reflect the homeostatic response to sleep deprivation. Furthermore, this increase was topographically more predominant in frontal regions in relation to the occipital regions. While all genotypes did not differ with respect to the effects of sleep deprivation on the occipital EEG, the effects on the frontal EEG were initially diminished in both mPer mutants. Differences between the genotypes were observed during the time course of the 24-h distribution of sleep, and reflected especially the phase advance of motor activity onset in m*Per2* mutants. Whereas the daily distribution of sleep was modulated by m*Per1* and m*Per2* genes, sleep homeostasis, as indexed by the increase in slow-wave activity after sleep deprivation, was relatively preserved in the mPer mutants.

Another evidence in support of a non-circadian role for clock genes in sleep homeostasis buildsup from a study focused on the time course of forebrain changes in the expression of *per1* and *per2* during sleep deprivation and subsequent recovery sleep in three inbred strains of mice [124]. Accordingly, in all mice strains *per1* and *per2* expression increased during sleep deprivation, particularly in the D2 mice strain for which the sleep rebound was lowest. Furthermore, while in the other two strains *per1* and *per2* reverted to control levels during recovery sleep, *per2* expression continued high in D2 mice. Taken together, this provides a functional role for clock genes in the homeostatic sleep regulation.

The non-circadian role of clock genes has also been described for humans. For instance, the human *period* 3, one of the several clock genes, is under current extensive investigation due to its implication in the circadian regulation and sleep homeostasis.

first day of the forced desynchronization protocol, immediately upon release from entrainment. Plasma melatonin data were expressed as Z-scores to correct for inter-individual differences in mean values. REM sleep is expressed as a percentage of total sleep time (TST). Low-frequency sleep spindle activity in NREM sleep and Alfa activity in REM sleep are expressed as percentage deviation from the mean. Data are double plotted, i.e., all data plotted left from the dashed vertical line are repeated to the right of this vertical line (with permission from [112])

Peak levels of *PER*1, 2 and 3 are known to happen during diurnal activity time in human peripheral blood cells, with greater levels of mRNA expression approximately 9 h after the melatonin peak [125]. Furthermore, the phase of peripheral clock gene expression in leukocytes can be strongly associated with the habitual sleep timing and *PER3* expression can be correlated with melatonin and cortisol phase [126]. While considering all these three clock genes, *PER*3 has been shown to be the most robust marker of the circadian gene expression, together with cortisol and melatonin. However, despite the fact that *PER3* expression can act as a marker for the entrained phase, this does not imply that it can have direct repercussions on entrained phase. It can be argued, for instance, that its effects on the diurnal preference can be mediated by the effects on the sleep–wake homeostat, and not through the timing of core circadian markers [127]. This idea is supported by a recent study focusing on the contribution of *PER*3 polymorphism on sleep–wake regulation [127]. Briefly, the coding region of the *PER*3 gene has a variable-number tandem-repeat polymorphism, in which a design encoding 18 amino acids is repeated either four ($PER3^4$) or five times ($PER3^5$), thus indicating that subjects can either be homozygous for $PER3^4$ or $PER3^5$ alleles or heterozygous $PER3^{4/5}$ [120]. In comparison to $PER3^{4/4}$ individuals, homozygosity for $PER3^5$ had a remarkable effect on several markers of sleep homeostasis: increase in slow-wave sleep and slow-wave activity in NREM sleep (baseline night), together with an increase in theta and alpha activity in REM sleep (baseline night), as illustrated in Fig. 9.8.

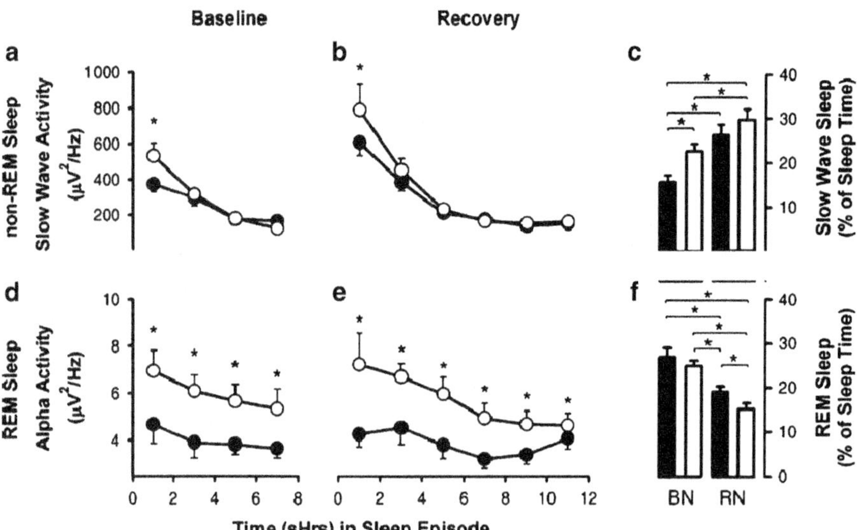

Fig. 9.8 *PER 3 polymorphism and sleep structure* – Dynamics of slow-wave activity in NREM Sleep, alpha activity in REM Sleep, and sleep-stage response to sleep loss differ between *PER*35/5 and *PER*34/4 (**a**, **b**, **d**, and **e**) Time courses of slow-wave activity in NREM sleep and alpha activity in REM sleep during baseline and recovery sleep episodes for *PER*35/5 (*open symbols*) and *PER*34/4 (*filled symbols*). (**c**) Slow-wave sleep time and (**f**) REM sleep time expressed as a percentage of total sleep time for baseline and recovery sleep in *PER*35/5 (*white bar*) and *PER*34/4 (*black bar*) (with permission from [127])

These differences in slow-wave and slow-wave activity were sustained during recovery sleep even after 40 h of sleep deprivation Taken together, *PER*3$^{5/5}$ individuals are likely to live under a comparatively higher sleep pressure and are more susceptible to the effects of sleep loss [127]. Interestingly, the circadian rhythms of melatonin and cortisol were not affected. In other words, the classical markers of amplitude and phase of circadian physiology are not affected by this polymorphism. Within the framework of the homeostatic and circadian regulation of sleep, these data imply that the *PER*3 polymorphism appears to contribute to the differences in sleep structure by its effects on the homeostatic facet of sleep regulation. The independence of these two processes has been a mainstay of our current models of sleep regulation [128]. Nevertheless, it might be that this concept does not extend to the molecular level, and that some genes previously described as clock genes can perform the noncircadian roles.

How can one see the interaction of these two oscillatory systems in healthy people? There are three examples of natural models: the interaction of these two systems in morning and evening types, in short and long sleepers and in healthy aging. In Sect. 9.4.5, we will focus on the first model.

9.4.5 Morningness and Eveningness: The Clock Ticks with the Homeostatic Hourglass

Morningness–eveningness, also described as "chronotype", comprises the individual characteristic related to the preference in sleep timing, according to which morning types habitually choose to sleep approximately 2 h earlier than the evening types [129]. An underlying hypothesis for this individual difference could be the internal circadian phase, with an earlier circadian phase in morning than in evening types [130]. However, it has been recently shown that morning and evening types have similar circadian phases [131]. Given that the homeostatic process plays a fundamental role in the regulation of sleep–wake behavior [132], one could argue that morningness–eveningness may result from the individual differences in the homeostatic sleep regulation. In line with this assumption, recent studies have questioned whether differences in the dynamics of nocturnal homeostatic sleep pressure can underlie differences in sleep timing preference [133, 134]. In the subjects with intermediate circadian phases (Fig. 9.9), both initial level and decay rate of slow-wave activity in the frontal derivation during NREM sleep were higher in morning than in evening types. Furthermore, no correlation was elicited between slow-wave activity decay and dim light melatonin onset (DLMO). This supports the idea that chronotypes can occur as a consequence of different dissipation levels of sleep pressure. All in all, it might be that, instead of "chronotype", the expression "homeotype" coins what really happens in the morningness and eveningness.

Recently, a fascinating study [135] using functional magnetic resonance imaging (fMRI) in extreme morning and evening types revealed that sustained attention in evening was intertwined to enhanced activity in evening than morning chronotypes

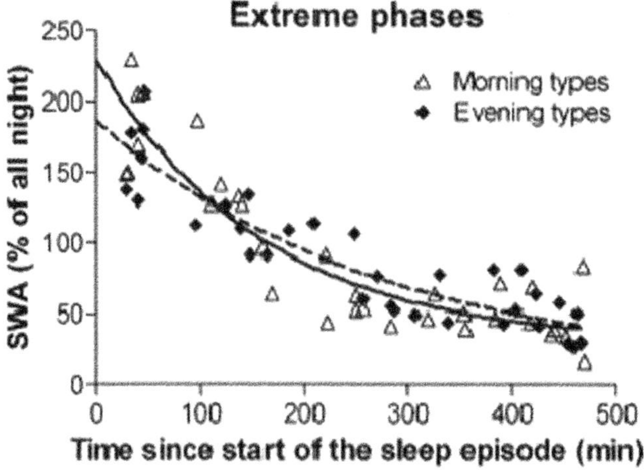

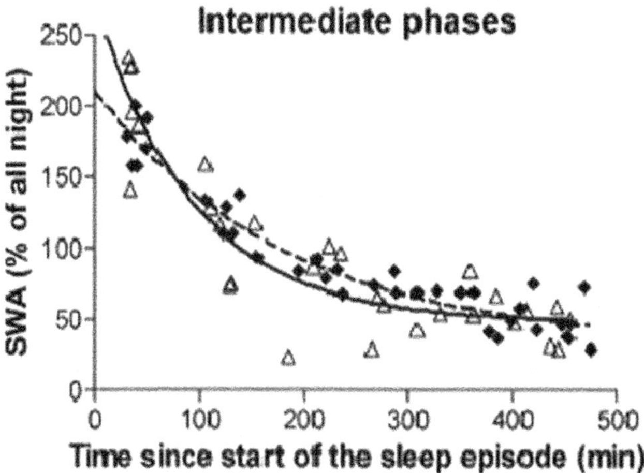

Fig. 9.9 *Slow-wave activity in morningness and eveningness* – Exponential decay functions adjusted on relative slow-wave activity in NREM sleep for all night EEG in frontal derivation. *Solid line*, morning types; *dashed line*, evening types. Subjects with extreme circadian phases are depicted in upper panel and subjects with intermediate circadian phases are depicted in lower panel (with permission from [134])

in the LC and in a suprachiasmatic area (SCA), both deeply involved in the circadian wake-promoting signal. In parallel, it was revealed that SCA activity was inversely related to slow-wave activity during the first NREM sleep episode. In other words, SCA activity decreases with increasing homeostatic sleep pressure, which provides strong evidence for the direct influence of the homeostatic and circadian interaction on the neural activity underpinning sleep. In Sect. 9.4.6, we focus on how these two sleep oscillators play a role in short and long sleepers.

9.4.6 *Short and Long Sleepers: How Do the Circadian Clock and the Sleep Homeostat Work?*

To look at the possible differences of these two oscillatory systems in habitual short and long sleepers, parameters of the buildup and dissipation of sleep pressure, as indexed by the time course of slow-wave activity during sleep before and after 24 h of sleep deprivation, were carried out [136]. Accordingly, total sleep time was more than 3 h longer in the long sleepers than in the short sleepers. The enhancement of EEG slow-wave activity in NREM sleep after sleep deprivation was larger in long sleepers than in short sleepers. This difference in slow-wave activity was predicted by the two-process model of sleep regulation on the basis of the different sleep durations for both groups. This suggests that short sleepers live under a higher NREM sleep pressure than long sleepers, even though they do not necessarily differ with respect to sleep homeostat mechanisms. On the circadian domain, long sleepers usually exhibit longer circadian rhythms of core body temperature, melatonin and cortisol in detriment to short sleepers [137]. These differences may be attributed as either the cause or consequence of differences in sleep duration, since sleep is tightly related to the end of light input to the circadian pacemaker.

Another natural model looking at the interaction of these two systems is healthy aging. As every dynamic system, it is clear that neither the circadian modulation of sleep nor the sleep homeostat remains constant throughout one's entire lifetime. Given that, in what manner do theses changes occur? Does the balance of changes tilt more towards one of these processes? Section 9.4.7 summarizes the current state-of-art about age-related changes within these two complex oscillatory systems.

9.4.7 *Circadian and Homeostatic Sleep Regulation: What Happens with Age?*

Advanced age implicates changes in numerous aspects of the sleep–wake cycle. Among these, evidence points towards shallower nocturnal sleep associated with increase number of arousals, a decrease in slow wave sleep and more daytime naps [138, 139]. Similarly, there appears to be attenuated amplitude of circadian markers, such as melatonin, core body temperature and cortisol [140]. Older individuals tend to present earlier sleep times with a concomitant advanced circadian phase in relation to core body temperature minimum [141, 142], although the endogenous circadian period is quite stable [4]. However, it is still a matter of debate as to whether it is the circadian or the homeostatic facet of sleep which undergoes more changes with aging. It is very likely that the sleep homeostat remains operational after sleep deprivation in older individuals [143]. On the circadian domain, although some aspects of circadian sleep regulation seem to be affected by age [144], it is unclear whether it is aging *per se* or the modified regulation of circadian signaling downstream or both that are the underlying reason for these changes [40].

In order to unravel these questions, it was hypothesized if age-related changes in sleep can result from an attenuated circadian arousal signal during the evening. The main assumption underlying this idea is that the human circadian pacemaker notoriously maintains timing and consolidation of sleep by opposing increased homeostatic sleep pressure, particularly in the evening during the "wake maintenance zone" [40]. If the circadian signal is dampened with age, this opposition would not be so clear-cut. As one might expect, quantitative evidence for a dampened circadian arousal signal in older individuals (Fig. 9.10) was observed through an increased sleep in the wake maintenance zone [145]. Additionally, older individuals had a comparatively decreased melatonin secretion. On the EEG domain, older participants exhibited a reduction in circadian modulation of REM sleep, together with less obvious day-night differences in the alpha and spindle range of sleep, both of which clearly undergo a circadian regulation. Taken together, this implies that the age-related changes in sleep propensity can be underpinned by a reduced circadian signal opposing the homeostatic sleep drive.

While considering the sleep homeostat, there is evidence for two types of situations: (1) under high sleep pressure, older subjects exhibit an attenuated frontal predominance of sleep EEG delta activity [143]; (2) under low sleep pressure, in which sleep pressure is maintained low by intermittent 75-min naps scheduled every 150 min for 40 h under constant routine conditions, there are slight age-related differences [146]. In this case, there appears to be a significant decline of EEG power density in the delta range and an increase in the sleep spindle range during recovery sleep. Nonetheless, the delta range response to low sleep pressure was more enduring in young individuals, given that it lasted for the first two NREM sleep episodes,

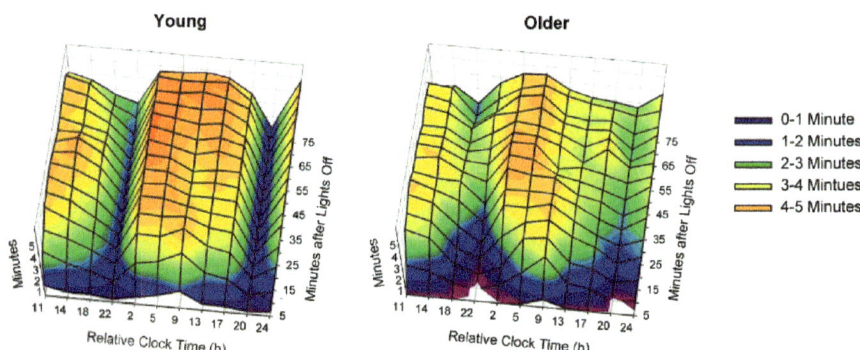

Fig. 9.10 *Dampened circadian arousal signal in older individuals* – Three-dimensional plots of total sleep time for young and older subjects. The *x*-axis represents the averaged mid-nap clock times for both age groups and the *y*-axis the time course within the respective naps (3–4 min). The *z*-axis specifies the amount of sleep per 5 min bin of each nap (min). Short-wavelength colors (*blue*, *green*) illustrate less sleep, longer wavelength colors (*yellow*, *orange*), more sleep. During the wake-maintenance zone (around 22 h – day 1 – and 20 h – day 2 – respectively) older individuals have comparatively more sleep (with permission from [145])

in comparison to the older participants, in which it lasted solely in the first NREM sleep episode. This dataset implies that the homeostatic sleep response to low sleep pressure is rather similar in older and young subjects.

What if these two processes go out of tune? Are there any clinical implications for a mal-functioning of these processes? Section 9.5 addresses some clinical scenarios in which the circadian and/or the homeostatic processes undergo some clear disruptions.

9.5 What Happens When the Two Sleep Processes Go Out of Sinc?

Changes in the circadian pacemaker may contribute to a wide array of dysfunctions, such as mood, sleep disturbances and cognitive performance [147, 148]. Given that the circadian timing system is extremely sensitive to environmental light and melatonin, it may not function in optimal levels in the absence or decrease of their synchronizing effects. For instance, elderly patients with dementia are very likely to experience functional deficits in the circadian timing system and attenuated synchronization if light exposure and melatonin production are reduced [149]. Recently, it was investigated whether cognitive and non-cognitive symptoms associated with dementia can decrease by bright light exposure [149]. Accordingly, light substantially attenuated cognitive deterioration, depressive symptoms and the increase in functional limitations. The possible underlying reasons may be that the long-term use of bright light improves the SCN ability to synchronize rhythms in hormones, metabolism and peripheral oscillators, with further repercussions on general functioning.

Alzheimer disease (AD) is another neurological condition associated with a wide range of changes in circadian rhythms [150]. For instance, rhythmic levels of several hormones, such as cortisol, melatonin, vasopressin, pulsatile luteinizing hormone and β-endorphine, are substantially modified in AD patients [151]. A significant increase of nocturnal serum cortisol levels in AD patients appears to be associated with a disturbance of sleep–wake rhythms [152]. Furthermore, the sensitivity of the hypothalamic–pituitary–adrenal axis to the steroid feedback appears to be highly impaired. In particular, pineal melatonin secretion and pineal clock gene oscillation can be disrupted in AD patients, and, surprisingly, even in non-demented controls with the earliest signs of AD neuropathology, in contrast to non-demented controls without AD neuropathology [151]. Furthermore, a functional disruption of the SCN was described from the earliest AD stages onwards, as shown by a decreased vasopressin mRNA, a clock-controlled major output of the SCN. This functional disconnection between the SCN and the pineal gland from the earliest AD stage onwards may be responsible for the pineal clock gene and melatonin changes, which can underlie the circadian rhythm disturbances in AD.

Circadian and homeostatic disruptions can play an essential role in the pathogenesis of mood disorders, particularly in depression. Sleep disturbances in depression include shortening of the REM sleep latency and an abnormal distribution of REM sleep during the night [153]. These REM sleep alterations, which exhibit a clear circadian pattern, appear to be more prominent in depressive states [154]. Abnormalities in the circadian rhythm of depressed patients may also include variations in the circadian phase and amplitude of core body temperature, with an increase in nocturnal body temperature amplitude [147, 155]. It is still unclear whether these observations represent functional changes of the circadian oscillator or are influenced by other factors. Similarly, there is some evidence in support of a decrease in nocturnal secretion of melatonin in depression [156], although this is still controversial.

Depression is habitually associated with sleep disturbances, like insomnia [157] and excessive daytime sleepiness [158, 159]. Interestingly, individuals genetically predisposed towards "eveningness" (a preference for the evening) in relation to "morningness" (a preference for the morning) are more prone to develop depression [160]. Genetic variations in the circadian genes have been found to associate with these sleep disorders and diurnal preference measures, and includes an association between certain variants of *PER1*, *PER2*, *PER3* and *Clock* [120, 161–163]. This suggests a connection between proper mood regulation and normal functioning circadian molecular machinery.

Together with the circadian system, depressed patients may also suffer a deficiency in the homeostatic buildup of sleep pressure during extended wakefulness. It is unclear whether alterations in sleep regulatory mechanisms are associated with the pathophysiology of depression, and it might be that the homeostatic facet of sleep remains functional in this clinical condition [164]. Nevertheless, in some types of depressions, such as seasonal affective disorder (SAD), patients do appear to have changes in the buildup of homeostatic sleep pressure during wakefulness, irrespective of clinical state or season [114]. It is possible that SAD patients may have a trait – rather than state – deficiency in the homeostatic build up of sleep pressure during an extended wakefulness, as indexed by subjective sleepiness and EEG theta-alpha activity.

9.6 Concluding Remarks

Day-to-day rhythms in sleep are underpinned by the interplay of the circadian clock and the sleep homeostat. Both these internal oscillators contribute in an independent and non-additive manner to numerous dimensions of sleep, such as sleep timing and duration, REM sleep, NREM sleep, and so forth. The relative contribution of these two systems cannot be predicted by either oscillator independently. Instead, they decisively depend on their phase relationship and amplitude. In other words, the amplitude of the most circadian variations in sleep crucially relies on the sleep homeostat, and vice-versa. Therefore, it is this opposing interaction between the circadian and homeostatic changes in sleep propensity that consolidates sleep. The implications of disruption in these two systems can be clearly evidenced in psychiatric and neurological disorders that are intertwined to circadian rhythms and to the sleep homeostat on a wide array of dimensions. Knowledge on these systems may help to unravel the pathophysiology of these disorders and offer potential therapeutic strategies. The theoretical framework for both processes is under intense investigation. While several neuroanatomical pathways have been described for the circadian timing system, the mechanisms responsible for the generation of the homeostatic sleep process are still under thorough debate. New data on the genetical and molecular expression, like clock gene polymorphisms, may shed new light on the inter-individual differences associated with sleep architecture, timing and duration. This promises a fascinating future for the circadian and homeostatic sleep research.

References

1. Dijk DJ, Von Schantz M (2005) Timing and consolidation of human sleep, wakefulness, and performance by a symphony of oscillators. J Biol Rhythms 20:279–290
2. Cajochen C, Dijk DJ (2003) Electroencephalographic activity during wakefulness, rapid eye movement and non-rapid eye movement sleep in humans: Comparison of their circadian and homeostatic modulation. Sleep Biol Rhythms 1:85–95
3. Moore RY, Speh JC, Leak RK (2002) Suprachiasmatic nucleus organization. Cell Tissue Res 309:89–98
4. Czeisler CA, Duffy JF, Shanahan TL, Brown EN, Mitchell JF, Rimmer DW, Ronda JM, Silva EJ, Allan JS, Emens JS, Dijk DJ, Kronauer RE (1999) Stability, precision, and near-24-hour period of the human circadian pacemaker. Science 284:2177–2181
5. Gachon F, Nagoshi E, Brown SA, Ripperger J, Schibler U (2004) The mammalian circadian timing system: from gene expression to physiology. Chromosoma 113:103–112
6. Schibler U, Sassone-Corsi P (2002) A web of circadian pacemakers. Cell 111:919–922
7. Hastings MH, Herzog ED (2004) Clock genes, oscillators, and cellular networks in the suprachiasmatic nuclei. J Biol Rhythms 19:400–413
8. Moore RY (1997) Circadian rhythms: basic neurobiology and clinical applications. Annu Rev Med 48:253–266
9. Borbély AA (1982) A two process model of sleep regulation. Hum Neurobiol 1:195–204
10. Tononi G, Cirelli C (2006) Sleep function and synaptic homeostasis. Sleep Med Rev 10:49–62
11. Tononi G, Cirelli C (2003) Sleep and synaptic homeostasis: a hypothesis. Brain Res Bull 62:143–150
12. Cajochen C, Wyatt JK, Czeisler CA, Dijk DJ (2002) Separation of circadian and wake duration-dependent modulation of EEG activation during wakefulness. Neuroscience 114:1047–1060
13. Dijk DJ (1999) Reply to technical note: nonlinear interactions between circadian and homeostatic processes: models or metrics? J Biol Rhythms 14:604–605
14. Czeisler CA, Buxton OM, Khalsa SBS (2005) The human circadian timing system and sleep–wake regulation. In: Kryger MH, Roth T, Dement W (eds) Principles and practice of sleep medicine, 4th edn. Philadelphia, Elsevier, pp 375–394
15. Reppert SM, Weaver DR (2002) Coordination of circadian timing in mammals. Nature 418:935–941
16. Moore RY (1983) Organization and function of a central nervous system circadian oscillator: the suprachiasmatic hypothalamic nucleus. Fed Proc 42:2783–2789
17. Klein DC, Moore RY, Reppert SM (1991) Suprachiasmatic nucleus: the mind's clock. Oxford University Press, New York

18. Berson DM, Dunn FA, Takao M (2002) Phototransduction by retinal ganglion cells that set the circadian clock. Science 295:1070–1073
19. Provencio I, Rodriguez IR, Jiang G, Hayes WP, Moreira EF, Rollag MD (2000) A novel human opsin in the inner retina. J Neurosci 20:600–605
20. Wirz-Justice A (2007) How to measure circadian rhythms in human. Medicographia 29:84–90
21. Lockley SW, Skene DJ, James K, Thapan K, Wright J, Arendt J (2000) Melatonin administration can entrain the free-running circadian system of blind subjects. J Endocrinol 164:R1–R6
22. Berson DM (2003) Strange vision: ganglion cells as circadian photoreceptors. Trends Neurosci 26:314–320
23. Provencio I, Rollag MD, Castrucci AM (2002) Photoreceptive net in the mammalian retina. Nature 415:493–494
24. Moore RY, Lenn NJ (1972) A retinohypothalamic projection in the rat. J Comp Neurol 146:1–9
25. Meyer-Bernstein EL, Morin LP (1996) Differential serotonergic innervation of the suprachiasmatic nucleus and the intergeniculate leaflet and its role in circadian rhythm modulation. J Neurosci 16:2097–2111
26. Cohen RA, Albers HE (1991) Disruption of human circadian and cognitive regulation following a discrete hypothalamic lesion: a case study. Neurology 41:726–729
27. Inouye ST, Kawamura H (1979) Persistence of circadian rhythmicity in a mamalian hypothalamic "island" containing the suprachiasmatic nucleus. Proc Natl Acad Sci U S A 76:5962–5966
28. Moore R (2007) Suprachiasmatic nucleus in sleep–wake regulation. Sleep Med 8:27–33
29. Saper CB, Chou TC, Scammell TE (2001) The sleep switch: hypothalamic control of sleep and wakefulness. Trends Neurosci 24:726–731
30. Mc Carley RW (2007) Neurobiology of REM and NREM sleep. Sleep Med 8:302–330
31. Llinas R, Steriade M (2006) Bursting of thalamic neurons and states of vigilance. J Neurophysiol 95:3297–3308
32. Abrahamson EE, Leak RK, Moore RY (2001) The suprachiasmatic nucleus projects to posterior hypothalamic arousal systems. Neuroreport 12:435–440
33. Deboer T, Overeem S, Visser NAH, Duindam H, Frölich M, Lammers GJ, Meijer JH (2004) Convergence of circadian and sleep regulatory mechanisms on hypocretin-1. Neuroscience 129:727–732
34. Yoshida Y, Fujiki N, Nakajima T, Ripley B, Matsumura H, Yoneda H, Mignot E, Nishino S (2001) Fluctuation of extracellular hypocretin-1 (orexin A) levels in the rat in relation to the light–dark cycle and sleep–wake activities. Eur J Neurosci 14:1075–1081
35. Zeitzer JM, Buckmaster CL, Parker KJ, Hauck CM, Lyons DM, Mignot E (2003) Circadian and homeostatic regulation of hypocretin in a primate model: implications for the consolidation of wakefulness. J Neurosci 23:3555–3560
36. Brisbare-Roch C, Dingemanse J, Koberstein R, Hoever P, Aissaoui H, Flores S (2007) Promotion of sleep by targeting the orexin system in rats, dogs and humans. Nature 12:150–155
37. Mistlberger RE (2005) Circadian regulation of sleep in mammals: role of the suprachiasmatic nucleus. Brain Res Rev 49:429–454
38. Moore RY (1996) Neural control of the pineal gland. Behav Brain Res 73:125–130
39. Czeisler CA, Gooley J (2007) Sleep and circadian rhythms in humans. Cold Spring Harb Symp Quant Biol 72:579–597
40. Cajochen C, Münch M, Knoblauch V, Blatter K, Wirz-Justice A (2006) Age-related changes in the circadian and homeostatic regulation of human sleep. Chronobiol Int 23:1–14
41. Dibner C, Sage D, Unser M, Bauer C, d'Esmond T, Naef F, Schibler U (2008) Circadian gene expression is resilient to large fluctuations in overall transcription rates. EMBO J. 21;28(2):123–134
42. Schibler U, Ripperger J, Brown SA (2003) Peripheral circadian oscillators in mammals: time and food. J Biol Rhythms 18:250–260
43. Mc Carley RW, Massaquoi SG (1992) Neurobiological structure of the revised limit cycle reciprocal interaction model of REM cycle control. J Sleep Res 1:132–137

44. Saper CB, Cano G, Scammell TE (2005) Homeostatic, circadian, and emotional regulation of sleep. J Comp Neurol 493:92–98
45. Saper CB, Scammell TE, Lu J (2005) Hypothalamic regulation of sleep and circadian rhythms. Nature 437:1257–1263
46. Chou TC, Scammell TE, Gooley JJ, Gaus SE, Saper CB, Lu J (2003) Critical role of dorsomedial hypothalamic nucleus in a wide range of behavioral circadian rhythms. J Neurosci 23:10691–10702
47. Dijk DJ, Duffy JF, Czeisler CA (1992) Circadian and sleep/wake dependent aspects of subjective alertness and cognitive performance. J Sleep Res 1:112–117
48. Dijk DJ (1995) EEG slow waves and sleep spindles: windows on the sleeping brain. Behav Brain Res 69:109–116
49. Jewett ME, Rimmer DW, Duffy JF, Klerman EB, Kronauer RE, Czeisler CA (1997) Human circadian pacemaker is sensitive to light throughout subjective day without evidence of transients. Am J Physiol Regul Integr Comp Physiol 273:R1800–R1809
50. Van Dongen HPA, Maislin G, Mullington JM, Dinges DF (2003) The cumulative cost of additional wakefulness: dose-response effects on neurobehavioral functions and sleep physiology from chronic sleep restriction and total sleep deprivation. Sleep 26:117–126
51. Lockley SW, Cronin JW, Evans EE, Cade BE, Lee CJ, Landrigan CP, Rothschild JM, Katz JT, Lilly CM, Stone PH, Aeschbach D, Czeisler CA, the Harvard Work Hours, H.A.S.G. (2004) Effect of reducing interns' weekly work hours on sleep and attentional failures. N Engl J Med 351:1829–1837
52. Brunner DP, Dijk DJ, Borbély AA (1993) Repeated partial sleep deprivation progressively changes the EEG during sleep and wakefulness. Sleep 16:100–113
53. Dijk DJ, Hayes B, Czeisler CA (1993) Dynamics of electroencephalographic sleep spindles and slow wave activity in men: effect of sleep deprivation. Brain Res 626:190–199
54. Cajochen C, Foy R, Dijk DJ (1999) Frontal predominance of a relative increase in sleep delta and theta EEG activity after sleep loss in humans. Sleep Res Online 2:65–69
55. Borbély AA, Baumann F, Brandeis D, Strauch I, Lehmann D (1981) Sleep deprivation: effect on sleep stages and EEG power density in man. Electroencephalogr Clin Neurophysiol 51:483–495
56. Cajochen C, Knoblauch V, Kräuchi K, Renz C, Wirz-Justice A (2001) Dynamics of frontal EEG activity, sleepiness and body temperature under high and low sleep pressure. Neuroreport 12:2277–2281
57. Werth E, Dijk DJ, Achermann P, Borbély AA (1996) Dynamics of the sleep EEG after an early evening nap: experimental data and simulations. Am J Physiol Regul Integr Comp Physiol 271:501–510
58. Werth E, Achermann P, Borbély AA (1996) Brain topography of the human sleep EEG: Antero-posterior shifts of spectral power. Neuroreport 8:123–127
59. Finelli LA, Baumann H, Borbély AA, Achermann P (2000) Dual electroencephalogram markers of human sleep homeostasis: correlation between theta activity in waking and slow-wave activity in sleep. Neuroscience 101:523–529
60. Knoblauch V, Kräuchi K, Renz C, Wirz-Justice A, Cajochen C (2002) Homeostatic control of slow-wave and spindle frequency activity during human sleep: effect of differential sleep pressure and brain topography. Cereb Cortex 12:1092–1100
61. Borbély AA, Achermann P (2000) Sleep homeostasis and models of sleep regulation. In: Kryger MH, Roth T, Dement WC (eds) Principles and practice of sleep medicine. W B Saunders Company, Philadelphia, pp 377–390
62. Rétey JV, Adam M, Honegger E, Khatami R, Luhmann UF, Jung HH, Berger W, Landolt HP (2005) A functional genetic variation of adenosine deaminase affects the duration and intensity of deep sleep in humans. Proc Natl Acad Sci U S A 102:15676–15681
63. Rétey JV, Adam M, Gottselig JM, Khatami R, Dürr R, Achermann P, Landolt HP (2006) Adenosinergic mechanisms contribute to individual differences in sleep deprivation-induced changes in neurobehavioral function and brain rhythmic activity. J Neurosci 26:10472–10479
64. Landolt HP (2008) Sleep homeostasis: A role for adenosine in humans? Biochem Pharmacol 75:2070–2079

65. Huston JP, Haas HL, Boix F, Pfister MUD, Schrader J (1996) Extracellular adenosine levels in neostriatum and hippocampus during rest and activity periods of rats. Neuroscience 73:99–107
66. Murillo-Rodriguez E, Blanco-Centurion C, Gerashchenko D, Salin-Pascual RJ, Shiromani P (2004) The diurnal rhythm of forebrain of young and adenosine levels in the basal old rats. Neuroscience 123:361–370
67. Strecker RE, Morairty S, Thakkar MM, Porkka-Heiskanen T, Basheer R, Dauphin LJ (2000) Adenosinergic modulation of basal forebrain and preoptic/anterior hypothalamic neuronal activity in the control of behavioral state. Behav Brain Res 115:183–204
68. Basheer R, Strecker RE, Thakkar MM, McCarley RW (2004) Adenosine and sleep–wake regulation. Prog Neurobiol 73:379–396
69. Chamberlin NL, Arrigoni E, Chou TC, Scammell TE, Greene RW, Saper CB (2003) Effects of adenosine on GABAergic synaptic inputs to identified ventrolateral preoptic neurons. Neuroscience 119:913–918
70. Morairty S, Rainnie D, McCarley R, Greene RW (2004) Disinhibition of ventrolateral preoptic area sleep-active neurons by adenosine: a new mechanism for sleep promotion. Neuroscience 123:451–457
71. Timofeev I, Grenier F, Steriade M (2001) Disfacilitation and active inhibition in the neocortex during the natural sleep–wake cycle: an intracellular study. Proc Natl Acad Sci U S A 98:1924–1929
72. Maret S, Dorsaz S, Gurcel L, Pradervand S, Petit B, Pfister C, Hagenbuchle O, O'Hara BF, Franken P, Tafti M (2007) Homer1a is a core brain molecular correlate of sleep loss. Proc Natl Acad Sci U S A 104:20090–20095
73. Compte A, Sanchez-Vives MV, McCormick DA, Wang XJ (2003) Cellular and network mechanisms of slow oscillatory activity (<1 Hz) and wave propagations in a cortical network model. J Neurophysiol 89:2707–2725
74. Bazhenov M, Timofeev I, Steriade M, Sejnowski TJ (2002) Model of thalamocortical slow-wave sleep oscillations and transitions to activated states. J Neurosci 22:8691–8704
75. Hill S, Tononi G (2005) Modeling sleep and wakefulness in the thalamocortical system. J Neurophysiol 93:1671–1698
76. Massimini M, Huber R, Ferrarelli F, Hill S, Tononi G (2004) The sleep slow oscillation as a travelling wave. J Neurosci 24:6862–6870
77. Cajochen C, Brunner DP, Kräuchi K, Graw P, Wirz-Justice A (1995) Power density in theta/alpha frequencies of the waking EEG progressively increases during sustained wakefulness. Sleep 18:890–894
78. Cajochen C, Kräuchi K, Knoblauch V, Renz C, Rössler A, Balestrieri G, Dattler MF, Graw P, Wirz-Justice A (2001) Dynamics of frontal low EEG-activity, subjective sleepiness and body temperature under high and low sleep pressure. Neuroreport 20;12(10):2277–2281
79. Aeschbach D, Matthews JR, Postolache TT, Jackson MA, Giesen HA, Wehr TA (1997) Dynamics of the human EEG during prolonged wakefulness: evidence for frequency-specific circadian and homeostatic influences. Neurosci Lett 239:121–124
80. Dijk DJ, Beersma DGM, Daan S (1987) EEG power density during nap sleep: reflection of an hourglass measuring the duration of prior wakefulness. J Biol Rhythms 2:207–219
81. Cirelli C, Tononi G (2000) Differential expression of plasticity-related genes in waking and sleep and their regulation by the noradrenergic system. J Neurosci 20:9187–9194
82. Cirelli C, Pompeiano M, Tononi G (1996) Neuronal gene expression in the waking state: a role for the locus coeruleus. Science 274:1211–1215
83. Kopp C, Longordo F, Nicholson JR, Lüthi A (2006) Insufficient sleep reversibly alters bidirectional synaptic plasticity and NMDA receptor function. J Neurosci 26:12456–12465
84. Krueger JM, Obàl F (1993) A neuronal group theory of sleep function. J Sleep Res 2:63–69
85. Krueger JM, Obàl F Jr, Kapàs L, Fang J (1995) Brain organization and sleep function. Behav Brain Res 69:177–185
86. Vyazovskiy VV, Cirelli C, Pfister-Genskow M, Faraguna U, Tononi G (2008) Molecular and electrophysiological evidence for net synaptic potentiation in wake and depression in sleep. Nat Neurosci 11:200–208

87. Vyazovskiy VV, Riedner BA, Cirelli C, Tononi G (2007) Sleep homeostasis and cortical synchronization: II. A local field potential study of sleep slow waves in the rat. Sleep 30:1631–1642
88. Grutzendler J, Kasthuri N, Gan WB (2002) Long-term dendritic spine stability in the adult cortex. Nature 420:812–816
89. Dijk DJ, Edgar DM (1999) Circadian and homeostatic control of wakefulness and sleep. In: Turek FW, Zee PC (eds) Regulation of sleep and circadian rhythms, vol 133. Marcel Dekker, Inc, New York Basel, pp 111–147
90. Dijk DJ, Czeisler CA (1995) Contribution of the circadian pacemaker and the sleep homeostat to sleep propensity, sleep structure, electroencephalographic slow waves, and sleep spindle activity in humans. J Neurosci 15:3526–3538
91. Kronauer RE, Czeisler CA, Pilato SF, Moore-Ede MC, Weitzman ED (1982) Mathematical model of the human circadian system with two interacting oscillators. Am J Physiol Regul Integr Comp Physiol 242:R3–R17
92. Daan S, Beersma DGM, Borbély AA (1984) Timing of human sleep: recovery process gated by a circadian pacemaker. Am J Physiol Regul Integr Comp Physiol 246:R161–R178
93. Edgar DM, Dement WC, Fuller CA (1993) Effect of SCN lesions on sleep in squirrel monkeys: evidence for opponent processes in sleep–wake regulation. J Neurosci 13:1065–1079
94. Cajochen C, Khalsa SBS, Wyatt JK, Czeisler CA, Dijk DJ (1999) EEG and ocular correlates of circadian melatonin phase and human performance decrements during sleep loss. Am J Physiol Regul Integr Comp Physiol 277:R640–R649
95. Johnson MP, Duffy JF, Dijk DJ, Ronda JM, Dyal CM, Czeisler CA (1992) Short-term memory, alertness and performance: a reappraisal of their relationship to body temperature. J Sleep Res 1:24–29
96. Dijk DJ, Czeisler CA (1994) Paradoxical timing of the circadian rhythm of sleep propensity serves to consolidate sleep and wakefulness in humans. Neurosci Lett 166:63–68
97. Liu C, Weaver D, Jin X, Shearman L, SM R (1997) Molecular dissection of two distinct actions of melatonin on the suprachiasmatic circadian clock. Neuron 19:91
98. Barinaga M (1997) How jet-lag hormone does double duty in the brain. Science 277:480
99. Czeisler CA, Zimmerman JC, Ronda JM, Moore-Ede MC, Weitzman ED (1980) Timing of REM sleep is coupled to the circadian rhythm of body temperature in man. Sleep 2:329–346
100. Tobler I, Borbély AA, Groos G (1983) The effect of sleep deprivation on sleep in rats with suprachiasmatic lesions. Neurosci Lett 21:49–54
101. Achermann P (1999) Technical note: a problem with identifying nonlinear interactions of circadian and homeostatic processes. J Biol Rhythms 14:602–603
102. Deboer T, Vansteensel MJ, Détari L, Meijer JH (2003) Sleep states alter activity of suprachiasmatic nucleus neurons. Nat Neurosci 6:1086–1090
103. Zulley J, Wever R, Aschoff J (1981) The dependence of onset and duration of sleep on the circadian rhythm of rectal temperature. Pflügers Arch 391:314–318
104. Aschoff J, Von Goetz C, Wildgruber C, Wever RA (1986) Meal timing in humans during isolation without time cues. J Biol Rhythms 1:151–162
105. Aschoff J, Gerecke U, Wever R (1967) Desynchronization of human circadian rhythms. Jpn J Physiol 17:450–457
106. Zulley J (1980) Distribution of REM sleep in entrained 24 hour and free-running sleep–wake cycles. Sleep 2:377–389
107. Zimmerman JC, Czeisler CA, Laxminarayan S, Knauer RS, Weitzman ED (1980) REM density is dissociated from REM sleep timing during free-running sleep episodes. Sleep 2:409–415
108. Weitzman ED, Czeisler CA, Zimmermann JC, Ronda JM (1980) Timing of REM and stages 3+4 sleep during temporal isolation in man. Sleep 2:391–407
109. Hiddinga AE, Beersma DGM, Van Den Hoofdakker RH (1997) Endogenous and exogenous components in the circadian variation of core body temperature in humans. J Sleep Res 6:156–163

110. Wyatt JK, Ritz-De Cecco A, Czeisler CA, Dijk DJ (1999) Circadian temperature and melatonin rhythms, sleep, and neurobehavioral function in humans living on a 20-h day. Am J Physiol Regul Integr Comp Physiol 277:R1152–R1163
111. Hull JT, Wright KP Jr, Czeisler CA (2003) The influence of subjective alertness and motivation on human performance independent of circadian and homeostatic regulation. J Biol Rhythms 18:329–338
112. Dijk DJ, Shanahan TL, Duffy JF, Ronda JM, Czeisler CA (1997) Variation of electroencephalographic activity during non-rapid eye movement and rapid eye movement sleep with phase of circadian melatonin rhythm in humans. J Physiol 505:851–858
113. Wyatt J, Dijk D, Ritz-De Cecco A, Ronda J, Czeisler C (2006) Sleep facilitating effect of exogenous melatonin in healthy young men and women is circadian-phase dependent. Sleep 29:609–610
114. Cajochen C, Brunner DP, Kräuchi K, Graw P, Wirz-Justice A (2000) EEG and subjective sleepiness during extended wakefulness in seasonal affective disorder: circadian and homeostatic influences. Biol Psychol 47:610–617
115. Steriade M, McCormick DA, Sejnowski TJ (1993) Thalamocortical oscillations in the sleeping and aroused brain. Science 262:679–685
116. Amzica F, Steriade M (1998) Electrophysiological correlates of sleep delta waves. Electroencephalogr Clin Neurophysiol 107:69–83
117. Siapas AG, Wilson MA (1998) Coordinated interactions between hippocampal ripples and cortical spindles during slow-wave sleep. Neuron 21:1123–1128
118. Schibler U (2005) The daily rhythms of genes, cells and organs. Biological clocks and circadian timing in cells. EMBO J 6:9–13
119. Schibler U, Naef F (2005) Cellular oscillators: rhythmic gene expression and metabolism. Curr Opin Cell Biol 17:223–229
120. Ebisawa T, Uchiyama M, Kajimura N, Mishima K, Kamei Y, Katoh M, Watanabe T, Sekimoto M, Shibui K, Kim K, Kudo Y, Ozeki Y, Sugishita M, Toyoshima R, Inoue Y, Yamada N, Nagase T, Ozaki N, Ohara O, Ishida N, Okawa M, Takahashi K, Yamauchi T (2001) Association of structural polymorphisms in the human period 3 gene with delayed sleep phase syndrome. EMBO report 2:342–346
121. Naylor E, Bergmann BM, Krauski K, Zee PC, Takahashi JS, Vitaterna MH, Turek FW (2000) The circadian clock mutation after sleep homeostasis in the mouse. J Neurosci 20:8138–8143
122. Antoch M, Song E, Chang A, Vitaterna M, Zhao Y, Wilsbacher L, Sangoram A, King D, Pinto L, Takahashi J (1997) Functional identification of the mouse circadian Clock gene by transgenic BAC rescue. Cell 89:655–667
123. Kopp C, Albrecht U, Zheng B, Tobler I (2002) Homeostatic sleep regulation is preserved in mPer1and mPer2 mutant mice. Eur J Neurosci 16:1099–1106
124. Franken P, Thomason R, Heller HC, O'Hara BF (2007) A non-circadian role for clock-genes in sleep homeostasis: a strain comparison. BMC Neurosci 8:87–97
125. Boivin DB, James FO, Wu A, Cho-Park F, Xiong H, Sun ZS (2003) Circadian clock genes oscillate in human peripheral blood mononuclear cells. Blood 102:4143–4145
126. Archer SN, Viola AU, Kyriakopoulou V, von Schantz M, Dijk D (2008) Inter-Individual differences in habitual sleep timing and entrained phase of endogenous circadian rhythms of BMAL1, PER2 and PER3 mRNA in human leucocytes. Sleep 5:608–617
127. Viola AU, Archer SN, James LM, Groeger JA, Lo JCY, Skene DJ, Von Schantz M, Dijk DJ (2007) PER3 polymorphism predicts sleep structure and waking performance. Curr Biol 17:613–618
128. Dijk DJ, Franken P (2005) Interaction of sleep homeostasis and circadian rhythmicity: Dependent or independent systems? In: Kryger MH, Roth T, Dement WC (eds) Principles and practice of sleep medicine, 4th edn. Elsevier, Philadelphia, pp 418–434
129. Mecacci L, Zani A (1983) Morningness-eveningness preferences and sleep-waking diary data of morning and evening types in student and worker samples. Ergonomics 26:1147–1153
130. Duffy JF, Dijk DJ, Hall E, Czeisler CA (1999) Relationship of endogenous circadian melatonin and temperature rhythms to self-reported preference for morning or evening activity in young and older people. J Invest Med 47:141–150

131. Mongrain V, Lavoie S, Selmaoui B, Paquet J, Dumont M (2004) Phase relationships between sleep – wake cycle and underlying circadian rhythms in morningness – eveningness. J Biol Rhythms 19:248–257
132. Dijk DJ, Lockley SW (2002) Integration of human sleep–wake regulation and circadian rhythmicity. J Appl Psysiol 92:852–862
133. Mongrain V, Carrier J, Dumont M (2004) Circadian and homeostatic sleep regulation in morningness-eveningness. J Sleep Res 15:162–166
134. Mongrain V, Carrier J, Dumont M (2006) Difference in sleep regulation between morning and evening circadian types as indexed by antero-posterior analyses of the sleep EEG. Eur J Neurosci 23:497–504
135. Schmidt C, Collette F, Leclercq Y, Sterpenich V, Vandewalle G, Berthomier P, Berthomier C, Phillips C, Tinguely G, Darsaud A, Gais S, Schabus M, Desseilles M, Dang-Vu T.T, Salmon E, Balteau E, Dequeldre C, Luxen A, Maquet P, Cajochen C, Peigneux P (2009) Homeostatic sleep pressure and responses to sustained attention in the suprachiasmatic area. Science, 324(5926):516–519
136. Aeschbach D, Cajochen C, Landolt H, Borbély AA (1996) Homeostatic sleep regulation in habitual short sleepers and long sleepers. Am J Physiol Regul Integr Comp Physiol 270:R41–R53
137. Aeschbach D, Sher L, TT P (2003) A longer biological night in long sleepers than in short sleepers. Am J Physiol Endocrinol Metab 88:26–30
138. Buysse DJ, Browman KE, Monk TH, Reynolds CF III, Fasiczka AL, Kupfer DJ (1992) Napping and 24 – hour sleep/wake patterns in healthy elderly and young adults. J Am Geriatr Soc 40:779–786
139. Bliwise DL (1993) Sleep in normal aging and dementia. Sleep 16:40–81
140. Czeisler CA, Dumont M, Duffy JF, Steinberg JD, Richardson GS, Brown EN, Sànchez R, Rios CD, Ronda JM (1992) Association of sleep–wake habits in older people with changes in output of circadian pacemaker. Lancet 340:933–936
141. Duffy JF, Dijk DJ, Klerman BE, Czeisler CA (1998) Later endogenous circadian temperature nadir relative to an earlier wake time in older people. Am J Physiol Regul Integr Comp Physiol 275:R1478–R1487
142. Duffy JF, Czeisler CA (2002) Age-related change in the relationship between circadian period, circadian phase, and diurnal preference in humans. Neurosci Lett 318:117–120
143. Münch M, Knoblauch V, Blatter K, Schröder C, Schnitzler C, Kräuchi K, Wirz-Justice A, Cajochen C (2004) The frontal predominance in human EEG delta activity after sleep loss decreases with age. Eur J Neurosci 20:1402–1410
144. Dijk DJ, Duffy JF, Riel E, Shanahan TL, Czeisler CA (1999) Ageing and the circadian and homeostatic regulation of human sleep during forced desynchrony of rest, melatonin and temperature rhythms. J Physiol 516:611–627
145. Münch M, Knoblauch V, Blatter K, Schroder C, Schnitzler C, Krauchi K, Wirz-Justice A, Cajochen C (2005) Age-related attenuation of the evening circadian arousal signal in humans. Neurobiol Aging 26:1307–1319
146. Münch M, Knoblauch V, Blatter K, Wirz-Justice A, Cajochen C (2007) Is homeostatic sleep regulation under low sleep pressure modified by age? Sleep 30:781–792
147. Wirz-Justice A (2003) Chronobiology and mood disorders. Dialogues Clin Neurosci 5:223–233
148. Van Someren EJW (2000) Circadian and sleep disturbances in the elderly. Exp Gerontol 35:1229–1237
149. Riemersma-van der Lek RF, Swaab DF, Twisk J, Hol EM, Hoogendijk WJG, Van Someren E (2008) Effect of bright light and melatonin on cognitive and noncognitive function in elderly residents of group care facilities. JAMA 299:2642–2655
150. Van Someren EJW, Hagebeuk EEO, Lijzenga C, Scheltens P, De Rooij SEJA, Jonker C, Pot AM, Mirmiran M, Swaab DF (1996) Circadian rest-activity rhythm disturbances in Alzheimer's disease. Biol Psychiatry 40:259–270
151. Wu YH, Swaab DF (2006) Disturbance and strategies for reactivation of the circadian rhythm system in aging and Alzheimer's disease. Sleep Med 8:623–636

152. Ferrari E, Arcaini A, Gornati R, Pelanconi L, Cravello L, Fioravanti M, Solerte SB, Magri F (2000) Pineal and pituitary-adrenocortical function in physiological aging and in senile dementia. Exp Gerontol 35:1239–1250
153. Perlis ML, Giles DE, Buysse DJ, Thase ME, Tu X, Kupfer DJ (1997) Which depressive symptoms are related to which sleep electroencephalographic variables? Biol Psychiatry 42:904–913
154. Riemann D, Berger M, Voderholzer U (2001) Sleep and depression – results from psychobiological studies: an overview. Biol Psychol 57:67–103
155. Boivin DB (2000) Influence of sleep–wake and circadian rhythm disturbances in psychiatric disorders. J Psychiatry Neurosci 25:446–458
156. Lewy AJ, Lefler BJ, Emens JS, Bauer VK (2006) The circadian basis of winter depression. Proc Natl Acad Sci U S A 103:7414–7419
157. Chellappa SL, Araujo JF (2007) Sleep disorders and suicidal ideation in patients with depressive disorder. Psychiatr Res 153:131–136
158. Chellappa SL, Araújo JF (2006) Excessive daytime sleepiness in patients with depressive disorder. Rev Bras Psiquiatr 28:126–130
159. Fava M (2004) Daytime sleepiness and Insomnia as correlates of depression. J Clin Psychiatry 65:27–32
160. Drennan MD, Klauber MR, Kripke DF, Goyette LM (1991) The effects of depression and age on the Horne–Ostberg morningness-eveningness score. J Affect Disord 23:93–98
161. Johansson C, Willeit M, Smedh C, Ekholm J, Paunio T, Kieseppä T, Lichtermann D, Praschak-Rieder N, Neumeister A, Nilsson LG, Kasper S, Peltonen L, Adolfsson R, Schalling M, Partonen T (2003) Circadian clock-related polymorphisms in seasonal affective disorder and their relevance to diurnal preference. Neuropsychopharmacology 28:734–739
162. Katzenberg D, Young T, Finn L, Lin L, King DP, Takahashi J (1998) A CLOCK polymorphism associated with human diurnal preference. Sleep 21:569–576
163. McClung CA (2007) Circadian genes, rhythms and the biology of mood disorders. Pharmacol Thera 114:222–232
164. Landolt HP, Raimo EB, Schnierow BJ, Kelsoe JR, Rapaport MH, Gillin JC (2001) Sleep and sleep electroencephalogram in depressed patients treated with phenelzine. Arch Gen Psychiatry 58:268–276
165. Cajochen C (2007) Alerting effects of light. Sleep Med Rev 11:453–464
166. Cajochen C, Wyatt JK, Bonikowska M, Czeisler C, Dijk DJ (2000) Non-linear interaction between circadian and homeostatic modulation of slow eye movements during wakefulness in humans. J Sleep Res 9(S1):29

Chapter 10
Clocks, Brain Function, and Dysfunction

Céline Feillet and Urs Albrecht

10.1 Introduction

All living organisms on our planet experience daily and seasonal changes. The physiology of plants and animals is dynamically controlled to respond to these recurring fluctuations. Daily and seasonal modifications of the environmental conditions are predictable as long as the organism possesses a system allowing time tracking. Clocks provide an estimation of time and thus permit to anticipate environmental changes by preparing the organism through the initiation of cellular and physiological mechanisms before they are actually required. At the time those processes are needed, they are already running at full capacity.

Clocks represent a highly advantageous evolutionary feature and have been widely implicated in normal brain and physiological functions. They are connected to both afferent and efferent pathways, which allow them to "listen" to their environment and to "speak" to the rest of the organism. When the clock is impaired, it impacts on both the brain and peripheral organ physiology. In return, a dysfunction in physiology can dysregulate the clock, thus giving rise to secondary symptoms.

Clocks control the so-called "biological rhythms", which can be characterized by their period (time separating two identical events) and separated into three groups: infradian rhythms, occurring less than once a day (seasonal rhythms are an example), circadian rhythms (from the latin *circa*: about and *dies*: a day) showing periods of about 24 h and ultradian rhythms with periods shorter than 24 h (e.g., sleep stages, heart beats). Here, we will essentially focus on circadian rhythms in the brain of rodents and humans.

U. Albrecht (✉)
Department of Medicine, Unit of Biochemistry, University of Fribourg, Fribourg 1700, Switzerland
e-mail: urs.albrecht@unifr.ch

U. Albrecht (ed.), *The Circadian Clock*, Protein Reviews 12,
DOI 10.1007/978-1-4419-1262-6_10, © Springer Science+Business Media, LLC 2010

10.2 Circadian Rhythms in the Brain

10.2.1 The Suprachiasmatic Nuclei (SCN): The Main Circadian Clock

In the manner of many other brain structures, the location of the main circadian clock was identified using a lesional approach. In 1972, Moore & Eichler [1] and Stephan & Zucker [2] demonstrated that the lesion of a paired structure in the hypothalamus just above the optic chiasm eliminates the rhythmicity in corticosterone secretion and locomotor activity. The so-called suprachiasmatic nuclei (SCN) are composed of about 16,000 small neurons in rats, organized in a dense network [3]. Strikingly, Lehman and colleagues [4] restored a free-running locomotor activity in animals placed in DD by transplanting the fetal SCN tissue directly into the third ventricle. Additionally, the transplantation of fetal SCN tissue coming from mutant hamsters restored rhythmicity in the SCN-lesioned wild type animals with a free running period characteristic of the mutant donor [5]. Since these pioneer studies, the existence of circadian clocks controlling rhythmic physiology and behavior has been demonstrated in numerous mammalian species, including humans (see sect. 10.3).

10.2.2 Clock Gene Expression Throughout the Brain

In the SCN, selfsustained oscillatory mechanisms rely on clock genes and clock proteins. The interactions between them constitute an autoregulatory feedback loop (see Chap. 2 for details). Since 1997, many groups have demonstrated the rhythmic expression of both clock genes and clock proteins in the brain. Many brain structures showing oscillating gene expression have been identified. Are these oscillations driven by the SCN clock or are they generated by independent "clocks"? (for peripheral oscillators see Chap. 5).

Rhythmic expression of clock genes has been observed in numerous brain structures (reviewed in [6]; see also Fig. 10.1). For instance, rhythmic expression of *Per1*, *Per2*, and/or *Bmal1* mRNA is observed in forebrain structures such as olfactory bulbs, piriform and cerebral cortices, hippocampus, various hypothalamic nuclei, pituitary, pineal gland, Raphe nuclei, cerebellum, and the nucleus of the solitary tract [7–15]. Using brain structure explants from transgenic *Per1*-Luciferase rats, Abe and collaborators demonstrated that bioluminescence of brain tissue in vitro either dampened after a few cycles, or explants were immediately arrhythmic when isolated from the SCN [10]. Bioluminescence studies with cultured fibroblasts revealed that each cell is capable of self-sustained oscillations, but they tend to desynchronize over time and as a consequence overall rhythmicity is lost. The observed amplitude of the rhythm at the level of a cell culture thus decreases although each cell continues to cycle independently with its own phase [16, 17]. This feature could be an explanation why oscillations in isolated brain structures tend to dampen over time.

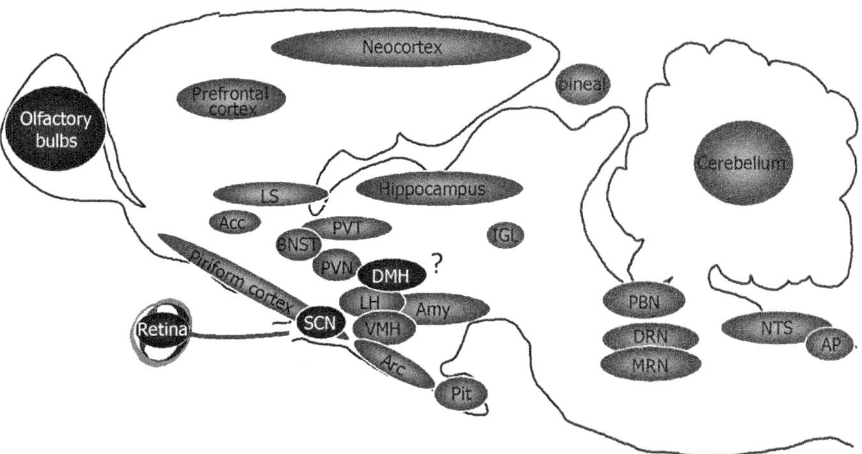

Fig. 10.1 Schematic representation of the localization of brain oscillators in the sagittal plane of a mouse brain. Structures in *light gray* are oscillators and structures in *dark gray* are known or possible brain clocks. *Acc* nucleus accumbens, *Amy* amygdala, *AP* area postrema, *Arc* arcuate nuclei, *BNST* bed nucleus of the stria terminalis, *DMH* dorsomedial hypothalamus, *DRN* dorsal raphe nuclei, *IGL* intergeniculate leaflets, *LH* lateral hypothalamus, *LS* lateral septum; *MRN* medial raphe nucleus, *NTS* nuclei of the tractus solitaris, *PBN* parabrachial nuclei, *Pit* pituitary, *PVN* paraventricular hypothalamic nuclei, *PVT* paraventricular thalamic nucleus, *SCN* suprachiasmatic nuclei, *VMH* ventromedial hypothalamus

For clock function, clock gene rhythmicity alone is not sufficient. Posttranscriptional and -translational modifications are essential to prevent the clock from getting into a steady state (see Chap. 3). Protein stability and cellular localization of clock proteins can be observed by immunohistochemistry. Oscillations of PER1 and/or PER2 proteins could be observed in numerous forebrain structures such as amygdala and bed nucleus of the stria terminalis, hippocampus, nucleus accumbens, lateral septum, and prefrontal cortex [18–20]. A study exploring the concomitant expression of both PER1 and PER2 in the forebrain of mice revealed that PER1 is detected alone in the DMH, anterior piriform cortex, and dorsal striatum, and concomitant with PER2 in the PVT, hippocampus central amygdala, and arcute nucleus. PER2 alone is detected in the PVN, basal amygdala, and VMH [21]. Consequently, it is possible that forebrain oscillators may not need all clock proteins to keep their clock going. This might also explain why they cannot sustain autonomous oscillations without the SCN. In addition, there may be oscillators ticking with different clock hands, depending on their function. This suggests that the PER proteins may have different functions in clocks in different brain nuclei or tissues. This observation already has a precedent as genes coding for nuclear orphan receptors (ROR) (see Chap. 2) are unevenly expressed in various peripheral tissues. In this context, the possibility of differential function for clock genes in various tissues has been suggested [22].

10.2.3 Requirements for Having a Clock

Rhythmic expression of clock genes or clock proteins apparently is not sufficient for defining a clock. Studying the SCN helped delineate the basics for having a clock: (1) the candidate structure (or network) should exhibit endogenous and self-sustained oscillations within the circadian range (i.e., 21–28 h) that persist in total isolation. (2) The oscillations should be temperature compensated, which means that the period of the oscillations should not be modified by variations in temperatures within a physiological range. (3) The structure should be able to communicate with the environment, i.e., receive inputs from the environment (via receptors or nervous pathways), be able to be synchronized by those cues, and send output signals to downstream targets to transmit a circadian message.

Any structure that has the machinery to generate autonomous oscillations for a few cycles only will be termed "semi-autonomous oscillator". If arrhythmic in the absence of a master clock, the structure will be a "slave oscillator". If selfsustained and independent from the environment, the structure will be called a "pacemaker". Only an entity that presents all criteria listed above will be termed a true "circadian clock".

10.3 True Brain Clocks

10.3.1 The SCN Clock as a Basis for Studying Brain Clocks

10.3.1.1 Receiving and Integrating Information from the Environment

For an integration center like the SCN, it is important to be able to stay in touch with both the outside world and the rest of the body; thus, the SCN receive multiple nervous afferents and express a variety of receptors connecting them to the sensory systems receiving environmental input. All cues from the environment can act as synchronizers, and influence the phase of the SCN. This in turn impacts on physiology and behavior.

Nervous Pathways

Based on the expression of neuropeptides, the SCN have been divided into a dorsomedial part (also called "shell", mainly expressing arginine vasopressine – AVP) and a ventrolateral part (also called "core", expressing essentially vasoactive intestinal peptide – VIP). The gastrin-releasing peptide (GRP)-expressing neurons are mostly localized in the central core as well [23, 24]. This "core" part of the SCN appears to integrate most afferents. Thus, VIP or "core" neurons receive three main nervous projections from the following tissues:

Retina: Perception of light is mainly achieved by a specific photopigment called melanopsin, which is expressed in specific ganglion cells of the retina [25–28] (see Chap. 4 for detailed information). The retinohypothalamic tract then conveys light information to the SCN as well as to the other brain areas such as the intergeniculate leaflets (IGL). These projections mainly release glutamate [29] and PACAP [30], and are the most important ones for synchronization to light.

The intergeniculate leaflets (IGL): Another afferent conveying light information to the SCN is an indirect pathway, passing through the IGL via the geniculohypothalamic tract. This pathway secretes Neuropeptide Y, enkephalin and GABA in the SCN core [31–36]. Knowing that the information arriving at both the SCN and the IGL emanate from the same groups of ganglion cells, it is possible that one of the roles of this pathways is to modulate photic information to the SCN. Note that the geniculohypothalamic tract can also convey nonphotic cues to the SCN [32].

Raphe nuclei: Serotonergic projection from the median raphe nucleus are the third direct entry to the SCN. Another projection derives from the dorsal raphe nuclei through the IGL. Application of agonists or chemical impairment of the serotonergic system affects the locomotor activity in both constant darkness (DD) and light/dark (LD) cycles [37].

Receptors

The SCN not only receive nervous projections but also express receptors that allow integration of humoral information from the rest of the body:

Melatonin receptors: The hormone melatonin is synthesized exclusively during the night by the pineal gland. This rhythm is under SCN control. Melatonin gives rise to cues encoding both the occurrence of the night and its length, thus playing a crucial role in seasonal rhythms. Characterization of melatonin receptors revealed 2 subtypes that have now been cloned: MT1 et MT2 [38, 39]. Their density and distribution varies among species, but they can often be seen in the SCN, suggesting that the clock can receive a melatonin feedback message capable of modulating the circadian system (see below).

Metabolic receptors: The clock can also integrate metabolic information from the periphery. Leptin is an adipose tissue-derived hormone that regulates body mass by lowering food intake and increasing energy expenditure. It acts through leptin receptors in the hypothalamus. An immunohistochemical study unravelled their presence in the SCN thus wiring the clock to adipose tissue [40].

In addition to leptin, the SCN also express ghrelin receptors: This hormone synthesized by the stomach mirrors the effect of leptin; it stimulates food intake and signals energy depletion. Ghrelin receptors have been identified by in situ hybridization in the whole hypothalamus [41]. Also insulin receptors have been detected in the SCN. It seems that binding of insulin to those receptors could modulate synaptic transmission and neurotransmitter release in the central nervous system [42].

VPAC2 receptors: VIP (vasoactive intestinal peptide) is a peptide released by the ventrolateral part of the SCN. The SCN itself possesses receptors for VIP.

The VPAC2 receptor in particular seems to play a role in circadian rhythms as VPAC2 antagonists block the electrical activity in the SCN [43]. A mutation of the Vipr2 gene (encoding for VPAC2 receptors) induces an arrhythmic locomotor activity in mice in both LD and DD conditions [43–45]. Nevertheless, real-time recording of *Per1* expression in individual SCN cells revealed that each cell does oscillate autonomously and independently of the others [46]. Thus, VPAC2 receptors are not directly important for synchronization to the environment, but for internal synchronization of SCN cells, which in turn stabilizes individual oscillators to produce a coherent self sustaining oscillator capable of transmitting circadian information.

10.3.1.2 Normal Functioning of the Clock

In this part, we will focus on biological rhythms from the behavioral point of view.

In the Absence of a Synchronizer

It is possible to create a controlled environment to study the clock functioning in isolation from the environment. Rodents are kept under constant temperature and lighting conditions (constant darkness in general). Food and water are maintained ad libitum. In this case, the animals stay rhythmic with a period that slightly differs from 24 h, called the free-running period. It is usually measured by recording locomotor activity, one of the major output of the SCN. The onset of activity marks the beginning of the subjective night (in nocturnal rodents) and defines circadian time 12 (CT12). CT0 is defined by default in antiphase to CT12 (Fig. 10.2).

Synchronization of the Clock to the Environment

Photic synchronization: Once the animal is exposed to a synchronizer which can influence the clock via nervous or metabolic pathways, rhythms are no longer free-running. For the SCN, light is the most powerful synchronizer. Exposing animals to light/dark (LD) cycles of 12 h light and 12 h dark changes their pattern of activity and physiology. Most laboratory rodents are active during the dark period of the day. In those conditions, times of the day are no longer expressed as CTs but as ZTs for zeitgeber time. A zeitgeber literally is a time giver ; under LD 12/12 conditions, ZT0 is the time of lights on and ZT12 the time of lights off (Fig. 10.2).

In addition to simple LD cycles, a single light pulse under DD conditions can shift the locomotor activity. A light pulse given around the time of activity onset elicits phase delays (i.e., activity starts later than expected the next day). If given around the end of the active phase, the light pulse gives rise to phase advances (i.e., activity starts earlier than expected) on the day following the light pulse. A light pulse delivered during the inactive phase has no effect [47] (Fig. 10.3).

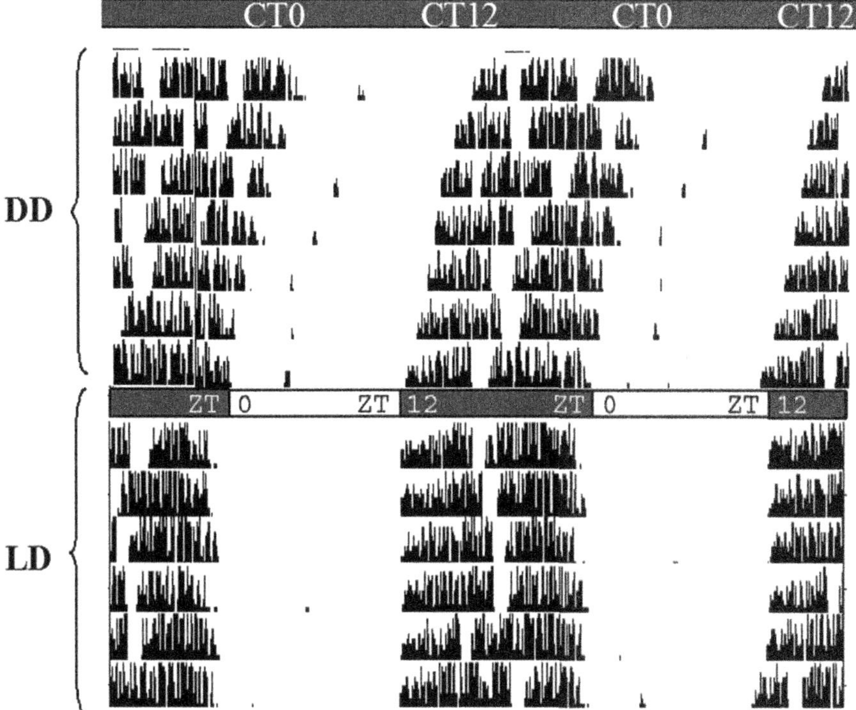

Fig. 10.2 Example of a double plotted actogram of a mouse. The mouse was placed in a cage with a running wheel. Wheel turns are recorded in 5-min blocks. The higher the number of wheel revolutions, the higher the black bar representing that block. The actogram has been double plotted, meaning two consecutive days are placed one after another both vertically and horizontally. In the first part of the actogram, the mouse is in constant darkness. Time is measured as circadian time (CT). Under these conditions, the mouse still shows a rhythm with a daily bout of locomotor activity (*black bars*) corresponding to the subjective night (CT12: beginning of activity). Its period of activity is inferior (but close) to 24 h. In the second part of the actogram, the animal is synchronized to the light/dark (LD) cycle. Under these conditions, time is expressed as zeitgeber time (ZT), with ZT0 = lights on and ZT12 = lights off. The period of activity is synchronized exactly to the zeitgeber and is equal to 24 h

When the LD cycle is shifted as a whole to mimic a jet lag such as the ones experienced during a transmeridian trip, for example, the locomotor activity observed is not directly synchronized to the new LD schedule. Transients, where the activity is progressively shifted, are observed instead and a few days are necessary before clock function is back to normal [48].

Nonphotic synchronization: Even if light is the most powerful synchronizer for the SCN, the other so-called nonphotic stimuli can influence and/or shift the SCN phase in rodents. These factors share the common feature not to be a light stimulation and usually induce phase advances of locomotor activity when applied during

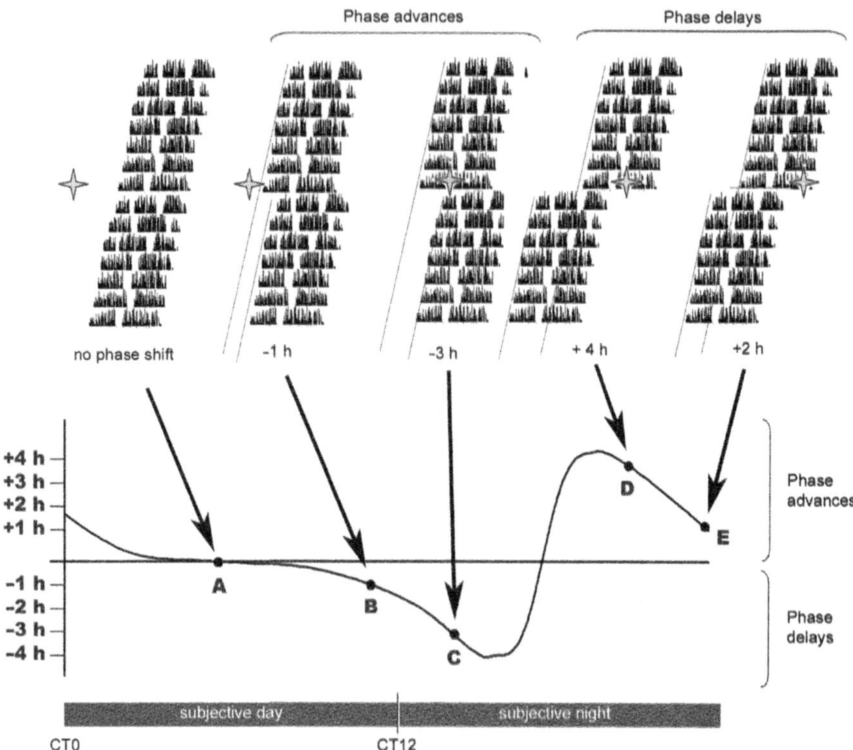

Fig. 10.3 Phase response curve corresponding to photic stimulation. Mice are placed in constant darkness. A light pulse (15 min) was applied at a certain circadian time (CT) (**A**) CT6, (**B**) CT11, (**C**) CT 14, (**D**) CT21, (**E**): CT24). (**A**) CT6: a light pulse during the subjective day does not elicit any modification of phase the next day; (**B**) CT11; (**C**) CT 14: a light pulse at the beginning of the subjective night provokes a phase delay the next day; (**D**) CT21, (**E**) CT24: a light pulse at the end of the subjective night provokes a phase advance the next day. (Adapted from [47])

the subjective day [49]. Most of the following results have been established in DD to eliminate the dominant influence of the LD cycle.

As mentioned above, the SCN express melatonin receptors and thus can receive information from this hormone. When administered at supra-physiological doses during the subjective day, melatonin can induce phase advances of locomotor activity [50]. Moreover, a daily subcutaneous infusion of melatonin in DD around the beginning of the subjective night can synchronize the SCN clock in nocturnal rodents [51, 52].

Other factors can induce a behavioral activation and thus are capable of modifying the period of the clock: Benzodiazepins such as Triazolam [53, 54] or morphine [55] can induce phase advances when injected during the subjective day.

In golden hamsters, it has been demonstrated that arousal and behavioral activation during the subjective day, elicited by a de novo access to a running wheel at an unexpected time (i.e., during the inactive phase of the animal), can also shift the

SCN clock [56, 57]. Note that except for melatonin, nonphotic effects on the clock are abolished if serotonergic inputs from the raphe nuclei to the SCN are destroyed [54, 58, 59]. Moreover, IGL lesions affect the phase changes induced by different nonphotic stimuli [58, 60, 61]. These results emphasize the importance of both IGL and raphe inputs to the SCN for normal integration of nonphotic messages from the environment.

The last nonphotic synchronizer we are going to talk about is food. It is crucial for an animal to be able to adapt to changes in food availability. In this respect, rodents are able to modify their pattern of activity in response to food access. Mice and rats for example spontaneously feed at night but if food access is restricted to the light period, they will become active during the day in the hours preceding the new food access, thus anticipating the new schedule. The so-called food anticipatory activity is a characteristic of changing food availability and is controlled by a clock situated outside the SCN (see sect. 10.3.4).

Even if restricting food access to the light part of the day greatly affects the peripheral tissues, the SCN are quite insensitive to this treatment [62, 63]. An experiment on hypothalamic slices in culture showed that a temporal restricted feeding schedule does not alter electrical activity in the SCN [64]. From a behavioral point of view, animals placed in DD or LL (constant light) and receiving food every day at the same time show both a free-running component and a food anticipatory activity [65]. Once again, the SCN cannot be synchronized by feeding schedules in normal conditions. This feature has nevertheless been questioned in some recent studies: in rats rendered arrhythmic by prolonged exposure to LL, rhythmicity can be restored by restricted feeding schedules, showing that the SCN can become sensitive to restricted feeding under some conditions [19]. Also, in some species of domestic mice, or in some strains like CS mice the SCN can be synchronized by food [66, 67]. In all those experiments, the animals have access to food at an unusual time of the day, but the amount of food is not limited in quantity.

If a temporal restriction is not capable of influencing the SCN, a calorie restriction on the contrary will have the same effect as other nonphotic factors: a daily ration of hypocaloric food will phase advance by 2 h the nocturnal peak of melatonin secretion, a well-known output of the SCN ([68, 69] but see also [70, 71]). Locomotor activity is also phase advanced and can start as soon as in the middle of the light phase (6 h phase advance; [59]). These results indicate that the SCN can be influenced by factors related to metabolism and feeding.

Limits of entrainment: The SCN have an endogenous period of about 24 h, that is different for each individual and expressed in constant conditions. A zeitgeber with a period T can synchronize the clock such that the rhythms controlled and expressed by the SCN will have a period of exactly T. This is possible only if T is close enough to the endogenous period. So, there are limits of entrainment for the SCN. In rats, for example, limits of entrainment to light have been defined by exposing the animals to days of increasing length (from 22 h: LD 11/11 corresponding to 11 h of light and 11 h of darkness, to 27 h: LD 13.5/13.5). Hence it was demonstrated that limits of entrainment were between 22.5 and 26 h in that species. When these limits are exceeded, behavioral rhythms neither have the same period as the synchronizer nor the free-running period [48].

10.3.1.3 The SCN: A Relay to Synchronize Physiology to the Environment

In 1972, Stephan and Zucker established that the SCN are necessary for expressing locomotor activity rhythms [2]. Unfortunately, a lesional approach alone could not account for the endogenous oscillatory nature of the SCN. It could have been "only" an obligatory relay in the control of rhythmic activities. In 1990, Ralph and collaborators demonstrated in arrhythmic SCN-X Syrian hamsters that transplanting fetal SCN into the third ventricle restores circadian behavioral rhythmicity, the period being imposed by the donor ([5]; Note however that grafts cannot restore rhythmicity in hormone secretion [72]). Consequently, the SCN really are the site of a true circadian clock [5, 73, 74]. Its numerous outputs impact on physiology in normal conditions.

Molecular Outputs (Clock-Controlled Genes=CCG)

AVP: Arginine-vasopressine (AVP) is a neuropeptide. Its mRNA expression has often been used as a molecular marker to determine the phase of the SCN output. It has an E-box in its promoter region, which means its transcription can be positively regulated by the CLOCK/BMAL1 hetero-dimer [75]. The precise role of AVP inside the SCN remains unclear. It is dispensable for expression of circadian rhythmicity within the SCN, but a selective elimination of AVP neurons renders rats arrhythmic in water intake. Thus, AVP appears to be a true molecular output of the clock [37]. AVP mRNA and protein cycle in LD as well as in DD, with a peak during the (subjective) day [75–79].

DBP: Albumin D-element binding protein (DBP) was first identified in the liver and other peripheral organs. It belongs to a family of transcription factors, which also comprises FOS and JUN [80]. DBP expression is rhythmic in the liver and most peripheral tissues in rats and mice. It oscillates with a high amplitude in the SCN with a maximum of expression at ZT5 [81]. Activation of DBP is achieved via E-boxes in its promoter, thus indicating that it belongs to the group of CCG [82]. Knock-out mice for *Dbp* show diminished locomotor activity. When released in DD, these mice are rhythmic and their free-running period is slightly shorter than that of WT mice [81].

Neurotransmitters: Most SCN neurons are GABAergic and present receptors for this neurotransmitter [33]. Moreover, SCN neurons releasing glutamate in the PVN have been identified [83].

Neuropeptides and SCN Regionalization

From a morphological point of view, the SCN can be subdivided in two regions: the dorsomedial SCN (dmSCN) expressing mainly AVP [84–86] and the ventrolateral SCN (vlSCN) synthesizing mainly VIP and GRP (see below). Other peptides will be mentioned briefly.

dmSCN: AVP-synthetizing neurons delimitate this region. They send projections to the vlSCN [87]. Applying AVP in vitro increases electrical activity in SCN neurons by 50% [88]. A vasopressinergic antagonist has the opposite effect [89].

vlSCN: Neurons in this region mainly synthetize VIP, Gastrin-releasing peptide (GRP) and peptide histidine isoleucine (PHI) [24, 90, 91]. RNA and protein levels of vasoactive intestinal peptide (VIP) show a daily rhythm peaking during the night in the SCN in LD conditions. Its expression is, however, constant in DD conditions. VIP is neither necessary for generating oscillations in the SCN nor for the transmission of rhythms to the outputs [37]. Recent studies established that VIP and its receptor (VIPR) rather act as internal synchronizers inside the SCN: Knock-out mice for VIPR become behaviorally arrhythmic although individual SCN neurons are still rhythmic [46]. This study shows that SCN neurons synchronize with each other via VIP communication.

Gastrin-releasing peptide (GRP) presents a daily diurnal peak in LD, in anti-phase to VIP. Since GRP neurons project outside the SCN, it is possible that they carry photic information to other brain areas [37].

Between these regions, a small population of neurons synthesizes somatostatin [91]. Somatostatin shows a rhythmic expression in both LD and DD in the SCN. Projections of somatostatin neurons are limited to the SCN area and their role is still unknown.

Other neuropeptides have been detected in rats such as angiotensin II [92–94], Calcitonin gene-related peptide [95], Galanin [96] and Substance P [36, 97]. Some other neuropeptides delineate subregions in the SCN, like Calbindin D28k expressing neurons in mice and hamsters [98, 99]. A description of all the neuropeptides expressed in the SCN, reveals a complex picture. For some of them like AVP or VIP, their role has been largely investigated and unraveled. Most of them, however, still have not revealed their secrets regarding their possible role in clock physiology.

Control of Locomotor Activity by Peptides

The SCN are responsible for generating and schedule processes at the physiological and behavioral levels. Among its most studied outputs, locomotor activity has seen increasing interest since it has been proposed that it could be controlled directly via factors secreted by the SCN.

Transforming growth factor (TGF): In 2001, Weitz's team identified TGFα (Transforming growth factor alpha) as a possible inhibiting factor of locomotor activity. It is rhythmically expressed in the SCN with a broad peak during the day. A TGFα infusion in the third ventricle reversibly inhibits the locomotor activity. This effect is mediated by the EGF receptor (epidermal growth factor) located in the subparaventricular zone of the hypothalamus. TGFα can then be implicated in the diurnal inhibition of locomotor activity by the SCN [100].

Prokineticin 2 (*PK2*): More recently, it has been shown that PK2 has the same effect as TGFα. Its peak of expression in the SCN occurs at the beginning of the day. PK2 receptors are mainly situated in SCN target structures. Once more, an intracerebroventricular administration of PK2 during the subjective night inhibits locomotor activity. Interestingly, PK2 might be a clock-controlled gene influencing at least in part the locomotor output [101].

Cardiotrophin-Like Cytokin (*CLC*): In 2006,Weitz's team published a new diffusible factor that is secreted by the SCN and capable of inhibiting diurnal locomotor activity: Cardiotrophin-Like Cytokin (CLC) seems to meet all conditions: It is expressed in the SCN in a subpopulation of AVP neurons. Its expression is rhythmic with a peak at the end of the day when nocturnal rodents are inactive. CLC receptors are located on the walls of the third ventricle and a CLC infusion transiently blocks the locomotor activity without influencing clock phase per se [102]. This experiment, however, does not exclude previous results. It is probable that the SCN control locomotor activity via multiple factors, and other peptides might still be discovered that can activate the locomotor activity during the night in nocturnal rodents.

Clock Genes and Behavior

Photic response: The SCN receive light information and thus it is not surprising that light has an effect on core clock mechanisms, and more precisely on clock genes. *Per1* [103], *Per2* [104] and *Dec1* [105] have been demonstrated to be light-inducible. *Per1* and *Dec1* are induced during the whole night [105], whereas *Per2* is strongly induced only at the beginning of the night [106, 107]. These genes would then be the main actors of phase shifts observed in response to light pulses.

Nonphotic response: MT1 and MT2 receptors are present in the SCN. In mice, ablating those receptors allowed the association of the MT1 receptor with a suppression of electrical activity in the SCN in response to melatonin application. MT2 receptors would be associated with phase shifts after melatonin application [108]. In normal conditions, melatonin exerts a retrocontrol on the clock. When applied at the day/night transition, it induces phase advances of electrical activity in the SCN in vitro [109–111]. However, its action on the SCN at the molecular level is still not well understood: it seems to have no effect on clock genes such as *Per*, *Cry*, *Bmal* or *Clock* [112] but a recent study showed a phase advance of *Rev-erba* gene expression [113].

Behavioral activation also has an opposite effect on the clock genes when compared to photic stimuli, i.e., it elicits a decrease in *Per1* and *Per2* transcription during the subjective day [114].

As previously mentioned, changes in food availability can efficiently synchronize the peripheral tissues [62, 63]. Nevertheless, it seems that the SCN (except in certain animal strains or under particular conditions) are quite impervious to temporal restriction. The expression of various clock genes remains unchanged when food access is limited to the light phase of the day (when nocturnal rodents are usually inactive) [62]. Conversely, as suggested by behavioral and physiological changes (activity and melatonin secretion) observed during a hypocaloric feeding, the SCN can be affected by certain feeding cues. Analyzing clock gene expression in these conditions reveals that *Per1* and *Cry2* peaks are phase advanced by 1–3 h. SCN outputs are also affected as AVP expression is advanced by 4 h [115].

Electrical Activity

Circadian variations in the frequency of electrical activity is one of the best known properties of the SCN; moreover, when isolated from the rest of the brain, SCN neurons continue to show rhythmic multiunit activity [116–119]. The highest electrical activity is observed during the light phase of the day. Additionally, this rhythm persists in constant darkness [120].

Electrical activity is an important feature of nervous tissues as it determines neurotransmitter and neuropeptide release. Since clock oscillations originate in feedback loops of clock gene expression, understanding what links the electrical activity and gene expression is crucial. A clear correlation between these two elements remains to be established, but some experiments have started to unravel their connection: Firstly, *Cry1*/*Cry2* double knock-out mice placed in DD do not show any circadian rhythmicity, neither in discharge frequency in the SCN nor in behavior [121]. Other experiments suggest that there is a correlation between *Per1* expression and electrical activity in SCN cells [122]. Thus, electrical activity can be linked to known circadian parameters. It could therefore also be a clock output, although this still remains to be demonstrated.

Nervous Outputs

The SCN send projections to numerous structures, mainly hypothalamic. Tracing neuronatomic pathways from the clock has been done extensively in rats [123, 124] and in Syrian hamster [125]. Projections are hardly different between species and four major pathways have been established (Fig. 10.4):

- A rostral projection, innervating the anterior median preoptic area, the lateral septum, the BNST (bed nucleus of the stria terminalis) and the anterior paraventricular thalamic nucleus.
- A periventricular projection, very dense, making connections with the SPVZ (subparaventricular zone), the PVN (paraventricular nuclei of the hypothalamus), the MPOA (median preoptic area), the VMH (ventromedial hypothalamus), the DMH (dorsomedial hypothalamus), the arcuate nuclei, and the premamillary nuclei
- A lateral projection to the IGL.
- A posterior projection innervating the posterior paraventricular thalamic nucleus, the ventral part of the VMH, the commissural nucleus, and the pretectal olivary nucleus.

Projection density is not the same in all structures; it is denser to the MPOA, the PVN, the DMH, the SPVZ, and the paraventricular thalamic nucleus.

The median preoptic area (*MPOA*): The MPOA plays a role in sleep and body temperature regulation and in reproductive behavior. Innervation from the SCN to the MPOA is essentially AVPergic [125]. Conversely, VIPergic innervation to this region is sparse (Fig. 10.4).

The paraventricular hypothalamic nuclei (*PVN*): The PVN is an important control and relay center for homeostasis. Neurons in the parvocellular area of the PVN

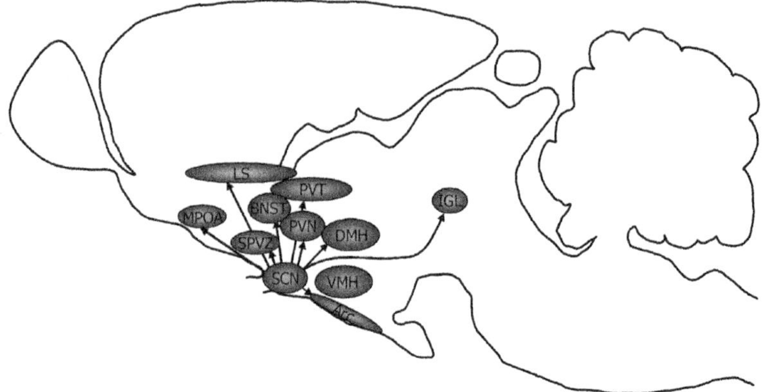

Fig. 10.4 Main SCN efferent structures. *Arc* arcuate nuclei, *BNST* bed nucleus of the stria terminalis, *DMH* dorsomedial hypothalamus, *IGL* intergeniculate leaflets, *LS* lateral septum, *MPOA* medial preoptic area, *PVN* paraventricular hypothalamic nuclei, *PVT* paraventricular thalamic nucleus, *SPVZ* subparaventricular zone, *VMH* ventromedial hypothalamus

project to the neurohypophysis; they contain factors eliciting release or inhibition of hormones (e.g., CRF; see also part below for melatonin and corticosterone secretion control). These neurons constitute the basis for circadian control of hormones by the SCN [126]. In the magnocellular area of the PVN, preautonomous neurons project through the spinal cord to sympathetic ganglions. This neuroanatomic pathway gives rise to a communication way from the SCN to peripheral tissues (reviewed in [127, 128]). The PVN are innervated by at least three populations of SCN neurons: AVP (magnocellular area and posterior part of the parvocellular area), VIP and GRP cells [125] (Fig. 10.4).

The dorsomedial hypothalamic nuclei of the hypothalamus (*DMH*) *and paraventricular thalamic nucleus PVT*): DMH and PVT are innervated by AVP, VIP, and GRP fibers [123, 125]. The DMH have a great number of connections with neuroendocrine and autonomous areas. It is thus probably an integration center which allows circadian information to participate in the regulation of a large number of physiological processes [129] (Fig. 10.4).

Through its connection to the nucleus accumbens, prefrontal cortex, and amygdala, the PVT allows the SCN to influence motor behavior and the reward system [130, 131].

Physiological Outputs

Melatonin: The hormone melatonin is secreted by the pineal gland exclusively during the night (or subjective night reviewed in [132]). Tracing studies have established the polysynaptic pathway allowing the SCN to control melatonin liberation. Briefly, through excitatory (glutamate) and inhibitory (GABA) projections to the

PVN, the SCN target the first relay [83]. The intermediolateral column of the spinal cord and superior cervical ganglions constitute further successive relays between the SCN and the pineal gland [133, 134]. Ultimately, sympathetic fibers release norepinephrin which elicits rhythmic transcription of aa-nat (Aryl-alkylamine N-acetyl transferase), the rate-limiting enzyme for melatonin synthesis. aa-nat mRNA and protein and melatonin secretion rhythms are abolished by a lesion of any structure of the polysynaptic pathway, thus confirming their implication in melatonin control [135–137].

Corticosterone: GABAergic SCN neurons target the PVN and dynamically inhibit their activity during the day. During the night, this inhibition is lifted and PVN neurons become active. This phenomenon is reinforced by an activation from glutamatergic SCN projections [83]. The SCN thus stimulate CRH (corticotropin-releasing hormone) release by PVN neurons in the hypothalamo-pituitary axis, which induces ACTH synthesis (adrenocorticotropin hormone) by the anterior pituitary. ACTH in turn stimulates the adrenal cortex, which synthesizes and liberates glucocorticoids. Depending on the species, the main glucocorticoid produced is corticosterone (rat and mouse), cortisol (human) or both (Siberian hamster). Other than the humoral control of glucocorticoid secretion via ACTH, the SCN modulate adrenal sensitivity to ACTH via sympathetic innervations of the adrenal gland [138]. In the end, the glucocorticoid peak tends to anticipate wakefulness (during the night in nocturnal animals) and thus prepares the organism for activity.

Glucose and plasmatic metabolites: Many enzymes such as insulin and glucagon, inducing hypo- and hyper-glycemia, respectively, are rhythmically observed in plasma [139, 140]. However, if food intake rhythmicity is artificially abolished by giving several meals a day, this rhythmicity disappears. Those rhythms therefore appear to be passively driven by food intake. Nevertheless, in normal feeding conditions, the SCN drive rhythmic food intake, thus creating indirectly a rhythmic secretion of a number of hormones. Insulin has receptors in many organs. As a consequence of rhythmic food intake, these organs will receive a rhythmic insuline message and will in turn be influenced by the clock.

Unlike insulin and glucagon, blood glucose rhythmicity persists even if food intake is not rhythmic in SCN intact animals. The SCN thus also control certain metabolic parameters. Like glucocorticoids, glucose is mobilized in anticipation of the beginning of the active phase [139]. More recently, Kalsbeek and collaborators have demonstrated how SCN projections releasing GABA and glutamate on preautonomic neurons in the PVN participate in the circadian control of plasma glucose concentration. The most important message emanates from inhibitory GABAergic SCN neurons that are mainly active during the light period. They contact sympathetic preautonomic neurons at the origin of hepatic glucose production and preautonomic parasympathetic neurons controlling ultimately insulin release from the pancreas. Both sympathetic and parasympathetic preautonomic PVN neurons also receive excitatory inputs, both from the SCN and noncircadian structures [141].

Daily rhythms in plasma can also be observed for a number of other metabolites such as triglycerides or ketone bodies [142].

Temperature: Core body temperature is controlled by two variables: homeostasis and the clock. Circadian regulation of temperature becomes evident when it is dissociated from locomotor activity. In rodents with partial lesions of the SCN, locomotor activity can be arrhythmic while core body temperature remains rhythmic [143]. Nevertheless, complete lesions of the SCN abolish temperature rhythms without impairing thermoregulation [144] controlled by sympathetic and parasympathetic systems. Lesions of the MPOA partially abolish temperature rhythms, so the SCN probably control this output at least in part via projections to the MPOA [145].

Behavioral Outputs

Locomotor activity and food and water intake are the typical behavioral rhythms studied as phase markers of the SCN output. They remain rhythmic as long as the SCN are intact [146, 147]. They show the same phase in regular conditions but can be forced out of phase by restriction protocols [65].

10.3.2 Retina

The retina was the first site, apart from the SCN, to be identified as a clock. The retina is a peripheral extension of the brain. Its physiology and characteristics have been explained in Chap. 4. Here, we briefly point out the characteristics making it a true clock.

(1) Retinal photoreceptors and ganglion cells receive and transmit light via the retinohypothalamic tract (RHT) to the SCN (reviewed in [148]). In the absence of both rods and cones, nonvisual photoreception is still possible in the ganglion cells due to the photosensitive pigment melanopsin residing in the ganglionic cell layer of the retina [25–28].
(2) Circadian synthesis of melatonin has been first detected in cultured retinas of hamsters. It persisted in DD and was entrainable by light [149].
(3) This rhythm is temperature compensated.

Clock gene expression and/or oscillations have also been reported in diverse cell types in the retina, but some discrepancies exist between different studies (*Clock*: [12, 150], *Bmal1*: [12, 151], *Per/Cry*: [152, 153]).

10.3.3 Olfactory Bulbs

The general assertion that all brain oscillators are phase locked to the SCN is not absolute: the olfactory bulbs have an endogenous rhythm of *Per1* expression in rhythmic as well as in behaviorally arrhythmic animals [11]. Moreover, cells of the olfactory bulb synchronize to each other as demonstrated by firing rate rhythms [11]. Furthermore, a rhythm of c-FOS induction has been detected in response to odor presentation.

Interestingly, the number of responsive cells is higher at the beginning of the night when compared to the rest of the day. This rhythm persists both in constant darkness and in SCN-lesioned rats. In contrast, a lesion of the olfactory bulbs abolished the rhythm in c-FOS induction in the primary olfactory cortex [154]. Taken together, the olfactory bulbs receive, integrate, generate, and distribute a rhythmic message. It appears that the olfactory bulb contains an independent circadian clock.

10.3.4 Food Entrainable Clock (FEC)

For about 30 years now, another putative clock has attracted the attention. First reports are even older: in the twenties, Behling and Wahl made the first observation that bees anticipated their daily breakfast, and came even if breakfast was not served (reviewed in [65]). From these empirical observations, it was nevertheless impossible to deduce that this behavior was not controlled by the clock.

In 1979, Stephan and collaborators established that in arrhythmic SCN-lesioned rats, it is still possible to restore a daily rhythm of locomotor activity by submitting the animals to a daily timed access to food. After a few days, the animals showed an increase in activity in the hours preceding food access. This food anticipatory activity (FAA) is one of the outputs of another clock, situated outside the SCN. It would correspond to foraging in natural conditions: when food is scarce, it is crucial for survival that animals can look for food sources in a temporal window different from their usual period of activity. Using a clock, it is then possible for the animal to anticipate food access, forage and prepare physiology for food intake. This food anticipation can be observed in many species such as mice and rats [55, 155], on which we will focus here (for a review of other species, see [65]).

10.3.4.1 Outputs of the FEC

When looking for a clock phenomenon it is easier to focus first on what is accessible and observable: the outputs. When submitting SCN-lesioned animals to restricted feeding schedules, a lot of physiological and behavioral functions become rhythmic again: wheel-running and general locomotor activity (FAA - [155, 156]), water intake [147, 157], unreinforced lever-pressing, anticipatory food-bin approach [65, 157], thermogenesis [68, 158, 159], corticosterone secretion (in SCN intact animals [158, 160]), and melatonin release by the pineal gland [161]. Leptin and ghrelin secretion is also affected by restricted feeding schedules [162, 163], as well as insulin and glucagon [164]. This is not astonishing since these hormones regulate metabolism and food intake (see also Chap. 5).

Note that FAA can also be observed in SCN intact animals [65]. If food is provided ad libitum after a period of restricted feeding, FAA persists for a few days at the expected time [155, 165–167]. Interestingly, after a long period (many weeks) under ad libitum feeding, FAA can reappear at the expected time if the animals are placed in fasting conditions, indicating that it is based on a self-sustained oscillator [147, 155, 157, 167, 168].

When food timing is shifted, a progressive resetting of FAA (transients) occurs [169] and entrainment of FAA to feeding T-cycles is only possible between 23 and 32 h (i.e., food is provided for a few hours every 23–32 h [170]). These phenomena are characteristic of clocks, indicating that the FEC is a true clock.

10.3.4.2 Receiving and Integrating Information from the Environment

To attest that the FEC is able to receive and integrate information from the environment, an alteration in one of its outputs should be observable in response to changes in the feeding environment. Locomotor activity is the most commonly measured parameter. In 1979, Stephan and collaborators [155] observed that changes in food availability elicited an adaptation of locomotor activity. Feeding schedules with or without calorie restriction can trigger FAA in both SCN intact and SCN-lesioned animals [65, 66, 115, 155, 167, 171–173]. Other studies have demonstrated that this is not the case for salt, water or single nutrients [174–177]. On the other hand, it seems that timed palatable meals without any other food restriction can induce anticipatory activity [178, 179].

Unlike the SCN, the FEC seems to be quite insensitive to light because FAA is independent of the lighting conditions [180]. Nevertheless, it is likely that under normal conditions, the FEC and SCN are coupled and that the FEC can have a time reference related to the LD cycle via the SCN [181].

10.3.4.3 Generating a Circadian Message

In the SCN, the core clock mechanisms originate in oscillations of clock genes. Based on studies in mutant and knock-out mice for various clock genes, the importance of known gene products for the functioning of the FEC has been established. For example, CLOCK seems dispensable [182] and would be replaced by its analog, NPAS2 [183, 184], that also dimerizes with BMAL1. CRY1 and CRY2 have also been implicated in the stability of the oscillations for the FEC [185]. Among all clock genes, *Per2* was the first one to drastically alter FAA when mutated. In *Per2* mutant mice, food anticipation is absent in both locomotor activity and temperature, whereas *Per1* mutant mice show normal food anticipation [186]. Recently, Fuller et al. demonstrated that knocking out *Bmal1* in mice also impairs food anticipation [187]. Consequently, it seems that the FEC works more or less like the SCN, at least concerning its molecular machinery.

10.3.4.4 A True New Clock, Without a Location!

Considering all the elements that have been previously cited, the FEC seems to be a true clock. Nonetheless, its localization has been investigated and debated for

30 years, but it still is not clear where the FEC resides: is it a unique brain or peripheral structure or a network comprising several brain nuclei?

Since numerous peripheral organs such as kidneys, intestine or liver are food entrainable [62, 63, 188], it was supposed that they were a part of the FEC. However, many studies indicate that they are not the primary location of the FEC [189–195]. According to these findings, peripheral organs would be either sensors or outputs of the FEC.

Therefore, the main hypothesis in the past years is that a unique brain structure is the site of the FEC. To investigate this hypothesis, many groups made lesions all around the brain without success. It was hypothesized that lesioning the FEC would suppress or impair its output(s). In the end, the olfactory bulbs [196], ventromedial hypothalamus [156, 197, 198], lateral hypothalamus [199], arcuate nuclei [200], paraventricular thalamic nucleus [201], hippocampus, nucleus accumbens ([176], but see also [69]) and the area postrema [202] seem to be dispensable for the FEC.

Other structures, when lesioned, selectively impair one output of the FEC: Lesions of the paraventricular nuclei of the hypothalamus (PVN) almost abolish the FAA but anticipation of feeding in a food-bin approach is still present [199]. Hence, the PVN would be essential for expression of the FAA but are not the primary site of the FEC. Lesions of the parabrachial nuclei also impair anticipatory food-bin approach and thermogenesis, but further investigation is required to clarify its role in the FEC [203]. Lesions of the pituitary [204] and infralimbic cortex [159] selectively suppress the food anticipatory thermogenesis.

Among all lesioned structures, one has attracted most attention in the past few years and generated a debate: the dorsomedial hypothalamus (DMH). Whereas, Saper's group claims that a lesion of the DMH abolishes FAA [205], Mistlberger's group finds that FAA persists after such a lesion [206]. Moreover, Mieda and collaborators found a de novo expression of *Per* genes in the DMH in response to restricted feeding, which would indicate that it contains a food-responsive oscillator [207]. More recently, Fuller [187] reported that injecting *Bmal1*-expressing adenoviruses in the DMH of *Bmal1* knock-out mice restores food anticipatory thermogenesis (that is otherwise not observed in these mice). This experiment would indicate that the DMH is part of the FEC but once again, this result is subject to great debate and will need to be clarified in the years to come [208].

The possibility remains that the FEC is not a unique structure (which would explain why single structure lesions do not completely suppress food anticipatory phenomena) and that many structures with partial redundancy cooperate to form the FEC. To investigate this possibility, various teams have looked for neuronal activation (c-FOS expression [20, 209–211]) alterations in clock protein rhythms (PER1 [20]), or variations of local glucose consumption (2-deoxyglucose imagery [212]) in response to restricted feeding. Many structures and possible networks have been identified by means of those techniques, but establishing a hierarchy between all those structures to precisely determine the site of the FEC is still impossible. Therefore, the FEC remains a mystery – a mysterious true circadian clock.

10.4 Environmental Influence on Brain Oscillators Outside the SCN

10.4.1 Photic Synchronization

Direct photic synchronization is only possible if a structure receives direct projections from the retina (see Chap. 4), which is the case, apart from the SCN, for the intergeniculate leaflet (IGL). Strong c-FOS induction is observed in this structure upon light exposure [213, 214], but the IGL also react to nonphotic stimuli such as novel access to a running wheel [215, 216].

Apart from the IGL, any brain structure that is connected to the SCN, directly or indirectly via nervous tracts, or bearing receptors to SCN humoral outputs (melatonin and corticosterone for example) can be indirectly influenced by light via the SCN. Hence, these structures can be viewed as SCN outputs, since they are not directly responsive to light.

10.4.2 Nonphotic Synchronization

In this section, we will focus on nonphotic synchronization, mainly by food, in central non-SCN oscillators. Note that individual cells in culture have been found to be sensitive to various nonphotic stimuli such as serum shock [217] or glucocorticoids [218]. Moreover, in all peripheral tissues, feeding schedules dominate light signaling in setting the phase of clock gene expression [62, 63, 188]. There is only one exception to date, the submaxillary salivary gland, where temporal restriction of feeding fails to entrain the clock gene expression [219].

In central oscillators, many studies report an influence of feeding on clock genes, clock proteins, and c-FOS expression. In particular, c-FOS is increased in anticipation to food access in the LH, perifornical area, DMH [209], and tuberomamillary nucleus [211]. In the NPB, NTS, AP, and the dorsal motor nucleus of the vagus nerve, c-FOS expression reacts to food intake itself [210]. Based on the expression of clock and clock-controlled genes, the cerebral cortex and hippocampus (*Per1* and *Per2*; [9]), as well as DMH, PVN and PVT (*Cry1* and *dbp*, [21]) strongly react to restricted feeding. Special attention has been drawn to the DMH, which express de novo rhythmic *Per1* and *Per2* in response to restricted feeding [207]. A daily injection of metamphetamine exerts strong effects on various clock genes (*Per1*, *Per2*, *Bmal1*, *Npas2*) in the striatum, showing that this structure can react to drugs [220].

In rats, PER1 protein expression is altered in the nucleus accumbens, basolateral (BLA) and central (CEA) amygdala, BNST, lateral septum, prefrontal cortex, and PVT in anticipation and during restricted feeding [20]. PER2 expression changes in the hippocampus and BLA [221], although in mice, Feillet et al. [21] report a milder sensitivity to feeding cues in these structures. In mice, alterations of PER1 and/or PER2 immunoreactivity were detected in CEA, DMH, PVN, PVT, VMH, and arcuate nuclei [21].

All these central structures that are implicated in processes as diverse as memory (hippocampus), energy metabolism (arcuate nuclei, VMH), reward system (nucleus accumbens, amygdala), HPA axis (PVN) or are even relay areas (PVT) not only harbor oscillators but also react to nonphotic environmental cues crucial for survival. They will in turn influence the whole physiology through their respective connections to other structures. This feature underlines the complexity of the circadian system that receives multiple cues from the environment, integrates them and adapts physiology to sometimes conflicting information.

10.5 The Reward System and Brain Clocks

The reward system is composed of brain structures that regulate and control behavior by inducing pleasurable effects. A reward, when presented more than once, causes a behavior to increase in intensity, which is termed reinforcement. Primary rewards are those necessary for survival, such as food, water, and sex. Secondary rewards derive their value from the primary rewards and include music, pleasant touch, money etc. Rewards modify behavior and emotions and induce learning and hence influence mood. Therefore, drugs of abuse, such as alcohol and cocaine, which positively influence the reward system, improve the subjective well-being, encouraging repetitive drug use, which will eventually lead to drug dependence.

10.5.1 Rhythms in Addiction

There is evidence that drug addiction is linked to diurnal rhythms. Drug addicts display disruptions in their sleep/wake and activity cycles as well as in their eating pattern and show abnormal body temperature rhythms, hormone levels, and blood pressure [222, 223]. These disruptions in behavioral and physiological processes persist long after drug use and often lead to relapse [223]. However, they are not only an indirect consequence of chronic drug use. Studies indicate that drugs can directly affect the diurnal rhythms. Cocaine exposure, for example, influences sleep [224], and sensitivity to drugs of abuse has been observed to be dependent on time of day. Interestingly, patients experiencing a drug overdose are mostly admitted to emergency rooms between 6 and 7 pm, suggesting a diurnal component in the body's response to drugs [225], although there might be environmental and social factors influencing this phenomenon as well.

Addiction may be more prevalent in people with a compromised circadian clock or with mood disorders, which might have a circadian basis, such as depressive disorders including bipolar depression and seasonal affective disorder [226–228]. Seasonal patterns appear to influence the use of addictive drugs such as alcohol predominantly during the winter when individuals are more susceptible to depression [229]. Furthermore, this seasonal incidence of affective illness is reflected in biochemical determinants [230] suggesting a cellular and molecular basis of such disorders. In support of this view are the observations that people with genetic sleep disorders are more prone to addiction [231] and that a rat strain with abnormal circadian rhythms shows high ethanol preference [232].

10.5.2 Influence of Addictive Drugs on the Circadian Clock

The major neurochemical pathways of the reward system in the brain are shown in Fig. 10.5. Of these signaling pathways, the mesolimbic pathway going from the ventral tegmental area (VTA) via the medial forebrain bundle to the nucleus accumbens (NAc) where dopamine is released, is thought to play an important role [233, 234]. Reward-related behavior exhibits a recurring pattern with a period of about 24 h and hence is circadian [235, 236]. This indicates interactions between the circadian and the reward system in the brain.

10.5.2.1 Drugs that Bind to Transporters of Biogenic Amines

Cocaine and Dopamine: First evidence for an involvement of circadian clock-associated genes in drug-induced behavior came from studies in *Drosophila*, which showed that cocaine sensitization is dependent on clock gene expression [237]. In mice lacking the *Per1* or the *Per2* genes, locomotor sensitization and conditioned preference for cocaine are abnormal [236]. Furthermore, expression of these genes is induced by cocaine in the dorsal striatum and the NAc, brain regions important for cocaine-mediated behavioral effects [238, 239]. Interestingly, cocaine differentially affects the expression of clock genes in the brain, depending on the treatment schedule (acute or chronic) and the brain area [240]. Cocaine and other drugs of abuse influence the reward system in part by modulating dopamine neurotransmission in the mesolimbic dopamine reward circuit, including the VTA and NAc of the striatum [233, 234]. Several interactions between dopamine and the circadian clock have been reported. Dopamine receptor responsiveness in *Drosophila* underlies a circadian modulation [241]. Furthermore, dopamine neurons in the retina regulate adaptation to light [242] and dopamine D2 receptor-null mice show impaired light masking [243]. It also appears that signaling via the dopamine D2 receptor potentiates circadian transcriptional regulation in the retina [244]. Mice with a point mutation in the gene *Clock* display increased excitability of dopamine neurons, cocaine reward and expression of tyrosine hydroxylase, the rate-limiting enzyme in dopamine synthesis [245]. Furthermore, these animals display a mania-like phenotype as observed in patients with bipolar disorder [246]. However, a molecular link between the circadian clock mechanism and dopamine metabolism has just recently been established [247]. It appears that, in mice, the clock proteins BMAL1, NPAS2, and PER2 regulate expression levels and activity of monoamine oxidase A (MaoA), an enzyme important in dopamine degradation. In this mechanism, PER2 appears to act as a coactivator and modulates dopamine levels in the mesolimbic dopaminergic system. This has an impact on mood-related behavior in mice leading to the conclusion that the clock can influence mood. In line with this conclusion is the finding that seasonal affective disorder (SAD) in humans correlates with the frequency of specific single nucleotide polymorphisms in *BMAL1*, *NPAS2*, and *PER2* [248].

Amphetamine and Methamphetamine: Amphetamine (1-phenylpropan-2-amine, also known as speed) and methamphetamine ((2S)-N-methyl-1-phenylpropan-2-amine) are both psychostimulants that can act on the mesolimbic reward pathway (VTA and NAc, see Fig. 10.5). The effects of these drugs, especially of methamphetamine, on circadian behavior have first been described in rats [249]. Methamphetamine reversibly increases the locomotor activity, the length of alpha (the duration between onset and offset of daily activity), and the period length of the activity rhythm. Even in constant darkness, these effects persist but disappear after withdrawal of methamphetamine. Interestingly, methamphetamine can induce robust activity rhythms in SCN-lesioned rats [250] as well as in *Clock* and *Cry1/Cry2* mutant mice that are arrhythmic under constant darkness conditions [251, 252]. All the behavioral responses observed in rats were confirmed in mice and it appears that these effects do not depend on rhythmic consumption of methamphetamine, indicating that a methamphetamine-sensitive circadian oscillator (MASCO) exists [253]. Since methamphetamine acts primarily on dopaminergic cells in the brain,

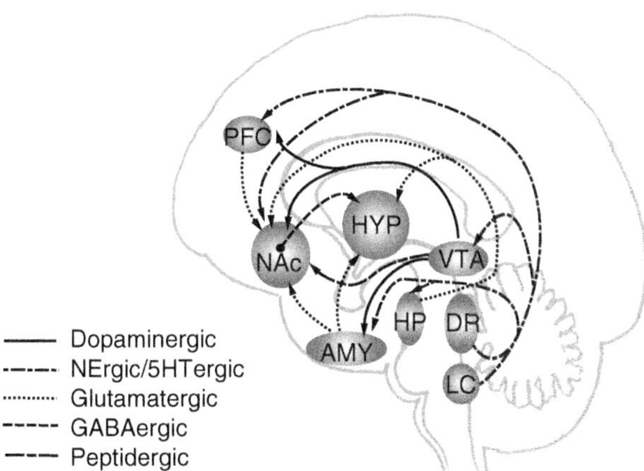

Fig. 10.5 Neural circuitry of mood. Several brain regions are implicated in the regulation of mood. Besides the hippocampus (HP) and the prefrontal cortex (PFC), several subcortical structures are involved in reward, fear, and motivation. These include the nucleus accumbens (NAc), amygdala (AMY), and hypothalamus (HYP). The figure shows only a subset of the many known interconnections between these various brain regions. The ventral tegmental area (VTA) provides dopaminergic input to the NAc, AMY, and PFC. *DR* dorsal raphe nuclei, *GABA* gamma-aminobutyric acid, *LC* locus coeruleus, *NE* norepinephrine, *5HT* serotonin

it is tempting to speculate that the MASCO resides in the mesolimbic dopaminergic system (see paragraph above and Fig. 10.5).

Methamphetamine modulation of gene expression in the brain has been studied. Repeated injections of methamphetamine cause a sensitized increase in *Per1* gene expression specifically in the mouse striatum without affecting *Per1* or *Per2* gene expression in the master clock of the SCN [254]. Furthermore, acute injection of methamphetamine increases the expression of *Per1*, *Bmal1*, and *Npas2* genes in the striatum. Additionally, chronic daytime methamphetamine treatment shifts rhythmic *Per1* and *Per2* expression in the striatum from a nocturnal to a diurnal rhythm, without affecting the rhythm in the SCN [220]. Among the genes coding for proteins involved in signal transduction, transcription, synaptic structure, and ion-channels, the gene most affected in its expression by methamphetamine treatment was the clock gene *Per2* [255]. These findings highlight the mutual interaction of the circadian clock with psychostimulants.

10.5.2.2 Drugs that Bind to Ionotropic Receptors and Ion Channels

Alcohol: In humans, chronic alcohol intake is associated with sleep disturbance and alterations in various daily physiological rhythms, including blood pressure, body temperature, metabolism, hormone secretion, and EEG topography [256–261]. Similar chronobiological disruptions were observed in patients with mood disorders [262]. Interestingly, mood dysregulation is a hallmark of alcohol withdrawal [263] and the persisting chronobiological disruptions over extended periods may promote relapse to drinking [264, 265]. In rats, chronic ethanol treatment and genetic ethanol preference are associated with alterations in free-running period and light response of the circadian clock, suggesting that the chronobiological disruptions seen in human alcoholics are partially caused by ethanol-induced disruption of the circadian system [266] and genetic parameters associated with alcoholism [232]. At the neuroendocrine and gene levels, ethanol administration in rats alters the circadian expression patterns of pro-opiomelanocortin (POMC) in the arcuate nucleus as well as *Per1* and *Per2* gene expression in this brain region and in the SCN [267].

Alcohol does not only affect the clock gene expression but also vice versa. Mice with a mutation in the *Per2* gene have a defective circadian clock [268] and show an increased ethanol intake [269]. This is probably caused by a decrease in the expression of a glutamate-aspartate transporter (Glast, Eaat1) on astrocytes which leads to a less effective clearance of glutamate from the synaptic cleft. As a consequence, glutamate levels in the striatum increase. This hyperglutamatergic state is thought to cause the elevated ethanol intake in these animals. Evidence for a genetic component being involved in hyperglutamatergic state and alcoholism in humans is provided by the differential response of alcoholics and controls to acamprosate [270]. Interestingly, increased alcohol intake in *Per2* mutant mice can be reversed by acamprosate leading to the hypothesis that genetic variations in human *PER2* or *EAAT1* might serve as predictors of a response to acamprosate. Although *PER2* was not associated with alcohol dependence, significant association of an SNP in the *PER2* gene with high versus low alcohol intake was observed [269]. These findings provide molecular evidence for the phenomenon of enhanced alcohol consumption in shift workers [271] and people suffering from jet lag as observed in aircraft personnel [272].

Nicotine: Effects of nicotine on the circadian system are not well studied. In rats, nicotine does not suppress the daily physiological rhythms but induces perturbations in daily rhythms of heart rate, body temperature, and locomotor activity [273]. In humans, transdermal nicotine affected rhythms in blood pressure and was associated with the impairment of endothelium-dependent vasodilation [274]. In humans, preferred timing of sleep and activity varies widely and gives rise to so-called "chronotypes". Chronotype is largely regulated by the circadian clock, which is influenced by genetic variations in clock genes and environmental influence. In a study associating chronotype, well-being, and stimulant consumption, the most striking correlation existed between chronotype and smoking [275]. Late chronotypes showed the strongest correlation indicating that social jetlag, i.e., the discrepancy between social and biological timing, is probably one reason for smoking. This suggests that misalignment of the circadian clock with the environment encourages nicotine consumption. However, molecular evidence for such a potential mechanism is lacking. Future studies will show how nicotine affects the circadian clock and vice versa.

Benzodiazepines: Benzodiazepines have sedative, but not anxiolytic, properties that are mediated by the GABA(A) receptor alpha1 subtype [276]. They are used in the treatment of insomnia and have been shown to phase-shift the mammalian circadian clock [277, 278]. Furthermore, daily injections of triazolam, a short-acting benzodiazepine, induced long-term changes in hamster circadian period [53]. In humans, triazolam administered at bedtime reduces sleep latency and duration of awakenings, which is accompanied by suppression of stages III and IV in late sleep [279]. Triazolam induces a transient hyperprolactinemia but does not abolish sleep-related hormone secretion and does not affect the timing of endocrine events controlled by the circadian clock [279]. At the molecular level, benzodiazepines reduce the mRNA levels of *Per1* in the mouse cerebellum [280]; however, it is not clear what the significance of this observation is. Interestingly, daily treatment of transgenic mice mimicking Huntington's disease with Alprazolam reversed dysregulated expression of *Per2* and *Prokineticin2*, an output factor of the SCN that controls behavioral rhythms, in these animals [281]. This raises hopes that benzodiazepines may improve the circadian-gene regulated functions that are impaired due to disease.

10.5.2.3 Drugs that Activate G Protein-Coupled Receptors

Opioids: Opioids have long been used to treat acute pain after operations and are invaluable in the treatment of chronic pain, as it is experienced e.g., in the terminal conditions of cancer. Opioids bind to opioid receptors, which are coupled to

G-proteins, to activate an intracellular signaling cascade. These receptors are found principally in the central nervous system and the gastrointestinal tract.

Opioid receptor activation appears to affect the clock in the hypothalamus of rats, which is generating pulsatile growth hormone secretion. The period of the growth hormone rhythm and the duration of the growth hormone burst are shortened by opioid mechanisms [282]. Interestingly, the synthetic opioid fentanyl can strongly modify the photic responsiveness of the circadian pacemaker in the hamster, probably via direct effects on the SCN electrical activity and the regulation of *Per* gene expression [283]. Furthermore, targeting *Per1* with DNAzyme affected the response of mice to morphine, but already formed morphine dependence was not affected by inactivation of *Per1* [284]. It appears that morphine-induced reward that affects *Per1* involves the ERK signaling pathway [285]. This signaling pathway also plays a mayor role in the photic response pathway [286, 287]. Taken together, these findings indicate that the photic signaling and opioid signaling pathways converge onto the circadian clock by ERK signaling.

Cannabinoids: Data on the influence of cannabinoids on the circadian system are scarce. It appears, however, that delta9-tetrahydrocannabinol (THC) increases brain temperature and inverts circadian rhythms in the rat [288]. In the clinic, THC has emerged to be useful in reducing nocturnal motor activity and agitation in severely demented patients [289]. Since nighttime agitation is thought to be the consequence of an abnormal circadian clock, the effects of THC on nighttime activity seem to be rooted in the circadian system. In line with this view is the finding that the cannabinoid type 1 receptor (CB1) is immunohistochemically detected in several brain nuclei of the hamster, including the SCN [290]. This same study also shows that the application of a CB1 agonist inhibits light-induced phase shifts, which can be reversed by CB1 antagonists [290]. Taken together, these studies indicate that the endocannabinoid system has the capability to modulate circadian rhythms.

Gammahydroxybutyric acid (GHB): GHB is used to treat narcolepsy in humans. It impairs nonrapid eye movement (NREM) sleep but improves rapid eye movement (REM) efficiency at night [291]. In mice, however, a clear sleep-promoting effect of GHB could not be detected, although at high doses, GHB causes electroencephaligraphic hypersynchronization together with a coma-like state [292]. In summary, it is not clear whether GHB affects the circadian system although effects on vigilance state are observed.

10.5.2.4 Conclusions

From the findings described above, it is clear that an interrelationship between drugs of abuse and the circadian system exists. Psychoactive drugs have the potential to influence clock gene expression in various brain areas. As a consequence circadian, clock function can become transiently or permanently altered, which might result in pathological conditions resembling aspects of drug addiction. Conversely, the circadian system influences the sensitivity of the organism to drugs of abuse because it can regulate enzymes involved in the metabolism of neurotransmitters [247].

For future studies, the identification of specific brain areas in which the clock influences drug-related behavior is of outstanding interest. Towards this goal, tissue-specific alteration of clock gene expression in mouse models combined with lentiviral rescue systems will uncover the different levels of interplay between drugs of abuse and clock genes.

10.6 Neurological and Behavioral Phenotypes in Mouse Clock Mutants

Most clock genes known to date have been genetically modified in the mouse. Investigations have primarily focused on circadian behavior by assessing activity and gene expression rhythms. The phenotypes observed can range from complete loss of behavioral and molecular rhythmicity in constant conditions to subtle alterations in phasing of these rhythms [107, 150, 293–295].

However, the circadian clock controls a great number of so-called clock-controlled genes (see Chaps. 1 and 7) that regulate biochemical and physiological processes. Therefore, systematic behavioral characterization of mouse clock mutants will uncover novel phenotypes associated with clock genes. Given the fact that clock genes are expressed in various brain areas with different phases [21], an impact of clock genes on brain function can be expected. Deficits in brain function could be due to a disturbance in SCN oscillator function or hampered oscillator function in discrete brain areas; alternatively, they might be a secondary consequence of neuronal dysfunction.

10.6.1 Sleep

Many neurological disorders are accompanied by a disturbance in sleep patterns (see Chap. 9). Only a small spectrum of these disorders is rooted in the circadian system itself. This is not astonishing since sleep is not exclusively regulated by the circadian clock but also by homeostatic parameters that are related to metabolism [296].

Mutations in the genes *Clock*, *Bmal1*, *Cry1,* and *Cry2* result in altered sleep time, sleep fragmentation, and atypical responses after sleep deprivation [297–299]. Elements of sleep homeostasis are also affected in mice mutant in a number of other clock-related genes including *Npas2*, *Dbp*, and *Prok2* [184, 300, 301]. However, evidence exists for a functional segregation of circadian and homeostatic parameters of sleep, because mutations in *Per1* and *Per2* cause no effects on the sleep homeostatic mechanism, although robust rhythms of sleep and wakefulness are affected under constant conditions [302, 303]. Taken together, these data indicate that clock genes have an effect on allocation of sleep time; however, only a subset of clock genes seems to affect the homeostat. This is probably due to indirect effects of clock genes on metabolism and neuronal function and might reflect brain region-specific effects of clock genes.

10.6.2 Memory

Memory is a process that allows an organism to recall events, locations, feelings, and more. If the matter that has to be recalled contains an aspect of time one can suppose that the circadian clock could be involved.

Expression of clock genes has been observed in brain regions relevant for memory regulation [106, 183]. Therefore, mice mutant for clock genes have been subjected to a variety of behavioral tasks testing different aspects of memory regulation. For example, mice lacking *Npas2* show deficits in the acquisition of cued and contextual fear-conditioning paradigms [304]. In contrast, mice mutant in the *Per1* or *Per2* genes show normal spatial and contextual learning in standardized tests for hippocampus-dependent learning [305]. This indicates that an association between clock genes and memory regulation is not universal and that a degree of functional segregation exists within the system. In this context, it is of note that the SCN is not required in the circadian modulation of conditioned place preference in hamsters [306], implying that memory containing a temporal aspect involves a circadian oscillator that is distinct from the SCN. However, hippocampus-dependent learning in the hamster appears to require a functional circadian system [307]. Interestingly, time-place associations, which are crucial for the survival and reproductive success of animals, appear to involve functional *Cry* genes in mice [308]. Taken together, these findings indicate that at least some aspects of memory regulation are modulated by the circadian clock or clock components.

10.6.3 Mood-Related Behavior

There is evidence that a defective circadian system affects the emotional behavior. Mice mutant in the gene *Clock* exhibit a spectrum of behavioral abnormalities, including low anxiety, mania, and hyperactivity [246, 309]. Furthermore, dysregulation of important neurotransmitters such as NPY or VIP, which are involved in signaling to and within the SCN, leads to alterations in anxiety-like behavior and aggression [310–312]. Lithium, a mood-stabilizing agent, can reverse behavioral disturbances observed in *Clock* mutant mice [246], indicating that the therapeutic effects of lithium may be partly mediated via the circadian system. This hypothesis is bolstered by the finding that lithium is a potent inhibitor of glycogen synthase kinase 3β (GSK3β see Fig. 10.6), a kinase that influences the nuclear expression and stability of a number of transcription factors and circadian clock proteins including REV-ERBα and PER2 [313–315] (see also Chap. 3).

Based on the finding that the circadian clock regulates dopamine metabolism [247] and in view of the fact that PER2 interacts with GSK-3β [313], the following model was proposed [316] (see Fig. 10.6).

PER2 phosphorylation by GSK-3β favors accumulation of PER2 in the nucleus where it enhances the NPAS2/BMAL1-mediated transcription of *Maoa*. More enzyme is generated and translocates to the inner mitochondrial membrane. MAOA degrades dopamine, producing 3,4-dihydroxyacetaldehyde, peroxide, and ammonia. Increased levels of PER2 might therefore lead to less dopamine and a more depressed mood state [317, 318]. The mood stabilizer lithium inhibits the action of GSK-3β [319] and less PER2 is phosphorylated [313]. As a consequence, PER2 becomes less abundant in the nucleus, which leads to less MAOA production. Consequently, dopamine will rise and mood status will improve. This might be a potential mechanism explaining the beneficial effects of lithium on mood.

An alternative to lithium in treating depression is bright light therapy, which has been successfully used for recurrent winter depression or seasonal affective disorder (SAD) [320]. Interestingly, clock genes, especially *Per1* and *Per2*, are inducible by light in the SCN [106, 321]. However, it is not known how this might relate to

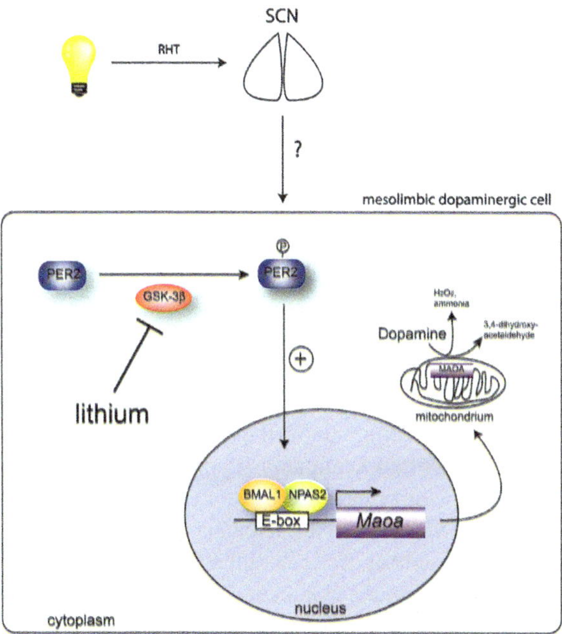

Fig. 10.6 Model how PER2 might affect monoamine oxidase A (Maoa) expression and dopamine levels in the mesolimbic dopaminergic system (for details see text). Light affects the molecular clock mechanism in the suprachiasmatic nuclei (SCN) via the retinohypothalamic tract (RHT), which directly connects the retina of the eye to the SCN. How this signal is transmitted to mesolimbic dopaminergic cells is not known. *BMAL1* brain and muscle arnt-like factor 1, *GSK-3β* glycogen synthase kinase-3β, *NPAS2* neuronal Period-Arndt-Single-minded-domain protein 2, *PER2* period 2

the relief of depressive episodes. One possibility is that light affects the nuclear/cytoplasmic distribution of PER2 in an unknown manner, thereby affecting dopamine metabolism (Fig. 10.6).

10.7 Human CNS Disorders and Clock Parameters

In many CNS disorders, disturbance of circadian rhythms and sleep is part of the clinical characterization of the disease (Table 10.1). Such clock and sleep disruptions are probably secondary to the compromised neuronal circuitry when brain regions regulating output rhythms are affected. In any case, it is difficult to dissect whether disruptions in the circadian clock are a cause or a consequence of CNS disorders. However, it is very likely that disruption of circadian oscillators can at least modify disease severity and that in some cases, it might play a more primary role in disease etiology.

10.7.1 Affective Disorders

Mice mutant in clock genes often display alterations in the period length of the circadian rhythm leading to alterations in sleep onset. Therefore, one would predict that also in the human population, such alterations in period length would be associated with sleep onset caused by variants of genes involved in the circadian clock mechanism. If such alterations were found in families, genetic analysis would allow identification of the genes involved. So far, the only familial syndrome hinting at a direct involvement of the circadian clock in early sleep onset is the familial advanced sleep phase syndrome (FASPS). This disorder is characterized by very early sleep on- and offset. Affected individuals show a profound phase advance of melatonin and temperature rhythms associated with a very short period length [322]. Since the trait segregates in an autosomal dominant manner with high penetrance, the genes affected in such families could be identified. In one family, the casein kinase 1 ε (CKIε) binding site in the gene product of *PER2* was mutated, which causes hypophosphorylation of the PER2 protein [323]. A mis-sense mutation in *CKIδ* that decreases the enzymatic activity of CKIδ also leads to hypophosphorylation of PER2 and FASPS [324] (see also Chap. 3).

Although an increasing number of people suffer from delayed sleep phase syndrome (DSPS) the mechanism for its development is obscure. Patients show psychiatric problems, longer sleep period and higher sensitivity to light [325]. A mis-sense variation in CKIε has been associated with DSPS [326], however, there are conflicting observations [327]. The lack of candidate genes and mouse models for this disorder indicates that it is probably caused by various mechanisms. Since blind people run on the internal human period of slightly more than 24 h and display DSPS, one can speculate that at least one mechanism involved in DSPS might be the light signal transduction pathway.

Table 10.1 Human CNS disorders and clock/sleep disturbance

Human disease/condition	Rhythm/sleep phenotype	Mouse model
Aging	Reduced sleep consolidation and earlier wake time. Reduction in amplitude and advanced phase [387, 388]	Lengthens period and reduces amplitude. Activity onset is delayed and more variable [345]
Alzheimer Disease (AD)	Fragmented sleep, activity distribution altered [334]	Alterations in sleep regulation and changes in activity profiles [335–338]
Angelman Syndrome (AS)	Sleep disturbance with alteration in melatonin metabolism [363]	Sleep disturbance [364]
Autism Spectrum Disorders (ASD)	Sleep disturbance, abnormal circadian rhythm, and altered melatonin levels [369]	*Pten* ko animals show longer period length [371]
Delayed Sleep Phase Syndrome (DSPS)	Evening preference, delayed phase of activity, sleep, and melatonin [325–327]	No model
Down Syndrome (DS)	Abnormal sleep [354–356]	Increased light phase activity, reduced amplitude and advance of phase [357, 358]
Familial Advanced Sleep Phase Syndrome (FASPS)	Early sleep and wake time, shortened circadian rhythm [322, 323]	*Per2* or *CKIδ* mis-sense mutation, advanced phase of activity, shortened period [324, 389]
Fragile X Syndrome (FXS)	Altered sleep and melatonin profiles [366]	Fragile X related proteins influence circadian activity rhythms [368]
Huntington Disease (HD)	Awakening during the night and disintegration of diurnal activity rhythm [343]	Disintegration of circadian rhythms [343]
Mood Disorders (unipolar, bipolar depression, schizophrenia)	Circadian phase disturbances in sleep, activity, temperature, and hormone levels [328, 329, 390]	Mice mutant in clock genes model aspects of mood disorders [236, 245–247, 269, 309]
Parkinson Disease (PD)	Sleep disorders, excessive daytime sleepiness [391]	No recorded circadian or sleep disturbance in genetic mouse models [392]
Prader–Willi Syndrome (PWS)	Sleep-related and behavioral disturbances [356, 360, 361]	Reduced circadian activity amplitude with more daytime activity [362]
Prion Disease	Sleep abnormalities, loss of circadian rest-activity and melatonin rhythms [339–341]	Sleep fragmentation, longer circadian period [342]

(continued)

Table 10.1 (continued)

Human disease/condition	Rhythm/sleep phenotype	Mouse model
Seasonal Affective Disorder (SAD)	Depressive symptoms in short winter days [331–333]	No model
Smith–Magenis Syndrome (SMS)	Inverted melatonin rhythm and sleep disorders [349–352]	Hypoactive and shorter circadian period [353]

Modified from [393]

Affective disorders are also believed to arise from conflicts between the internal circadian clock and environmental and social rhythms. Patients suffering from unipolar depression, bipolar disorder, and schizophrenia all display circadian phase disturbances in sleep, activity, temperature, and hormone regulation (reviewed in [328, 329]). Manipulation of the sleep-wake cycle and circadian phase have proven beneficial for some patients; for example, sleep deprivation can give a temporary remission from a depressive episode [330] and morning bright light therapy is currently the treatment of choice for recurrent winter depression or seasonal affective disorder (SAD) [320], a common condition where depressive symptoms occur during shorter winter days [331–333].

10.7.2 Neurodegenerative Diseases and Aging

In neurodegenerative diseases and aging, circadian and sleep disturbances seem to be secondary consequences of the deteriorating neuronal circuits in the brain. Common symptomatic features are circadian disturbances in phase and amplitude of activity and problems in temperature regulation. Circadian disturbances have been described in Alzheimer Disease (AD) in humans [334] as well as in mouse models for this disease [335–338]. Interestingly, in mouse models, circadian disturbances sometimes precede the onset of typical AD pathologies. Prion disease, also known as Creuzfeldt-Jakob or mad cow disease, can be accompanied by a spectrum of sleep and circadian clock disturbances [339–341]. These disturbances are mirrored in mouse models that lack expression of the gene encoding the prion protein [342]. The mouse transgenic line that mimics Huntington disease shows disintegration of activity rhythms similar to what is observed in Huntington patients. These behavioral disturbances are accompanied by altered clock gene rhythms in the SCN, motor cortex, and striatum [343]. Similarly, aging alters the precision of activity onset, lengthens the period and reduces the amplitude of circadian rhythms and alters SCN electrophysiology in mice [344, 345]. Conversely, mice mutant for both *Per1* and *Cry2* genes show accelerated aging in that they loose circadian activity rhythms after 6 months of age [346] and become infertile [347]. Alterations in brain histology, however, are not observed by that time, implying that circadian clock defects might precede onset of brain pathologies. However, comparable effects of aging are seen in wild-type and *Clock* mutant mice [348] indicating that age-related circadian changes may also be symptomatic of brain circuitry deterioration.

10.7.3 Syndromic Disorders

Syndromic disorders are the result of multiple defective biological pathways. Since many biological pathways are oscillating in a diurnal fashion it is likely that circadian disturbances contribute to the etiology of a number of syndromic disorders (see Table 10.1). Smith–Magenis syndrome (SMS) is associated with heterozygous deletions on chromosome 17, an inversed secretion rhythm of melatonin and sleep disturbance [349–352]. Although the human *PER1* gene is close to the deletion on chromosome 17 there is no evidence that alterations in the expression of this gene are the cause for the circadian rhythm disturbances. However, mice carrying a heterozygous deletion of the SMS syntenic region display a hypoactive phenotype and abnormal circadian rhythms [353].

Down syndrome (DS) patients display features of sleep disturbance such as reduced sleep maintenance, sleep fragmentation, sleep apnea, and reduction in REM sleep [354–356]. In a mouse model of DS, increased activity in the light phase (rest-phase), reduction in rhythm amplitude, and advanced activity phase are observed [357, 358]. These findings indicate that trisomy 21 interferes with the circadian, and probably also with the homeostatic, systems.

Uniparental dysomy disorders bring about a spectrum of sleep-related phenotypes. Patients with Prader–Willi syndrome (PWS), which is caused by a loss of paternal expression of a cluster of genes at human chromosome 15q11-q13, display hyperphagia, sleep problems, hypogonadism, and growth hormone deficiency indicating abnormalities in hypothalamic function [356, 359–361]. The clock-controlled output gene with unknown function *Magel2*, whose transcripts are imprinted and paternally expressed, maps to the PWS region. Mice deficient in *Magel2* entrain to normal light-dark cycles and show normal wheel-running rhythms, but with a reduced amplitude of activity and increased daytime activity [362]. This indicates that at least sleep-related phenotypes characteristic of PWS could be caused by the lack of this gene. Loss of maternal expression of a cluster of genes at 15q11-q13 causes Angelman syndrome (AS), which results in EEG abnormalities and sleep disturbance. This seems to be partly due to problems in metabolizing melatonin [363]. It appears that AS results from a lack of functional expression of the maternal allele of *UBE3A*, an ubiquitin-protein ligase. A mouse model for AS lacking maternal *Ube3a* expression shows sleep disturbances similar to those observed in the human phenotype [364].

Fragile X syndrome (FXS) is a common form of inherited mental retardation. Patients display physical and behavioral abnormalities including hyperactivity, cognitive impairment, and sleep disorders [365, 366]. The syndrome is caused by an expansion mutation of a CGG triplet repeat in the 5′ UTR of the *FMR1* gene.

This silences the transcription of the gene and leads to a loss of the encoded protein (FMRP) [367]. Mice lacking FMRP or FXR2P display a shorter free-running period of activity. Absence of both proteins leads to complete arrhythmicity even in a light-dark cycle [368] although clock gene expression and electrophysiological activity in the SCN are still normal. This indicates that the central clock mechanism is still functional but that the output pathway regulating clock-controlled genes is somehow derailed. This parallels the observations in the PWS model (see above) in which the output pathway of the clock seems to be affected.

Patients with autism spectrum disorders (ASD) exhibit circadian and sleep-related disturbances [369]. One of the genes implicated in ASD codes for acetylserotonin-O-methyltransferase, an enzyme critical in the synthesis of melatonin. This finding correlates with the low melatonin levels observed in patients [370] and might explain behavioral disturbances in ASD, which may arise from the inability of an individual's clock to entrain to environmental and social cues. Mutations in the phosphatase and tensin homologue on chromosome ten (PTEN) have been reported in autistic patients with macrocephaly. Mice with a conditional deletion of the homologous gene show macrocephaly, increased susceptibility to seizures, deficits in social interaction, anxiety, and a significantly longer free-running period [371].

Taken together, it appears that in syndromic disorders, the output pathways from the clock are disturbed.

10.8 Polymorphisms in Human Clock Genes

The circadian clock has an exclusive and central position in linking information from the environment with genetic information in the organism. From this exclusive position, it becomes clear that the clock, at least in part, regulates the plasticity of genetic information and enables an organism to optimally adapt to recurring events. Therefore, it might be possible that natural selection has genetically optimized adaptiveness along the latitudinal clines. In one study, latitude-driven selective effects on *PER2* genetic variability have been observed [372]. However, the other studies indicate that differences in the frequency of clock gene alleles are specific to ethnic group rather than latitude and that these differences arise from genetic drift rather than from natural selection [373, 374].

Since individuals in a population do not go to bed at exactly the same time, and because circadian rhythmicity is part of sleep, many laboratories investigated whether extreme diurnal or nocturnal preference is rooted in polymorphisms of clock genes. As we have seen in Sect. 10.6.1, this is the case for FASPS; however, for DSPS the case is less clear although a polymorphism in the gene *CLOCK* has been associated for this syndrome [375]. A silent polymorphism in the *PER1* gene also did associate with extreme diurnal preference [376], as did a structural polymorphism in the *PER3* gene [377]. A shorter 4-repeat allele of a 54 bp coding-region polymorphism in PER3 is associated with DSPS [378] and this polymorphism predicts sleep structure and waking performance [379]. Interestingly, polymorphisms in various clock genes do not only affect the sleep-associated phenotypes. A single nucleotide polymorphism in *CLOCK* appears to bias neural correlates of moral valence decision in depressed patients [380] and the other genetic variants of this gene are associated with individual susceptibility to obesity [381]. The *NPAS2* gene has been associated with SAD (see 10.6.1) [382] as well as the genes *BMAL1* and *PER2* [248] (see 10.4.2.1). Furthermore, *BMAL1* variants appear to be associated with susceptibility to hypertension and type 2 diabetes [383]. For *CRY1*, an involvement in schizophrenia has been suggested, because it is close to a linkage hotspot of this disease and interacts with antipsychotic drugs and the dopamine system [384]. However, another study could not find an association of *CRY1* with bipolar disorder [385].

In the future, more systematic studies in both humans and animals will uncover the contribution of clock gene variants to the onset and severity of CNS disorders.

10.9 Perspectives

In the last few years, evidence has accumulated pointing to an involvement of the circadian clock in brain disorders that reach from abnormal feeding behavior to mood disorders. It appears that SCN-independent oscillators exist in the brain that regulate feeding behavior and sensitivity to drugs of abuse. These oscillators might be connected to each other since feeding signals and signaling induced by drugs of abuse share neurobiological mechanisms that act in the midbrain (reviewed in [386]). Therefore, future studies will have to concentrate on tissue- or brain region-specific inactivation of clock genes to decipher the relative contribution of the clock to specific neurobiological systems. This will help uncover the localization of the food entrainable clock (FEC) and the methamphetamine-sensitive clock (MASCO). Lentiviral rescue systems will be used to rescue the clock defects in brain regions affecting the FEC and MASCO and will be a valuable tool to support genetic experiments. Furthermore, in vivo inducible and reversible tetracycline-controlled transactivator systems associated with specific clock genes will be necessary to exclude the effects of clock genes during development.

Approaches using systems biology technology will establish relationships between genetic networks, protein activity/protein interaction networks, and metabolic profiles (see Chap. 11). With the help of bioinformatics, we will uncover the biological circuits that interact with each other defining a signature for a "normal" state. Changes in metabolic equilibrium caused for example by a mutation in a gene will alter the signature indicating an abnormal state. However, a lot of data has to be gathered and properly annotated until we will see the fruits of these efforts and before we can dream of an instant biological profile-based medicine.

Genome-wide studies in humans will help us understand the mechanisms underlying "chronotype". This will also lead to improvements in treatments for many diseases which are not caused by defects in the circadian system but the symptoms of which are strongly modulated by the clock. This will hopefully lead to optimized chrono-pharmacological interventions. The insights into the genetics behind chronotype may lead to political changes facilitating more appropriate, individualized working and school schedules.

Acknowledgments We thank Dr. Sonja Langmesser for editing and critically reading the manuscript. Our work is supported by the European Molecular Biology Organization (EMBO) (CF), the Swiss National Science Foundation and the European Union Project EUCLOCK (UA).

References

1. Moore RY, Eichler VB (1972) Loss of a circadian adrenal corticosterone rhythm following suprachiasmatic lesions in the rat. Brain Res 42:201–206
2. Stephan FK, Zucker I (1972) Circadian rhythms in drinking behavior and locomotor activity of rats are eliminated by hypothalamic lesions. Proc Natl Acad Sci U S A 69:1583–1586
3. Van den Pol AN (1980) The hypothalamic suprachiasmatic nucleus of rat: intrinsic anatomy. J Comp Neurol 191:661–702
4. Lehman MN, Silver R, Gladstone WR, Kahn RM, Gibson M, Bittman EL (1987) Circadian rhythmicity restored by neural transplant. Immunocytochemical characterization of the graft and its integration with the host brain. J Neurosci 7:1626–1638
5. Ralph MR, Foster RG, Davis FC, Menaker M (1990) Transplanted suprachiasmatic nucleus determines circadian period. Science 247:975–978
6. Guilding C, Piggins HD (2007) Challenging the omnipotence of the suprachiasmatic time-keeper: are circadian oscillators present throughout the mammalian brain? Eur J NeuroSci 25:3195–3216
7. Abraham U, Prior JL, Granados-Fuentes D, Piwnica-Worms DR, Herzog ED (2005) Independent circadian oscillations of Period1 in specific brain areas in vivo and in vitro. J Neurosci 25:8620–8626
8. Asai M, Yoshinobu Y, Kaneko S, Mori A, Nikaido T, Moriya T, Akiyama M, Shibata S (2001) Circadian profile of Per gene mRNA expression in the suprachiasmatic nucleus, paraventricular nucleus, and pineal body of aged rats. J Neurosci Res 66:1133–1139
9. Wakamatsu H, Yoshinobu Y, Aida R, Moriya T, Akiyama M, Shibata S (2001) Restricted-feeding-induced anticipatory activity rhythm is associated with a phase-shift of the expression of mPer1 and mPer2 mRNA in the cerebral cortex and hippocampus but not in the suprachiasmatic nucleus of mice. Eur J NeuroSci 13:1190–1196
10. Abe M, Herzog ED, Yamazaki S, Straume M, Tei H, Sakaki Y, Menaker M, Block GD (2002) Circadian rhythms in isolated brain regions. J Neurosci 22:350–356
11. Granados-Fuentes D, Prolo LM, Abraham U, Herzog ED (2004) The suprachiasmatic nucleus entrains, but does not sustain, circadian rhythmicity in the olfactory bulb. J Neurosci 24:615–619
12. Namihira M, Honma S, Abe H, Tanahashi Y, Ikeda M, Honma K (1999) Daily variation and light responsiveness of mammalian clock gene, Clock and BMAL1, transcripts in the pineal body and different areas of brain in rats. Neurosci Lett 267:69–72
13. Shieh KR, Yang SC, Lu XY, Akil H, Watson SJ (2005) Diurnal rhythmic expression of the rhythm-related genes, rPeriod1, rPeriod2, and rClock, in the rat brain. J Biomed Sci 12:209–217
14. Sun ZS, Albrecht U, Zhuchenko O, Bailey J, Eichele G, Lee CC (1997) RIGUI, a putative mammalian ortholog of the Drosophila period gene. Cell 90:1003–1011
15. Herichova I, Mravec B, Stebelova K, Krizanova O, Jurkovicova D, Kvetnansky R, Zeman M (2007) Rhythmic clock gene expression in heart, kidney and some brain nuclei involved in blood pressure control in hypertensive TGR(mREN-2)27 rats. Mol Cell Biochem 296:25–34
16. Nagoshi E, Saini C, Bauer C, Laroche T, Naef F, Schibler U (2004) Circadian gene expression in individual fibroblasts: cell-autonomous and self-sustained oscillators pass time to daughter cells. Cell 119:693–705

17. Welsh DK, Yoo SH, Liu AC, Takahashi JS, Kay SA (2004) Bioluminescence imaging of individual fibroblasts reveals persistent, independently phased circadian rhythms of clock gene expression. Curr Biol 14:2289–2295
18. Amir S, Lamont EW, Robinson B, Stewart J (2004) A circadian rhythm in the expression of PERIOD2 protein reveals a novel SCN-controlled oscillator in the oval nucleus of the bed nucleus of the stria terminalis. J Neurosci 24:781–790
19. Lamont EW, Robinson B, Stewart J, Amir S (2005) The central and basolateral nuclei of the amygdala exhibit opposite diurnal rhythms of expression of the clock protein Period2. Proc Natl Acad Sci U S A 102:4180–4184
20. Angeles-Castellanos M, Mendoza J, Escobar C (2007) Restricted feeding schedules phase shift daily rhythms of c-Fos and protein Per1 immunoreactivity in corticolimbic regions in rats. Neuroscience 144:344–355
21. Feillet CA, Mendoza J, Albrecht U, Pevet P, Challet E (2008) Forebrain oscillators ticking with different clock hands. Mol Cell Neurosci 37:209–221
22. Guillaumond F, Dardente H, Giguere V, Cermakian N (2005) Differential control of Bmal1 circadian transcription by REV-ERB and ROR nuclear receptors. J Biol Rhythms 20:391–403
23. Abrahamson EE, Moore RY (2001) Suprachiasmatic nucleus in the mouse: retinal innervation, intrinsic organization and efferent projections. Brain Res 916:172–191
24. Mikkelsen JD, Larsen PJ, O'Hare MM, Wiegand SJ (1991) Gastrin releasing peptide in the rat suprachiasmatic nucleus: an immunohistochemical, chromatographic and radioimmunological study. Neuroscience 40:55–66
25. Provencio I, Cooper HM, Foster RG (1998) Retinal projections in mice with inherited retinal degeneration: implications for circadian photoentrainment. J Comp Neurol 395:417–439
26. Panda S, Antoch MP, Miller BH, Su AI, Schook AB, Straume M, Schultz PG, Kay SA, Takahashi JS, Hogenesch JB (2002) Coordinated transcription of key pathways in the mouse by the circadian clock. Cell 109:307–320
27. Ruby NF, Brennan TJ, Xie X, Cao V, Franken P, Heller HC, O'Hara BF (2002) Role of melanopsin in circadian responses to light. Science 298:2211–2213
28. Semo M, Lupi D, Peirson SN, Butler JN, Foster RG (2003) Light-induced c-fos in melanopsin retinal ganglion cells of young and aged rodless/coneless (rd/rdcl) mice. Eur J NeuroSci 18:3007–3017
29. Castel M, Belenky M, Cohen S, Ottersen OP, Storm-Mathisen J (1993) Glutamate-like immunoreactivity in retinal terminals of the mouse suprachiasmatic nucleus. Eur J NeuroSci 5:368–381
30. Hannibal J, Ding JM, Chen D, Fahrenkrug J, Larsen PJ, Gillette MU, Mikkelsen JD (1997) Pituitary adenylate cyclase-activating peptide (PACAP) in the retinohypothalamic tract: a potential daytime regulator of the biological clock. J Neurosci 17:2637–2644
31. Harrington ME, Nance DM, Rusak B (1985) Neuropeptide Y immunoreactivity in the hamster geniculo-suprachiasmatic tract. Brain Res Bull 15:465–472
32. Harrington ME (1997) The ventral lateral geniculate nucleus and the intergeniculate leaflet: interrelated structures in the visual and circadian systems. Neurosci Biobehav Rev 21:705–727
33. Moore RY, Speh JC (1993) GABA is the principal neurotransmitter of the circadian system. Neurosci Lett 150:112–116
34. Card JP, Moore RY (1989) Organization of lateral geniculate-hypothalamic connections in the rat. J Comp Neurol 284:135–147
35. Moore RY, Card JP (1994) Intergeniculate leaflet: an anatomically and functionally distinct subdivision of the lateral geniculate complex. J Comp Neurol 344:403–430
36. Morin LP, Blanchard J, Moore RY (1992) Intergeniculate leaflet and suprachiasmatic nucleus organization and connections in the golden hamster. Vis Neurosci 8:219–230
37. van Esseveldt KE, Lehman MN, Boer GJ (2000) The suprachiasmatic nucleus and the circadian time-keeping system revisited. Brain Res Brain Res Rev 33:34–77
38. Reppert SM, Godson C, Mahle CD, Weaver DR, Slaugenhaupt SA, Gusella JF (1995) Molecular characterization of a second melatonin receptor expressed in human retina and brain: the Mel1b melatonin receptor. Proc Natl Acad Sci U S A 92:8734–8738
39. Masana MI, Dubocovich ML (2001) Melatonin receptor signaling: finding the path through the dark. Sci STKE 2001:PE39

40. Hakansson ML, Brown H, Ghilardi N, Skoda RC, Meister B (1998) Leptin receptor immunoreactivity in chemically defined target neurons of the hypothalamus. J Neurosci 18:559–572
41. Zigman JM, Jones JE, Lee CE, Saper CB, Elmquist JK (2006) Expression of ghrelin receptor mRNA in the rat and the mouse brain. J Comp Neurol 494:528–548
42. Unger J, McNeill TH, Moxley RT 3 rd, White M, Moss A, Livingston JN (1989) Distribution of insulin receptor-like immunoreactivity in the rat forebrain. Neuroscience 31:143–157
43. Cutler DJ, Haraura M, Reed HE, Shen S, Sheward WJ, Morrison CF, Marston HM, Harmar AJ, Piggins HD (2003) The mouse VPAC2 receptor confers suprachiasmatic nuclei cellular rhythmicity and responsiveness to vasoactive intestinal polypeptide in vitro. Eur J NeuroSci 17:197–204
44. Harmar AJ, Marston HM, Shen S, Spratt C, West KM, Sheward WJ, Morrison CF, Dorin JR, Piggins HD, Reubi JC, Kelly JS, Maywood ES, Hastings MH (2002) The VPAC(2) receptor is essential for circadian function in the mouse suprachiasmatic nuclei. Cell 109:497–508
45. Aton SJ, Colwell CS, Harmar AJ, Waschek J, Herzog ED (2005) Vasoactive intestinal polypeptide mediates circadian rhythmicity and synchrony in mammalian clock neurons. Nat Neurosci 8:476–483
46. Maywood ES, Reddy AB, Wong GK, O'Neill JS, O'Brien JA, McMahon DG, Harmar AJ, Okamura H, Hastings MH (2006) Synchronization and maintenance of timekeeping in suprachiasmatic circadian clock cells by neuropeptidergic signaling. Curr Biol 16:599–605
47. Meijer JH, Rietveld WJ (1989) Neurophysiology of the suprachiasmatic circadian pacemaker in rodents. Physiol Rev 69:671–707
48. Daan S, Aschoff J (2001) The entrainment of circadian systems. In Handbook of behavioural neurobiology, Volume 12 Circadian clocks, J.S. Takahashi, F.W. Turek and R.Y. Moore, eds. Kluwer academics / Plenum publishers, New York
49. Mrosovsky N (1996) Locomotor activity and non-photic influences on circadian clocks. Biol Rev Camb Philos Soc 71:343–372
50. Armstrong SM, Cassone VM, Chesworth MJ, Redman JR, Short RV (1986) Synchronization of mammalian circadian rhythms by melatonin. J Neural Transm Suppl 21:375–394
51. Redman J, Armstrong S, Ng KT (1983) Free-running activity rhythms in the rat: entrainment by melatonin. Science 219:1089–1091
52. Pitrosky B, Kirsch R, Malan A, Mocaer E, Pevet P (1999) Organization of rat circadian rhythms during daily infusion of melatonin or S20098, a melatonin agonist. Am J Physiol 277:R812–R828
53. Van Reeth O, Turek FW (1990) Daily injections of triazolam induce long-term changes in hamster circadian period. Am J Physiol 259:R514–R520
54. Cutrera RA, Kalsbeek A, Pevet P (1994) Specific destruction of the serotonergic afferents to the suprachiasmatic nuclei prevents triazolam-induced phase advances of hamster activity rhythms. Behav Brain Res 62:21–28
55. Marchant EG, Mistlberger RE (1997) Anticipation and entrainment to feeding time in intact and SCN-ablated C57BL/6j mice. Brain Res 765:273–282
56. Reebs SG, Mrosovsky N (1989) Effects of induced wheel running on the circadian activity rhythms of Syrian hamsters: entrainment and phase response curve. J Biol Rhythms 4:39–48
57. Bobrzynska KJ, Mrosovsky N (1998) Phase shifting by novelty-induced running: activity dose-response curves at different circadian times. J Comp Physiol [A] 182:251–258
58. Miller JD, Morin LP, Schwartz WJ, Moore RY (1996) New insights into the mammalian circadian clock. Sleep 19:641–667
59. Challet E, Pevet P, Lakhdar-Ghazal N, Malan A (1997) Ventromedial nuclei of the hypothalamus are involved in the phase advance of temperature and activity rhythms in food-restricted rats fed during daytime. Brain Res Bull 43:209–218
60. Challet E, Pevet P, Malan A (1996) Intergeniculate leaflet lesion and daily rhythms in food-restricted rats fed during daytime. Neurosci Lett 216:214–218
61. Schuhler S, Pitrosky B, Saboureau M, Lakhdar-Ghazal N, Pevet P (1999) Role of the thalamic intergeniculate leaflet and its 5-HT afferences in the chronobiological properties of 8-OH-DPAT and triazolam in syrian hamster. Brain Res 849:16–24

62. Damiola F, Le Minh N, Preitner N, Kornmann B, Fleury-Olela F, Schibler U (2000) Restricted feeding uncouples circadian oscillators in peripheral tissues from the central pacemaker in the suprachiasmatic nucleus. Genes Dev 14:2950–2961
63. Stokkan KA, Yamazaki S, Tei H, Sakaki Y, Menaker M (2001) Entrainment of the circadian clock in the liver by feeding. Science 291:490–493
64. Shibata S, Liou SY, Ueki S, Oomura Y (1983) Effects of restricted feeding on single neuron activity of suprachiasmatic neurons in rat hypothalamic slice preparation. Physiol Behav 31:523–528
65. Mistlberger RE (1994) Circadian food-anticipatory activity: formal models and physiological mechanisms. Neurosci Biobehav Rev 18:171–195
66. Castillo MR, Hochstetler KJ, Tavernier RJ Jr, Greene DM, Bult-Ito A (2004) Entrainment of the master circadian clock by scheduled feeding. Am J Physiol Regul Integr Comp Physiol 287:R551–R555
67. Abe H, Kida M, Tsuji K, Mano T (1989) Feeding cycles entrain circadian rhythms of locomotor activity in CS mice but not in C57BL/6J mice. Physiol Behav 45:397–401
68. Challet E, Malan A, Pevet P (1996) Daily hypocaloric feeding entrains circadian rhythms of wheel-running and body temperature in rats kept in constant darkness. Neurosci Lett 211:1–4
69. Mendoza J, Angeles-Castellanos M, Escobar C (2005) Differential role of the accumbens Shell and Core subterritories in food-entrained rhythms of rats. Behav Brain Res 158:133–142
70. Holloway WR Jr, Tsui HW, Grota LJ, Brown GM (1979) Melatonin and corticosterone regulation: feeding time or the light:dark cycle? Life Sci 25:1837–1842
71. Chik CL, Ho AK, Brown GM (1987) Effect of food restriction on 24-h serum and pineal melatonin content in male rats. Acta Endocrinol (Copenh) 115:507–513
72. Meyer-Bernstein EL, Jetton AE, Matsumoto SI, Markuns JF, Lehman MN, Bittman EL (1999) Effects of suprachiasmatic transplants on circadian rhythms of neuroendocrine function in golden hamsters. Endocrinology 140:207–218
73. Sollars PJ, Kimble DP, Pickard GE (1995) Restoration of circadian behavior by anterior hypothalamic heterografts. J Neurosci 15:2109–2122
74. Boer GJ, van Esseveldt LE, Rietveld WJ (1998) Cellular requirements of suprachiasmatic nucleus transplants for restoration of circadian rhythm. Chronobiol Int 15:551–566
75. Jin X, Shearman LP, Weaver DR, Zylka MJ, de Vries GJ, Reppert SM (1999) A molecular mechanism regulating rhythmic output from the suprachiasmatic circadian clock. Cell 96:57–68
76. Uhl GR, Reppert SM (1986) Suprachiasmatic nucleus vasopressin messenger RNA: circadian variation in normal and Brattleboro rats. Science 232:390–393
77. Cagampang FR, Yang J, Nakayama Y, Fukuhara C, Inouye ST (1994) Circadian variation of arginine-vasopressin messenger RNA in the rat suprachiasmatic nucleus. Brain Res Mol Brain Res 24:179–184
78. Tominaga K, Shinohara K, Otori Y, Fukuhara C, Inouye ST (1992) Circadian rhythms of vasopressin content in the suprachiasmatic nucleus of the rat. NeuroReport 3:809–812
79. Yamase K, Takahashi S, Nomura K, Haruta K, Kawashima S (1991) Circadian changes in arginine vasopressin level in the suprachiasmatic nuclei in the rat. Neurosci Lett 130:255–258
80. Mueller CR, Maire P, Schibler U (1990) DBP, a liver-enriched transcriptional activator, is expressed late in ontogeny and its tissue specificity is determined posttranscriptionally. Cell 61:279–291
81. Lopez-Molina L, Conquet F, Dubois-Dauphin M, Schibler U (1997) The DBP gene is expressed according to a circadian rhythm in the suprachiasmatic nucleus and influences circadian behavior. Embo J 16:6762–6771
82. Ripperger JA, Shearman LP, Reppert SM, Schibler U (2000) CLOCK, an essential pacemaker component, controls expression of the circadian transcription factor DBP. Genes Dev 14:679–689
83. Perreau-Lenz S, Pevet P, Buijs RM, Kalsbeek A (2004) The biological clock: the bodyguard of temporal homeostasis. Chronobiol Int 21:1–25
84. Swaab DF, Nijveldt F, Pool CW (1975) Distribution of oxytocin and vasopressin in the rat supraoptic and paraventricular nucleus. J Endocrinol 67:461–462

85. Vandesande F, Dierickx K, DeMey J (1975) Identification of the vasopressin-neurophysin producing neurons of the rat suprachiasmatic nuclei. Cell Tissue Res 156:377–380
86. van den Pol AN, Tsujimoto KL (1985) Neurotransmitters of the hypothalamic suprachiasmatic nucleus: immunocytochemical analysis of 25 neuronal antigens. Neuroscience 15:1049–1086
87. Romijn HJ, Sluiter AA, Pool CW, Wortel J, Buijs RM (1996) Differences in colocalization between Fos and PHI, GRP, VIP and VP in neurons of the rat suprachiasmatic nucleus after a light stimulus during the phase delay versus the phase advance period of the night. J Comp Neurol 372:1–8
88. Ingram CD, Snowball RK, Mihai R (1996) Circadian rhythm of neuronal activity in suprachiasmatic nucleus slices from the vasopressin-deficient Brattleboro rat. Neuroscience 75:635–641
89. Mihai R, Juss TS, Ingram CD (1994) Suppression of suprachiasmatic nucleus neurone activity with a vasopressin receptor antagonist: possible role for endogenous vasopressin in circadian activity cycles in vitro. Neurosci Lett 179:95–99
90. Card JP, Brecha N, Karten HJ, Moore RY (1981) Immunocytochemical localization of vasoactive intestinal polypeptide-containing cells and processes in the suprachiasmatic nucleus of the rat: light and electron microscopic analysis. J Neurosci 1:1289–1303
91. Card JP, Fitzpatrick-McElligott S, Gozes I, Baldino F Jr (1988) Localization of vasopressin-, vasoactive intestinal polypeptide-, peptide histidine isoleucine- and somatostatin-mRNA in rat suprachiasmatic nucleus. Cell Tissue Res 252:307–315
92. Kilcoyne MM, Hoffman DL, Zimmerman EA (1980) Immunocytochemical localization of angiotensin II and vasopressin in rat hypothalamus: evidence for production in the same neuron. Clin Sci (Lond) 59(Suppl. 6):57s–60s
93. Lind RW, Swanson LW, Ganten D (1985) Organization of angiotensin II immunoreactive cells and fibers in the rat central nervous system. An immunohistochemical study. Neuroendocrinology 40:2–24
94. Block CH, Santos RA, Brosnihan KB, Ferrario CM (1988) Immunocytochemical localization of angiotensin-(1–7) in the rat forebrain. Peptides 9:1395–1401
95. Park HT, Baek SY, Kim BS, Kim JB, Kim JJ (1993) Calcitonin gene-related peptide-like immunoreactive (CGRPI) elements in the circadian system of the mouse: an immunohistochemistry combined with retrograde transport study. Brain Res 629:335–341
96. Skofitsch G, Jacobowitz DM (1985) Immunohistochemical mapping of galanin-like neurons in the rat central nervous system. Peptides 6:509–546
97. Mai JK, Kedziora O, Teckhaus L, Sofroniew MV (1991) Evidence for subdivisions in the human suprachiasmatic nucleus. J Comp Neurol 305:508–525
98. Silver R, Romero MT, Besmer HR, Leak R, Nunez JM, LeSauter J (1996) Calbindin-D28K cells in the hamster SCN express light-induced Fos. NeuroReport 7:1224–1228
99. Ikeda M, Allen CN (2003) Developmental changes in calbindin-D28k and calretinin expression in the mouse suprachiasmatic nucleus. Eur J NeuroSci 17:1111–1118
100. Kramer A, Yang FC, Snodgrass P, Li X, Scammell TE, Davis FC, Weitz CJ (2001) Regulation of daily locomotor activity and sleep by hypothalamic EGF receptor signaling. Science 294:2511–2515
101. Cheng MY, Bullock CM, Li C, Lee AG, Bermak JC, Belluzzi J, Weaver DR, Leslie FM, Zhou QY (2002) Prokineticin 2 transmits the behavioural circadian rhythm of the suprachiasmatic nucleus. Nature 417:405–410
102. Kraves S, Weitz CJ (2006) A role for cardiotrophin-like cytokine in the circadian control of mammalian locomotor activity. Nat Neurosci 9:212–219
103. Shigeyoshi Y, Taguchi K, Yamamoto S, Takekida S, Yan L, Tei H, Moriya T, Shibata S, Loros JJ, Dunlap JC, Okamura H (1997) Light-induced resetting of a mammalian circadian clock is associated with rapid induction of the mPer1 transcript. Cell 91:1043–1053
104. Shearman LP, Zylka MJ, Weaver DR, Kolakowski LF Jr, Reppert SM (1997) Two period homologs: circadian expression and photic regulation in the suprachiasmatic nuclei. Neuron 19:1261–1269
105. Honma S, Kawamoto T, Takagi Y, Fujimoto K, Sato F, Noshiro M, Kato Y, Honma K (2002) Dec1 and Dec2 are regulators of the mammalian molecular clock. Nature 419:841–844

106. Albrecht U, Sun ZS, Eichele G, Lee CC (1997) A differential response of two putative mammalian circadian regulators, mper1 and mper2, to light. Cell 91:1055–1064
107. Zheng B, Albrecht U, Kaasik K, Sage M, Lu W, Vaishnav S, Li Q, Sun ZS, Eichele G, Bradley A, Lee CC (2001) Nonredundant roles of the mPer1 and mPer2 genes in the mammalian circadian clock. Cell 105:683–694
108. Liu C, Weaver DR, Jin X, Shearman LP, Pieschl RL, Gribkoff VK, Reppert SM (1997) Molecular dissection of two distinct actions of melatonin on the suprachiasmatic circadian clock. Neuron 19:91–102
109. Shibata S, Cassone VM, Moore RY (1989) Effects of melatonin on neuronal activity in the rat suprachiasmatic nucleus in vitro. Neurosci Lett 97:140–144
110. Stehle J, Vanecek J, Vollrath L (1989) Effects of melatonin on spontaneous electrical activity of neurons in rat suprachiasmatic nuclei: an in vitro iontophoretic study. J Neural Transm 78:173–177
111. McArthur AJ, Gillette MU, Prosser RA (1991) Melatonin directly resets the rat suprachiasmatic circadian clock in vitro. Brain Res 565:158–161
112. Poirel VJ, Boggio V, Dardente H, Pevet P, Masson-Pevet M, Gauer F (2003) Contrary to other non-photic cues, acute melatonin injection does not induce immediate changes of clock gene mRNA expression in the rat suprachiasmatic nuclei. Neuroscience 120:745–755
113. Agez L, Laurent V, Pevet P, Masson-Pevet M, Gauer F (2007) Melatonin affects nuclear orphan receptors mRNA in the rat suprachiasmatic nuclei. Neuroscience 144:522–530
114. Maywood ES, Mrosovsky N, Field MD, Hastings MH (1999) Rapid down-regulation of mammalian period genes during behavioral resetting of the circadian clock. Proc Natl Acad Sci U S A 96:15211–15216
115. Mendoza J, Graff C, Dardente H, Pevet P, Challet E (2005) Feeding cues alter clock gene oscillations and photic responses in the suprachiasmatic nuclei of mice exposed to a light/dark cycle. J Neurosci 25:1514–1522
116. Inouye ST, Kawamura H (1979) Persistence of circadian rhythmicity in a mammalian hypothalamic "island" containing the suprachiasmatic nucleus. Proc Natl Acad Sci U S A 76:5962–5966
117. Green DJ, Gillette R (1982) Circadian rhythm of firing rate recorded from single cells in the rat suprachiasmatic brain slice. Brain Res 245:198–200
118. Shibata S, Oomura Y, Kita H, Hattori K (1982) Circadian rhythmic changes of neuronal activity in the suprachiasmatic nucleus of the rat hypothalamic slice. Brain Res 247:154–158
119. Gillette MU, Reppert SM (1987) The hypothalamic suprachiasmatic nuclei: circadian patterns of vasopressin secretion and neuronal activity in vitro. Brain Res Bull 19:135–139
120. Nakamura W, Honma S, Shirakawa T, Honma K (2001) Regional pacemakers composed of multiple oscillator neurons in the rat suprachiasmatic nucleus. Eur J NeuroSci 14:666–674
121. Albus H, Bonnefont X, Chaves I, Yasui A, Doczy J, van der Horst GT, Meijer JH (2002) Cryptochrome-deficient mice lack circadian electrical activity in the suprachiasmatic nuclei. Curr Biol 12:1130–1133
122. Quintero JE, Kuhlman SJ, McMahon DG (2003) The biological clock nucleus: a multiphasic oscillator network regulated by light. J Neurosci 23:8070–8076
123. Watts AG, Swanson LW (1987) Efferent projections of the suprachiasmatic nucleus: II. Studies using retrograde transport of fluorescent dyes and simultaneous peptide immunohistochemistry in the rat. J Comp Neurol 258:230–252
124. Watts AG, Swanson LW, Sanchez-Watts G (1987) Efferent projections of the suprachiasmatic nucleus: I. Studies using anterograde transport of Phaseolus vulgaris leucoagglutinin in the rat. J Comp Neurol 258:204–229
125. Kalsbeek A, Rikkers M, Vivien-Roels B, Pevet P (1993) Vasopressin and vasoactive intestinal peptide infused in the paraventricular nucleus of the hypothalamus elevate plasma melatonin levels. J Pineal Res 15:46–52
126. Saeb-Parsy K, Lombardelli S, Khan FZ, McDowall K, Au-Yong IT, Dyball RE (2000) Neural connections of hypothalamic neuroendocrine nuclei in the rat. J Neuroendocrinol 12:635–648
127. Buijs RM, Hermes MH, Kalsbeek A (1998) The suprachiasmatic nucleus–paraventricular nucleus interactions: a bridge to the neuroendocrine and autonomic nervous system. Prog Brain Res 119:365–382

128. Buijs RM, van Eden CG, Goncharuk VD, Kalsbeek A (2003) The biological clock tunes the organs of the body: timing by hormones and the autonomic nervous system. J Endocrinol 177:17–26
129. Luiten PG, ter Horst GJ, Steffens AB (1987) The hypothalamus, intrinsic connections and outflow pathways to the endocrine system in relation to the control of feeding and metabolism. Prog Neurobiol 28:1–54
130. Phillipson OT, Griffiths AC (1985) The topographic order of inputs to nucleus accumbens in the rat. Neuroscience 16:275–296
131. Berendse HW, Groenewegen HJ (1990) Organization of the thalamostriatal projections in the rat, with special emphasis on the ventral striatum. J Comp Neurol 299:187–228
132. Simonneaux V, Ribelayga C (2003) Generation of the melatonin endocrine message in mammals: a review of the complex regulation of melatonin synthesis by norepinephrine, peptides, and other pineal transmitters. Pharmacol Rev 55:325–395
133. Larsen PJ (1999) Tracing autonomic innervation of the rat pineal gland using viral transneuronal tracing. Microsc Res Tech 46:296–304
134. Teclemariam-Mesbah R, Ter Horst GJ, Postema F, Wortel J, Buijs RM (1999) Anatomical demonstration of the suprachiasmatic nucleus-pineal pathway. J Comp Neurol 406:171–182
135. Kalsbeek A, Garidou ML, Palm IF, Van Der Vliet J, Simonneaux V, Pevet P, Buijs RM (2000) Melatonin sees the light: blocking GABA-ergic transmission in the paraventricular nucleus induces daytime secretion of melatonin. Eur J NeuroSci 12:3146–3154
136. Garidou ML, Bartol I, Calgari C, Pevet P, Simonneaux V (2001) In vivo observation of a non-noradrenergic regulation of arylalkylamine N-acetyltransferase gene expression in the rat pineal complex. Neuroscience 105:721–729
137. Perreau-Lenz S, Kalsbeek A, Garidou ML, Wortel J, van der Vliet J, van Heijningen C, Simonneaux V, Pevet P, Buijs RM (2003) Suprachiasmatic control of melatonin synthesis in rats: inhibitory and stimulatory mechanisms. Eur J NeuroSci 17:221–228
138. Kalsbeek A, Buijs RM (2002) Output pathways of the mammalian suprachiasmatic nucleus: coding circadian time by transmitter selection and specific targeting. Cell Tissue Res 309:109–118
139. La Fleur SE, Kalsbeek A, Wortel J, Buijs RM (1999) A suprachiasmatic nucleus generated rhythm in basal glucose concentrations. J Neuroendocrinol 11:643–652
140. Kalsbeek A, Ruiter M, La Fleur SE, Cailotto C, Kreier F, Buijs RM (2006) The hypothalamic clock and its control of glucose homeostasis. Prog Brain Res 153:283–307
141. Kalsbeek A, Foppen E, Schalij I, Van Heijningen C, van der Vliet J, Fliers E, Buijs RM (2008) Circadian control of the daily plasma glucose rhythm: an interplay of GABA and glutamate. PLoS One 3:e3194
142. Escobar C, Diaz-Munoz M, Encinas F, Aguilar-Roblero R (1998) Persistence of metabolic rhythmicity during fasting and its entrainment by restricted feeding schedules in rats. Am J Physiol 274:R1309–R1316
143. Satinoff E, Prosser RA (1988) Suprachiasmatic nuclear lesions eliminate circadian rhythms of drinking and activity, but not of body temperature, in male rats. J Biol Rhythms 3:1–22
144. Wachulec M, Li H, Tanaka H, Peloso E, Satinoff E (1997) Suprachiasmatic nuclei lesions do not eliminate homeostatic thermoregulatory responses in rats. J Biol Rhythms 12:226–234
145. Osborne AR, Refinetti R (1995) Effects of hypothalamic lesions on the body temperature rhythm of the golden hamster. NeuroReport 6:2187–2192
146. Rosenwasser AM, Boulos Z, Terman M (1981) Circadian organization of food intake and meal patterns in the rat. Physiol Behav 27:33–39
147. Clarke JD, Coleman GJ (1986) Persistent meal-associated rhythms in SCN-lesioned rats. Physiol Behav 36:105–113
148. Meijer JH (2001) Photic entrainment in mammals. In Handbook of behavioural neurobiology, Volume 12 Circadian clocks, J.S. Takahashi, F.W. Turek and R.Y. Moore, eds. Kluwer academic / Plenum publishers, New York
149. Tosini G, Menaker M (1996) Circadian rhythms in cultured mammalian retina. Science 272:419–421

150. King DP, Zhao Y, Sangoram AM, Wilsbacher LD, Tanaka M, Antoch MP, Steeves TD, Vitaterna MH, Kornhauser JM, Lowrey PL, Turek FW, Takahashi JS (1997) Positional cloning of the mouse circadian clock gene. Cell 89:641–653
151. Oishi K, Sakamoto K, Okada T, Nagase T, Ishida N (1998) Antiphase circadian expression between BMAL1 and period homologue mRNA in the suprachiasmatic nucleus and peripheral tissues of rats. Biochem Biophys Res Commun 253:199–203
152. Kamphuis W, Cailotto C, Dijk F, Bergen A, Buijs RM (2005) Circadian expression of clock genes and clock-controlled genes in the rat retina. Biochem Biophys Res Commun 330:18–26
153. Ruan GX, Zhang DQ, Zhou T, Yamazaki S, McMahon DG (2006) Circadian organization of the mammalian retina. Proc Natl Acad Sci U S A 103:9703–9708
154. Granados-Fuentes D, Tseng A, Herzog ED (2006) A circadian clock in the olfactory bulb controls olfactory responsivity. J Neurosci 26:12219–12225
155. Stephan FK, Swann JM, Sisk CL (1979) Entrainment of circadian rhythms by feeding schedules in rats with suprachiasmatic lesions. Behav Neural Biol 25:545–554
156. Mistlberger RE, Rechtschaffen A (1984) Recovery of anticipatory activity to restricted feeding in rats with ventromedial hypothalamic lesions. Physiol Behav 33:227–235
157. Boulos Z, Rosenwasser AM, Terman M (1980) Feeding schedules and the circadian organization of behavior in the rat. Behav Brain Res 1:39–65
158. Nelson W, Scheving L, Halberg F (1975) Circadian rhythms in mice fed a single daily meal at different stages of lighting regimen. J Nutr 105:171–184
159. Recabarren MP, Valdes JL, Farias P, Seron-Ferre M, Torrealba F (2005) Differential effects of infralimbic cortical lesions on temperature and locomotor activity responses to feeding in rats. Neuroscience 134:1413–1422
160. Krieger DT (1974) New studies on the experimental alteration of the circadian periodicity of plasma corticosteroid levels in the rat. Chronobiologia 1(Suppl 1):82–90
161. Feillet CA, Mendoza J, Pevet P, Challet E (2008) Restricted feeding restores rhythmicity in the pineal gland of arrhythmic suprachiasmatic-lesioned rats. Eur J NeuroSci 28:2451–2458
162. Bodosi B, Gardi J, Hajdu I, Szentirmai E, Obal F Jr, Krueger JM (2004) Rhythms of ghrelin, leptin, and sleep in rats: effects of the normal diurnal cycle, restricted feeding, and sleep deprivation. Am J Physiol Regul Integr Comp Physiol 287:R1071–R1079
163. Martinez-Merlos MT, Angeles-Castellanos M, Diaz-Munoz M, Aguilar-Roblero R, Mendoza J, Escobar C (2004) Dissociation between adipose tissue signals, behavior and the food-entrained oscillator. J Endocrinol 181:53–63
164. Diaz-Munoz M, Vazquez-Martinez O, Aguilar-Roblero R, Escobar C (2000) Anticipatory changes in liver metabolism and entrainment of insulin, glucagon, and corticosterone in food-restricted rats. Am J Physiol Regul Integr Comp Physiol 279:R2048–R2056
165. Honma KI, Honma S, Hiroshige T (1983) Critical role of food amount for prefeeding corticosterone peak in rats. Am J Physiol 245:R339–R344
166. Stephan FK (1983) Circadian rhythms in the rat: constant darkness, entrainment to T cycles and to skeleton photoperiods. Physiol Behav 30:451–462
167. Coleman GJ, Harper S, Clarke JD, Armstrong S (1982) Evidence for a separate meal-associated oscillator in the rat. Physiol Behav 29:107–115
168. Ruis JF, Talamini LM, Buys JP, Rietveld WJ (1989) Effects of time of feeding on recovery of food-entrained rhythms during subsequent fasting in SCN-lesioned rats. Physiol Behav 46:857–866
169. Stephan FK (1984) Phase shifts of circadian rhythms in activity entrained to food access. Physiol Behav 32:663–671
170. Stephan FK (1981) Limits of entrainment to periodic feeding in rats with suprachiasmatic lesions. J Comp Physiol [A] 143:401–410
171. Challet E, Pevet P, Vivien-Roels B, Malan A (1997) Phase-advanced daily rhythms of melatonin, body temperature, and locomotor activity in food-restricted rats fed during daytime. J Biol Rhythms 12:65–79

172. Abe H, Honma S, Honma K (2007) Daily restricted feeding resets the circadian clock in the suprachiasmatic nucleus of CS mice. Am J Physiol Regul Integr Comp Physiol 292:R607–R615
173. Stephan FK (2002) The "other" circadian system: food as a Zeitgeber. J Biol Rhythms 17:284–292
174. Rosenwasser AM, Schulkin J, Adler NT (1985) Circadian wheel-running activity of rats under schedules of limited daily access to salt. Chronobiol Int 2:115–119
175. Mistlberger RE, Houpt TA, Moore-Ede MC (1990) Food-anticipatory rhythms under 24-hour schedules of limited access to single macronutrients. J Biol Rhythms 5:35–46
176. Mistlberger RE, Mumby DG (1992) The limbic system and food-anticipatory circadian rhythms in the rat: ablation and dopamine blocking studies. Behav Brain Res 47:159–168
177. Mistlberger RE (1993) Effects of scheduled food and water access on circadian rhythms of hamsters in constant light, dark, and light:dark. Physiol Behav 53:509–516
178. Mistlberger R, Rusak B (1987) Palatable daily meals entrain anticipatory activity rhythms in free-feeding rats: dependence on meal size and nutrient content. Physiol Behav 41:219–226
179. Mendoza J, Angeles-Castellanos M, Escobar C (2005) A daily palatable meal without food deprivation entrains the suprachiasmatic nucleus of rats. Eur J NeuroSci 22:2855–2862
180. Yoshihara T, Honma S, Mitome M, Honma K (1997) Independence of feeding-associated circadian rhythm from light conditions and meal intervals in SCN lesioned rats. Neurosci Lett 222:95–98
181. Stephan FK (1986) Coupling between feeding- and light-entrainable circadian pacemakers in the rat. Physiol Behav 38:537–544
182. Pitts S, Perone E, Silver R (2003) Food-entrained circadian rhythms are sustained in arrhythmic Clk/Clk mutant mice. Am J Physiol Regul Integr Comp Physiol 285:R57–R67
183. Reick M, Garcia JA, Dudley C, McKnight SL (2001) NPAS2: an analog of clock operative in the mammalian forebrain. Science 293:506–509
184. Dudley CA, Erbel-Sieler C, Estill SJ, Reick M, Franken P, Pitts S, McKnight SL (2003) Altered patterns of sleep and behavioral adaptability in NPAS2-deficient mice. Science 301:379–383
185. Iijima M, Yamaguchi S, van der Horst GT, Bonnefont X, Okamura H, Shibata S (2005) Altered food-anticipatory activity rhythm in Cryptochrome-deficient mice. Neurosci Res 52:166–173
186. Feillet CA, Ripperger JA, Magnone MC, Dulloo A, Albrecht U, Challet E (2006) Lack of food anticipation in Per2 mutant mice. Curr Biol 16:2016–2022
187. Fuller PM, Lu J, Saper CB (2008) Differential rescue of light- and food-entrainable circadian rhythms. Science 320:1074–1077
188. Hara R, Wan K, Wakamatsu H, Aida R, Moriya T, Akiyama M, Shibata S (2001) Restricted feeding entrains liver clock without participation of the suprachiasmatic nucleus. Genes Cells 6:269–278
189. Escobar C, Mendoza JY, Salazar-Juarez A, Avila J, Hernandez-Munoz R, Diaz-Munoz M, Aguilar-Roblero R (2002) Rats made cirrhotic by chronic CCl4 treatment still exhibit anticipatory activity to a restricted feeding schedule. Chronobiol Int 19:1073–1086
190. Davidson AJ, Stokkan KA, Yamazaki S, Menaker M (2002) Food-anticipatory activity and liver per1-luc activity in diabetic transgenic rats. Physiol Behav 76:21–26
191. Davidson AJ, Poole AS, Yamazaki S, Menaker M (2003) Is the food-entrainable circadian oscillator in the digestive system? Genes Brain Behav 2:32–39
192. Miki H, Yano M, Iwanaga H, Tsujinaka T, Nakayama M, Kobayashi M, Oishi K, Shiozaki H, Ishida N, Nagai K, Monden M (2003) Total parenteral nutrition entrains the central and peripheral circadian clocks. NeuroReport 14:1457–1461
193. Moreira AC, Krieger DT (1982) The effects of subdiaphragmatic vagotomy on circadian corticosterone rhythmicity in rats with continuous or restricted food access. Physiol Behav 28:787–790
194. Comperatore CA, Stephan FK (1990) Effects of vagotomy on entrainment of activity rhythms to food access. Physiol Behav 47:671–678

195. Davidson AJ, Stephan FK (1998) Circadian food anticipation persists in capsaicin deafferented rats. J Biol Rhythms 13:422–429
196. Davidson AJ, Aragona BJ, Werner RM, Schroeder E, Smith JC, Stephan FK (2001) Food-anticipatory activity persists after olfactory bulb ablation in the rat. Physiol Behav 72:231–235
197. Honma S, Honma K, Nagasaka T, Hiroshige T (1987) The ventromedial hypothalamic nucleus is not essential for the prefeeding corticosterone peak in rats under restricted daily feeding. Physiol Behav 39:211–215
198. Challet E, Pevet P, Malan A (1997) Lesion of the serotonergic terminals in the suprachiasmatic nuclei limits the phase advance of body temperature rhythm in food-restricted rats fed during daytime. J Biol Rhythms 12:235–244
199. Mistlberger RE, Rusak B (1988) Food anticipatory circadian rhythms in rats with paraventricular and lateral hypothalamic ablations. J Biol Rhythms 3:277–291
200. Mistlberger RE, Antle MC (1999) Neonatal monosodium glutamate alters circadian organization of feeding, food anticipatory activity and photic masking in the rat. Brain Res 842:73–83
201. Landry GJ, Yamakawa GR, Mistlberger RE (2007) Robust food anticipatory circadian rhythms in rats with complete ablation of the thalamic paraventricular nucleus. Brain Res 1141:108–118
202. Davidson AJ, Aragona BJ, Houpt TA, Stephan FK (2001) Persistence of meal-entrained circadian rhythms following area postrema lesions in the rat. Physiol Behav 74:349–354
203. Davidson AJ, Cappendijk SL, Stephan FK (2000) Feeding-entrained circadian rhythms are attenuated by lesions of the parabrachial region in rats. Am J Physiol Regul Integr Comp Physiol 278:R1296–R1304
204. Davidson AJ, Stephan FK (1999) Feeding-entrained circadian rhythms in hypophysectomized rats with suprachiasmatic nucleus lesions. Am J Physiol 277:R1376–R1384
205. Gooley JJ, Schomer A, Saper CB (2006) The dorsomedial hypothalamic nucleus is critical for the expression of food-entrainable circadian rhythms. Nat Neurosci 9:398–407
206. Landry GJ, Simon MM, Webb IC, Mistlberger RE (2006) Persistence of a behavioral food-anticipatory circadian rhythm following dorsomedial hypothalamic ablation in rats. Am J Physiol Regul Integr Comp Physiol 290:R1527–R1534
207. Mieda M, Williams SC, Richardson JA, Tanaka K, Yanagisawa M (2006) The dorsomedial hypothalamic nucleus as a putative food-entrainable circadian pacemaker. Proc Natl Acad Sci U S A 103:12150–12155
208. Mistlberger RE, Yamazaki S, Pendergast JS, Landry GJ, Takumi T, Nakamura W (2008) Comment on "Differential rescue of light- and food-entrainable circadian rhythms". Science 322:675 author reply 675
209. Angeles-Castellanos M, Aguilar-Roblero R, Escobar C (2004) c-Fos expression in hypothalamic nuclei of food-entrained rats. Am J Physiol Regul Integr Comp Physiol 286:R158–R165
210. Angeles-Castellanos M, Mendoza J, Diaz-Munoz M, Escobar C (2005) Food entrainment modifies the c-Fos expression pattern in brain stem nuclei of rats. Am J Physiol Regul Integr Comp Physiol 288:R678–R684
211. Inzunza O, Seron-Ferre MJ, Bravo H, Torrealba F (2000) Tuberomammillary nucleus activation anticipates feeding under a restricted schedule in rats. Neurosci Lett 293:139–142
212. de Vasconcelos AP, Bartol-Munier I, Feillet CA, Gourmelen S, Pevet P, Challet E (2006) Modifications of local cerebral glucose utilization during circadian food-anticipatory activity. Neuroscience 139:741–748
213. Beaule C, Amir S (1999) Photic entrainment and induction of immediate-early genes within the rat circadian system. Brain Res 821:95–100
214. Edelstein K, Beaule C, D'Abramo R, Amir S (2000) Expression profiles of JunB and c-Fos proteins in the rat circadian system. Brain Res 870:54–65
215. Janik D, Mrosovsky N (1992) Gene expression in the geniculate induced by a nonphotic circadian phase shifting stimulus. NeuroReport 3:575–578

216. Mikkelsen JD, Vrang N, Mrosovsky N (1998) Expression of Fos in the circadian system following nonphotic stimulation. Brain Res Bull 47:367–376
217. Balsalobre A, Damiola F, Schibler U (1998) A serum shock induces circadian gene expression in mammalian tissue culture cells. Cell 93:929–937
218. Balsalobre A, Brown SA, Marcacci L, Tronche F, Kellendonk C, Reichardt HM, Schutz G, Schibler U (2000) Resetting of circadian time in peripheral tissues by glucocorticoid signaling. Science 289:2344–2347
219. Vujovic N, Davidson AJ, Menaker M (2008) Sympathetic input modulates, but does not determine, phase of peripheral circadian oscillators. Am J Physiol Regul Integr Comp Physiol 295:R355–R360
220. Iijima M, Nikaido T, Akiyama M, Moriya T, Shibata S (2002) Methamphetamine-induced, suprachiasmatic nucleus-independent circadian rhythms of activity and mPer gene expression in the striatum of the mouse. Eur J NeuroSci 16:921–929
221. Waddington Lamont E, Harbour VL, Barry-Shaw J, Renteria Diaz L, Robinson B, Stewart J, Amir S (2007) Restricted access to food, but not sucrose, saccharine, or salt, synchronizes the expression of Period2 protein in the limbic forebrain. Neuroscience 144:402–411
222. Wasielewski JA, Holloway FA (2001) Alcohol's interactions with circadian rhythms. A focus on body temperature. Alcohol Res Health 25:94–100
223. Jones EM, Knutson D, Haines D (2003) Common problems in patients recovering from chemical dependency. Am Fam Physician 68:1971–1978
224. Morgan PT, Pace-Schott EF, Sahul ZH, Coric V, Stickgold R, Malison RT (2006) Sleep, sleep-dependent procedural learning and vigilance in chronic cocaine users: Evidence for occult insomnia. Drug Alcohol Depend 82:238–249
225. Raymond RC, Warren M, Morris RW, Leikin JB (1992) Periodicity of presentations of drugs of abuse and overdose in an emergency department. J Toxicol Clin Toxicol 30:467–478
226. Bunney WE, Bunney BG (2000) Molecular clock genes in man and lower animals: possible implications for circadian abnormalities in depression. Neuropsychopharmacology 22:335–345
227. Kandel DB, Huang FY, Davies M (2001) Comorbidity between patterns of substance use dependence and psychiatric syndromes. Drug Alcohol Depend 64:233–241
228. Grandin LD, Alloy LB, Abramson LY (2006) The social zeitgeber theory, circadian rhythms, and mood disorders: review and evaluation. Clin Psychol Rev 26:679–694
229. McGrath RE, Yahia M (1993) Preliminary data on seasonally related alcohol dependence. J Clin Psychiatry 54:260–262
230. Wirz-Justice A, Richter R (1979) Seasonality in biochemical determinations: a source of variance and a clue to the temporal incidence of affective illness. Psychiatry Res 1:53–60
231. Shibley HL, Malcolm RJ, Veatch LM (2008) Adolescents with insomnia and substance abuse: consequences and comorbidities. J Psychiatr Pract 14:146–153
232. Rosenwasser AM, Fecteau ME, Logan RW, Reed JD, Cotter SJ, Seggio JA (2005) Circadian activity rhythms in selectively bred ethanol-preferring and nonpreferring rats. Alcohol 36:69–81
233. Nestler EJ, Carlezon WA Jr (2006) The mesolimbic dopamine reward circuit in depression. Biol Psychiatry 59:1151–1159
234. LŸscher C (2007) Drugs of abuse. In: Katzung BG (ed) Basic and clinical pharmacology, 10th edn. New York, McGraw Hill, pp 511–525
235. Baird TJ, Gauvin D (2000) Characterization of cocaine self-administration and pharmacokinetics as a function of time of day in the rat. Pharmacol Biochem Behav 65:289–299
236. Abarca C, Albrecht U, Spanagel R (2002) Cocaine sensitization and reward are under the influence of circadian genes and rhythm. Proc Natl Acad Sci U S A 99:9026–9030
237. Andretic R, Chaney S, Hirsh J (1999) Requirement of circadian genes for cocaine sensitization in Drosophila. Science 285:1066–1068
238. Yuferov V, Kroslak T, Laforge KS, Zhou Y, Ho A, Kreek MJ (2003) Differential gene expression in the rat caudate putamen after "binge" cocaine administration: advantage of triplicate microarray analysis. Synapse 48:157–169
239. Sanchis-Segura C, Spanagel R (2006) Behavioural assessment of drug reinforcement and addictive features in rodents: an overview. Addict Biol 11:2–38

240. Uz T, Ahmed R, Akhisaroglu M, Kurtuncu M, Imbesi M, Dirim Arslan A, Manev H (2005) Effect of fluoxetine and cocaine on the expression of clock genes in the mouse hippocampus and striatum. Neuroscience 134:1309–1316
241. Andretic R, Hirsh J (2000) Circadian modulation of dopamine receptor responsiveness in *Drosophila melanogaster*. Proc Natl Acad Sci U S A 97:1873–1878
242. Witkovsky P (2004) Dopamine and retinal function. Doc Ophthalmol 108:17–40
243. Doi M, Yujnovsky I, Hirayama J, Malerba M, Tirotta E, Sassone-Corsi P, Borrelli E (2006) Impaired light masking in dopamine D2 receptor-null mice. Nat Neurosci 9:732–734
244. Yujnovsky I, Hirayama J, Doi M, Borrelli E, Sassone-Corsi P (2006) Signaling mediated by the dopamine D2 receptor potentiates circadian regulation by CLOCK:BMAL1. Proc Natl Acad Sci U S A 103:6386–6391
245. McClung CA, Sidiropoulou K, Vitaterna M, Takahashi JS, White FJ, Cooper DC, Nestler EJ (2005) Regulation of dopaminergic transmission and cocaine reward by the Clock gene. Proc Natl Acad Sci U S A 102:9377–9381
246. Roybal K, Theobold D, Graham A, Dinieri JA, Russo SJ, Krishnan V, Chakravarty S, Peevey J, Oehrlein N, Birnbaum S, Vitaterna MH, Orsulak P, Takahashi JS, Nestler EJ, Carlezon WA Jr, McClung CA (2007) Mania-like behavior induced by disruption of CLOCK. Proc Natl Acad Sci U S A 104(15):6406–6411
247. Hampp G, Ripperger JA, Houben T, Schmutz I, Blex C, Perreau-Lenz S, Brunk I, Spanagel R, Ahnert-Hilger G, Meijer JH, Albrecht U (2008) Regulation of monoamine oxidase a by circadian-clock components implies clock influence on mood. Curr Biol 18:678–683
248. Partonen T, Treutlein J, Alpman A, Frank J, Johansson C, Depner M, Aron L, Rietschel M, Wellek S, Soronen P, Paunio T, Koch A, Chen P, Lathrop M, Adolfsson R, Persson ML, Kasper S, Schalling M, Peltonen L, Schumann G (2007) Three circadian clock genes Per2, Arntl, and Npas2 contribute to winter depression. Ann Med 39:229–238
249. Honma K, Honma S, Hiroshige T (1986) Disorganization of the rat activity rhythm by chronic treatment with methamphetamine. Physiol Behav 38:687–695
250. Honma K, Honma S, Hiroshige T (1987) Activity rhythms in the circadian domain appear in suprachiasmatic nuclei lesioned rats given methamphetamine. Physiol Behav 40:767–774
251. Masubuchi S, Honma S, Abe H, Nakamura W, Honma K (2001) Circadian activity rhythm in methamphetamine-treated Clock mutant mice. Eur J NeuroSci 14:1177–1180
252. Honma S, Yasuda T, Yasui A, van der Horst GT, Honma K (2008) Circadian behavioral rhythms in Cry1/Cry2 double-deficient mice induced by methamphetamine. J Biol Rhythms 23:91–94
253. Tataroglu O, Davidson AJ, Benvenuto LJ, Menaker M (2006) The methamphetamine-sensitive circadian oscillator (MASCO) in mice. J Biol Rhythms 21:185–194
254. Nikaido T, Akiyama M, Moriya T, Shibata S (2001) Sensitized increase of period gene expression in the mouse caudate/putamen caused by repeated injection of methamphetamine. Mol Pharmacol 59:894–900
255. Yamamoto H, Imai K, Takamatsu Y, Kamegaya E, Kishida M, Hagino Y, Hara Y, Shimada K, Yamamoto T, Sora I, Koga H, Ikeda K (2005) Methamphetamine modulation of gene expression in the brain: analysis using customized cDNA microarray system with the mouse homologues of KIAA genes. Brain Res Mol Brain Res 137:40–46
256. Brower KJ (2001) Alcohol's effects on sleep in alcoholics. Alcohol Res Health 25:110–125
257. Devaney M, Graham D, Greeley J (2003) Circadian variation of the acute and delayed response to alcohol: investigation of core body temperature variations in humans. Pharmacol Biochem Behav 75:881–887
258. Fonzi S, Solinas GP, Costelli P, Parodi C, Murialdo G, Bo P, Albergati A, Montalbetti L, Savoldi F, Polleri A (1994) Melatonin and cortisol circadian secretion during ethanol withdrawal in chronic alcoholics. Chronobiologia 21:109–112
259. Imatoh N, Nakazawa Y, Ohshima H, Ishibashi M, Yokoyama T (1986) Circadian rhythm of REM sleep of chronic alcoholics during alcohol withdrawal. Drug Alcohol Depend 18:77–85
260. Kawano Y, Pontes CS, Abe H, Takishita S, Omae T (2002) Effects of alcohol consumption and restriction on home blood pressure in hypertensive patients: serial changes in the morning and evening records. Clin Exp Hypertens 24:33–39

261. Liu Y, Higuchi S, Motohashi Y (2000) Time-of-day effects of ethanol consumption on EEG topography and cognitive event-related potential in adult males. J Physiol Anthropol Appl Human Sci 19:249–254
262. Rosenwasser AM, Wirz-Justice A (1997) Circadian rhythms and depression: clinical and experimental models. In: Redfern PH, Lemmer B (eds) Physiology and Pharmacology of Biological Rhythms. Berlin, Springer, pp 457–486
263. Driessen M, Meier S, Hill A, Wetterling T, Lange W, Junghanns K (2001) The course of anxiety, depression and drinking behaviours after completed detoxification in alcoholics with and without comorbid anxiety and depressive disorders. Alcohol Alcohol 36:249–255
264. Drummond SP, Gillin JC, Smith TL, DeModena A (1998) The sleep of abstinent pure primary alcoholic patients: natural course and relationship to relapse. Alcohol Clin Exp Res 22:1796–1802
265. Landolt HP, Gillin JC (2001) Sleep abnormalities during abstinence in alcohol-dependent patients. Aetiology and management. CNS Drugs 15:413–425
266. Rosenwasser AM, Fecteau ME, Logan RW (2005) Effects of ethanol intake and ethanol withdrawal on free-running circadian activity rhythms in rats. Physiol Behav 84:537–542
267. Chen CP, Kuhn P, Advis JP, Sarkar DK (2004) Chronic ethanol consumption impairs the circadian rhythm of pro-opiomelanocortin and period genes mRNA expression in the hypothalamus of the male rat. J Neurochem 88:1547–1554
268. Zheng B, Larkin DW, Albrecht U, Sun ZS, Sage M, Eichele G, Lee CC, Bradley A (1999) The mPer2 gene encodes a functional component of the mammalian circadian clock. Nature 400:169–173
269. Spanagel R, Pendyala G, Abarca C, Zghoul T, Sanchis-Segura C, Magnone MC, Lascorz J, Depner M, Holzberg D, Soyka M, Schreiber S, Matsuda F, Lathrop M, Schumann G, Albrecht U (2005) The clock gene Per2 influences the glutamatergic system and modulates alcohol consumption. Nat Med 11:35–42
270. Mann K, Lehert P, Morgan MY (2004) The efficacy of acamprosate in the maintenance of abstinence in alcohol-dependent individuals: results of a meta-analysis. Alcohol Clin Exp Res 28:51–63
271. Trinkoff AM, Storr CL (1998) Work schedule characteristics and substance use in nurses. Am J Ind Med 34:266–271
272. Rogers HL, Reilly SM (2002) A survey of the health experiences of international business travelers. Part One – physiological aspects. Aaohn J 50:449–459
273. Pelissier AL, Gantenbein M, Bruguerolle B (1998) Nicotine-induced perturbations on heart rate, body temperature and locomotor activity daily rhythms in rats. J Pharm Pharmacol 50:929–934
274. Yugar-Toledo JC, Ferreira-Melo SE, Sabha M, Nogueira EA, Coelho OR, Consolin Colombo FM, Irigoyen MC, Moreno H Jr (2005) Blood pressure circadian rhythm and endothelial function in heavy smokers: acute effects of transdermal nicotine. J Clin Hypertens (Greenwich) 7:721–728
275. Wittmann M, Dinich J, Merrow M, Roenneberg T (2006) Social jetlag: misalignment of biological and social time. Chronobiol Int 23:497–509
276. McKernan RM, Rosahl TW, Reynolds DS, Sur C, Wafford KA, Atack JR, Farrar S, Myers J, Cook G, Ferris P, Garrett L, Bristow L, Marshall G, Macaulay A, Brown N, Howell O, Moore KW, Carling RW, Street LJ, Castro JL, Ragan CI, Dawson GR, Whiting PJ (2000) Sedative but not anxiolytic properties of benzodiazepines are mediated by the GABA(A) receptor alpha1 subtype. Nat Neurosci 3:587–592
277. Turek FW, Losee-Olson S (1986) A benzodiazepine used in the treatment of insomnia phase-shifts the mammalian circadian clock. Nature 321:167–168
278. Wee BE, Turek FW (1989) Midazolam, a short-acting benzodiazepine, resets the circadian clock of the hamster. Pharmacol Biochem Behav 32:901–906
279. Copinschi G, Van Onderbergen A, L'Hermite-Baleriaux M, Szyper M, Caufriez A, Bosson D, L'Hermite M, Robyn C, Turek FW, Van Cauter E (1990) Effects of the short-acting benzodiazepine triazolam, taken at bedtime, on circadian and sleep-related hormonal profiles in normal men. Sleep 13:232–244

280. Akiyama M, Kirihara T, Takahashi S, Minami Y, Yoshinobu Y, Moriya T, Shibata S (1999) Modulation of mPer1 gene expression by anxiolytic drugs in mouse cerebellum. Br J Pharmacol 128:1616–1622
281. Pallier PN, Maywood ES, Zheng Z, Chesham JE, Inyushkin AN, Dyball R, Hastings MH, Morton AJ (2007) Pharmacological imposition of sleep slows cognitive decline and reverses dysregulation of circadian gene expression in a transgenic mouse model of Huntington's disease. J Neurosci 27:7869–7878
282. Willoughby JO, Medvedev A (1996) Opioid receptor activation resets the hypothalamic clock generating growth hormone secretory bursts in the rat. J Endocrinol 148:149–155
283. Vansteensel MJ, Magnone MC, van Oosterhout F, Baeriswyl S, Albrecht U, Albus H, Dahan A, Meijer JH (2005) The opioid fentanyl affects light input, electrical activity and Per gene expression in the hamster suprachiasmatic nuclei. Eur J NeuroSci 21:2958–2966
284. Liu Y, Wang Y, Wan C, Zhou W, Peng T, Wang Z, Li G, Cornelisson G, Halberg F (2005) The role of mPer1 in morphine dependence in mice. Neuroscience 130:383–388
285. Liu Y, Wang Y, Jiang Z, Wan C, Zhou W, Wang Z (2007) The extracellular signal-regulated kinase signaling pathway is involved in the modulation of morphine-induced reward by mPer1. Neuroscience 146:265–271
286. Motzkus D, Maronde E, Grunenberg U, Lee CC, Forssmann W, Albrecht U (2000) The human PER1 gene is transcriptionally regulated by multiple signaling pathways. FEBS Lett 486:315–319
287. Akashi M, Hayasaka N, Yamazaki S, Node K (2008) Mitogen-activated protein kinase is a functional component of the autonomous circadian system in the suprachiasmatic nucleus. J Neurosci 28:4619–4623
288. Perron RR, Tyson RL, Sutherland GR (2001) Delta9-tetrahydrocannabinol increases brain temperature and inverts circadian rhythms. NeuroReport 12:3791–3794
289. Walther S, Mahlberg R, Eichmann U, Kunz D (2006) Delta-9-tetrahydrocannabinol for nighttime agitation in severe dementia. Psychopharmacology (Berl) 185:524–528
290. Sanford AE, Castillo E, Gannon RL (2008) Cannabinoids and hamster circadian activity rhythms. Brain Res 1222:141–148
291. Lapierre O, Montplaisir J, Lamarre M, Bedard MA (1990) The effect of gamma-hydroxybutyrate on nocturnal and diurnal sleep of normal subjects: further considerations on REM sleep-triggering mechanisms. Sleep 13:24–30
292. Meerlo P, Westerveld P, Turek FW, Koehl M (2004) Effects of gamma-hydroxybutyrate (GHB) on vigilance states and EEG in mice. Sleep 27:899–904
293. Bunger MK, Wilsbacher LD, Moran SM, Clendenin C, Radcliffe LA, Hogenesch JB, Simon MC, Takahashi JS, Bradfield CA (2000) Mop3 is an essential component of the master circadian pacemaker in mammals. Cell 103:1009–1017
294. Albrecht U, Zheng B, Larkin D, Sun ZS, Lee CC (2001) MPer1 and mper2 are essential for normal resetting of the circadian clock. J Biol Rhythms 16:100–104
295. van der Horst GT, Muijtjens M, Kobayashi K, Takano R, Kanno S, Takao M, de Wit J, Verkerk A, Eker AP, van Leenen D, Buijs R, Bootsma D, Hoeijmakers JH, Yasui A (1999) Mammalian Cry1 and Cry2 are essential for maintenance of circadian rhythms. Nature 398:627–630
296. Borbely AA (1982) A two process model of sleep regulation. Hum Neurobiol 1:195–204
297. Naylor E, Bergmann BM, Krauski K, Zee PC, Takahashi JS, Vitaterna MH, Turek FW (2000) The circadian clock mutation alters sleep homeostasis in the mouse. J Neurosci 20:8138–8143
298. Laposky A, Easton A, Dugovic C, Walisser J, Bradfield C, Turek F (2005) Deletion of the mammalian circadian clock gene BMAL1/Mop3 alters baseline sleep architecture and the response to sleep deprivation. Sleep 28:395–409
299. Wisor JP, O'Hara BF, Terao A, Selby CP, Kilduff TS, Sancar A, Edgar DM, Franken P (2002) A role for cryptochromes in sleep regulation. BMC Neurosci 3:20
300. Franken P, Lopez-Molina L, Marcacci L, Schibler U, Tafti M (2000) The transcription factor DBP affects circadian sleep consolidation and rhythmic EEG activity. J Neurosci 20:617–625

301. Hu WP, Li JD, Zhang C, Boehmer L, Siegel JM, Zhou QY (2007) Altered circadian and homeostatic sleep regulation in prokineticin 2-deficient mice. Sleep 30:247–256
302. Kopp C, Albrecht U, Zheng B, Tobler I (2002) Homeostatic sleep regulation is preserved in mPer1 and mPer2 mutant mice. Eur J NeuroSci 16:1099–1106
303. Shiromani PJ, Xu M, Winston EM, Shiromani SN, Gerashchenko D, Weaver DR (2004) Sleep rhythmicity and homeostasis in mice with targeted disruption of mPeriod genes. Am J Physiol Regul Integr Comp Physiol 287:R47–R57
304. Garcia JA, Zhang D, Estill SJ, Michnoff C, Rutter J, Reick M, Scott K, Diaz-Arrastia R, McKnight SL (2000) Impaired cued and contextual memory in NPAS2-deficient mice. Science 288:2226–2230
305. Zueger M, Urani A, Chourbaji S, Zacher C, Lipp HP, Albrecht U, Spanagel R, Wolfer DP, Gass P (2006) mPer1 and mPer2 mutant mice show regular spatial and contextual learning in standardized tests for hippocampus-dependent learning. J Neural Transm 113:347–356
306. Cain SW, Ralph MR (2008) Circadian modulation of conditioned place preference in hamsters does not require the suprachiasmatic nucleus. Neurobiol Learn Mem 91:81–84
307. Ruby NF, Hwang CE, Wessells C, Fernandez F, Zhang P, Sapolsky R, Heller HC (2008) Hippocampal-dependent learning requires a functional circadian system. Proc Natl Acad Sci U S A 105:15593–15598
308. Van der Zee EA, Havekes R, Barf RP, Hut RA, Nijholt IM, Jacobs EH, Gerkema MP (2008) Circadian time-place learning in mice depends on Cry genes. Curr Biol 18:844–848
309. Easton A, Arbuzova J, Turek FW (2003) The circadian Clock mutation increases exploratory activity and escape-seeking behavior. Genes Brain Behav 2:11–19
310. Karl T, Burne TH, Herzog H (2006) Effect of Y1 receptor deficiency on motor activity, exploration, and anxiety. Behav Brain Res 167:87–93
311. Wersinger SR, Caldwell HK, Christiansen M, Young WS III (2007) Disruption of the vasopressin 1b receptor gene impairs the attack component of aggressive behavior in mice. Genes Brain Behav 6:653–660
312. Karl T, Duffy L, Herzog H (2008) Behavioural profile of a new mouse model for NPY deficiency. Eur J NeuroSci 28:173–180
313. Iitaka C, Miyazaki K, Akaike T, Ishida N (2005) A role for glycogen synthase kinase-3beta in the mammalian circadian clock. J Biol Chem 280:29397–29402
314. Yin L, Wang J, Klein PS, Lazar MA (2006) Nuclear receptor Rev-erbalpha is a critical lithium-sensitive component of the circadian clock. Science 311:1002–1005
315. Kaladchibachi SA, Doble B, Anthopoulos N, Woodgett JR, Manoukian AS (2007) Glycogen synthase kinase 3, circadian rhythms, and bipolar disorder: a molecular link in the therapeutic action of lithium. J Circadian Rhythms 5:3
316. Hampp G, Albrecht U (2008) The circadian clock and mood-related behavior. Commun Integr Biol 1:1–3
317. Randrup A, Braestrup C (1977) Uptake inhibition of biogenic amines by newer antidepressant drugs: relevance to the dopamine hypothesis of depression. Psychopharmacology (Berl) 53:309–314
318. Berton O, Nestler EJ (2006) New approaches to antidepressant drug discovery: beyond monoamines. Nat Rev Neurosci 7:137–151
319. Klemfuss H (1992) Rhythms and the pharmacology of lithium. Pharmacol Ther 56:53–78
320. Rosenthal NE, Sack DA, Gillin JC, Lewy AJ, Goodwin FK, Davenport Y, Mueller PS, Newsome DA, Wehr TA (1984) Seasonal affective disorder. A description of the syndrome and preliminary findings with light therapy. Arch Gen Psychiatry 41:72–80
321. Yan L, Silver R (2002) Differential induction and localization of mPer1 and mPer2 during advancing and delaying phase shifts. Eur J NeuroSci 16:1531–1540
322. Jones CR, Campbell SS, Zone SE, Cooper F, DeSano A, Murphy PJ, Jones B, Czajkowski L, Ptacek LJ (1999) Familial advanced sleep-phase syndrome: a short-period circadian rhythm variant in humans. Nat Med 5:1062–1065
323. Toh KL, Jones CR, He Y, Eide EJ, Hinz WA, Virshup DM, Ptacek LJ, Fu YH (2001) An hPer2 phosphorylation site mutation in familial advanced sleep phase syndrome. Science 291:1040–1043

324. Xu Y, Padiath QS, Shapiro RE, Jones CR, Wu SC, Saigoh N, Saigoh K, Ptacek LJ, Fu YH (2005) Functional consequences of a CKIdelta mutation causing familial advanced sleep phase syndrome. Nature 434:640–644
325. Okawa M, Uchiyama M (2007) Circadian rhythm sleep disorders: characteristics and entrainment pathology in delayed sleep phase and non-24-h sleep-wake syndrome. Sleep Med Rev 11:485–496
326. Takano A, Uchiyama M, Kajimura N, Mishima K, Inoue Y, Kamei Y, Kitajima T, Shibui K, Katoh M, Watanabe T, Hashimotodani Y, Nakajima T, Ozeki Y, Hori T, Yamada N, Toyoshima R, Ozaki N, Okawa M, Nagai K, Takahashi K, Isojima Y, Yamauchi T, Ebisawa T (2004) A missense variation in human casein kinase I epsilon gene that induces functional alteration and shows an inverse association with circadian rhythm sleep disorders. Neuropsychopharmacology 29:1901–1909
327. Castro RM, Barbosa AA, Pedrazzoli M, Tufik S (2008) Casein kinase I epsilon (CKIvarepsilon) N408 allele is very rare in the Brazilian population and is not involved in susceptibility to circadian rhythm sleep disorders. Behav Brain Res 193:156–157
328. Boivin DB (2000) Influence of sleep-wake and circadian rhythm disturbances in psychiatric disorders. J Psychiatry Neurosci 25:446–458
329. Wirz-Justice A (2006) Biological rhythm disturbances in mood disorders. Int Clin Psychopharmacol 21(Suppl 1):S11–S15
330. Wirz-Justice A, Van den Hoofdakker RH (1999) Sleep deprivation in depression: what do we know, where do we go? Biol Psychiatry 46:445–453
331. Lam RW, Levitan RD (2000) Pathophysiology of seasonal affective disorder: a review. J Psychiatry Neurosci 25:469–480
332. Levitan RD (2007) The chronobiology and neurobiology of winter seasonal affective disorder. Dialogues Clin Neurosci 9:315–324
333. Magnusson A, Boivin D (2003) Seasonal affective disorder: an overview. Chronobiol Int 20:189–207
334. Volicer L, Harper DG, Manning BC, Goldstein R, Satlin A (2001) Sundowning and circadian rhythms in Alzheimer's disease. Am J Psychiatry 158:704–711
335. Huitron-Resendiz S, Sanchez-Alavez M, Gallegos R, Berg G, Crawford E, Giacchino JL, Games D, Henriksen SJ, Criado JR (2002) Age-independent and age-related deficits in visuospatial learning, sleep-wake states, thermoregulation and motor activity in PDAPP mice. Brain Res 928:126–137
336. Wisor JP, Edgar DM, Yesavage J, Ryan HS, McCormick CM, Lapustea N, Murphy GM Jr (2005) Sleep and circadian abnormalities in a transgenic mouse model of Alzheimer's disease: a role for cholinergic transmission. Neuroscience 131:375–385
337. Ambree O, Touma C, Gortz N, Keyvani K, Paulus W, Palme R, Sachser N (2006) Activity changes and marked stereotypic behavior precede Abeta pathology in TgCRND8 Alzheimer mice. Neurobiol Aging 27:955–964
338. Van Dam D, D'Hooge R, Staufenbiel M, Van Ginneken C, Van Meir F, De Deyn PP (2003) Age-dependent cognitive decline in the APP23 model precedes amyloid deposition. Eur J NeuroSci 17:388–396
339. Landolt HP, Glatzel M, Blattler T, Achermann P, Roth C, Mathis J, Weis J, Tobler I, Aguzzi A, Bassetti CL (2006) Sleep-wake disturbances in sporadic Creutzfeldt-Jakob disease. Neurology 66:1418–1424
340. Plazzi G, Schutz Y, Cortelli P, Provini F, Avoni P, Heikkila E, Tinuper P, Solieri L, Lugaresi E, Montagna P (1997) Motor overactivity and loss of motor circadian rhythm in fatal familial insomnia: an actigraphic study. Sleep 20:739–742
341. Portaluppi F, Cortelli P, Avoni P, Vergnani L, Maltoni P, Pavani A, Sforza E, Degli Uberti EC, Gambetti P, Lugaresi E (1994) Progressive disruption of the circadian rhythm of melatonin in fatal familial insomnia. J Clin Endocrinol Metab 78:1075–1078
342. Tobler I, Gaus SE, Deboer T, Achermann P, Fischer M, Rulicke T, Moser M, Oesch B, McBride PA, Manson JC (1996) Altered circadian activity rhythms and sleep in mice devoid of prion protein. Nature 380:639–642

343. Morton AJ, Wood NI, Hastings MH, Hurelbrink C, Barker RA, Maywood ES (2005) Disintegration of the sleep-wake cycle and circadian timing in Huntington's disease. J Neurosci 25:157–163
344. Nygard M, Hill RH, Wikstrom MA, Kristensson K (2005) Age-related changes in electrophysiological properties of the mouse suprachiasmatic nucleus in vitro. Brain Res Bull 65:149–154
345. Valentinuzzi VS, Scarbrough K, Takahashi JS, Turek FW (1997) Effects of aging on the circadian rhythm of wheel-running activity in C57BL/6 mice. Am J Physiol 273:R1957–R1964
346. Oster H, Baeriswyl S, Van Der Horst GT, Albrecht U (2003) Loss of circadian rhythmicity in aging mPer1–/–mCry2–/– mutant mice. Genes Dev 17:1366–1379
347. Oster H, van der Horst GT, Albrecht U (2003) Daily variation of clock output gene activation in behaviorally arrhythmic mPer/mCry triple mutant mice. Chronobiol Int 20:683–695
348. Kolker DE, Vitaterna MH, Fruechte EM, Takahashi JS, Turek FW (2004) Effects of age on circadian rhythms are similar in wild-type and heterozygous Clock mutant mice. Neurobiol Aging 25:517–523
349. De Leersnyder H, De Blois MC, Claustrat B, Romana S, Albrecht U, Von Kleist-Retzow JC, Delobel B, Viot G, Lyonnet S, Vekemans M, Munnich A (2001) Inversion of the circadian rhythm of melatonin in the Smith–Magenis syndrome. J Pediatr 139:111–116
350. De Leersnyder H, Bresson JL, de Blois MC, Souberbielle JC, Mogenet A, Delhotal-Landes B, Salefranque F, Munnich A (2003) Beta 1-adrenergic antagonists and melatonin reset the clock and restore sleep in a circadian disorder, Smith-Magenis syndrome. J Med Genet 40:74–78
351. De Leersnyder H, Claustrat B, Munnich A, Verloes A (2006) Circadian rhythm disorder in a rare disease: Smith–Magenis syndrome. Mol Cell Endocrinol 252:88–91
352. Potocki L, Glaze D, Tan DX, Park SS, Kashork CD, Shaffer LG, Reiter RJ, Lupski JR (2000) Circadian rhythm abnormalities of melatonin in Smith–Magenis syndrome. J Med Genet 37:428–433
353. Walz K, Spencer C, Kaasik K, Lee CC, Lupski JR, Paylor R (2004) Behavioral characterization of mouse models for Smith–Magenis syndrome and dup(17)(p11.2p11.2). Hum Mol Genet 13:367–378
354. Harvey MT, Kennedy CH (2002) Polysomnographic phenotypes in developmental disabilities. Int J Dev Neurosci 20:443–448
355. McKay SM, Angulo-Barroso RM (2006) Longitudinal assessment of leg motor activity and sleep patterns in infants with and without Down syndrome. Infant Behav Dev 29:153–168
356. Cotton S, Richdale A (2006) Brief report: parental descriptions of sleep problems in children with autism, Down syndrome, and Prader–Willi syndrome. Res Dev Disabil 27:151–161
357. Colas D, London J, Gharib A, Cespuglio R, Sarda N (2004) Sleep-wake architecture in mouse models for Down syndrome. Neurobiol Dis 16:291–299
358. Stewart LS, Persinger MA, Cortez MA, Snead OC 3 rd (2007) Chronobiometry of behavioral activity in the Ts65Dn model of Down syndrome. Behav Genet 37:388–398
359. Goldstone AP (2004) Prader–Willi syndrome: advances in genetics, pathophysiology and treatment. Trends Endocrinol Metab 15:12–20
360. Nixon GM, Brouillette RT (2002) Sleep and breathing in Prader–Willi syndrome. Pediatr Pulmonol 34:209–217
361. Vgontzas AN, Kales A, Seip J, Mascari MJ, Bixler EO, Myers DC, Vela-Bueno AV, Rogan PK (1996) Relationship of sleep abnormalities to patient genotypes in Prader–Willi syndrome. Am J Med Genet 67:478–482
362. Kozlov SV, Bogenpohl JW, Howell MP, Wevrick R, Panda S, Hogenesch JB, Muglia LJ, Van Gelder RN, Herzog ED, Stewart CL (2007) The imprinted gene Magel2 regulates normal circadian output. Nat Genet 39:1266–1272
363. Braam W, Smits MG, Didden R, Curfs LM (2008) Melatonin is effective in treating sleep problems in Angelman syndrome but problems in metabolising melatonin may be part of the Angelman phenotype. J Intellect Disabil Res 52:814
364. Colas D, Wagstaff J, Fort P, Salvert D, Sarda N (2005) Sleep disturbances in Ube3a maternal-deficient mice modeling Angelman syndrome. Neurobiol Dis 20:471–478
365. Hagerman RJ (1996) Fragile X syndrome: diagnosis, treatment, and research, 2nd edn. Johns Hopkins University Press, Baltimore, pp 3–87

366. Gould EL, Loesch DZ, Martin MJ, Hagerman RJ, Armstrong SM, Huggins RM (2000) Melatonin profiles and sleep characteristics in boys with fragile X syndrome: a preliminary study. Am J Med Genet 95:307–315
367. Verkerk AJ, Pieretti M, Sutcliffe JS, Fu YH, Kuhl DP, Pizzuti A, Reiner O, Richards S, Victoria MF, Zhang FP et al (1991) Identification of a gene (FMR-1) containing a CGG repeat coincident with a breakpoint cluster region exhibiting length variation in fragile X syndrome. Cell 65:905–914
368. Zhang J, Fang Z, Jud C, Vansteensel MJ, Kaasik K, Lee CC, Albrecht U, Tamanini F, Meijer JH, Oostra BA, Nelson DL (2008) Fragile X-related proteins regulate mammalian circadian behavioral rhythms. Am J Hum Genet 83:43–52
369. Malow BA (2004) Sleep disorders, epilepsy, and autism. Ment Retard Dev Disabil Res Rev 10:122–125
370. Melke J, Goubran Botros H, Chaste P, Betancur C, Nygren G, Anckarsater H, Rastam M, Stahlberg O, Gillberg IC, Delorme R, Chabane N, Mouren-Simeoni MC, Fauchereau F, Durand CM, Chevalier F, Drouot X, Collet C, Launay JM, Leboyer M, Gillberg C, Bourgeron T (2008) Abnormal melatonin synthesis in autism spectrum disorders. Mol Psychiatry 13:90–98
371. Ogawa S, Kwon CH, Zhou J, Koovakkattu D, Parada LF, Sinton CM (2007) A seizure-prone phenotype is associated with altered free-running rhythm in Pten mutant mice. Brain Res 1168:112–123
372. Cruciani F, Trombetta B, Labuda D, Modiano D, Torroni A, Costa R, Scozzari R (2008) Genetic diversity patterns at the human clock gene period 2 are suggestive of population-specific positive selection. Eur J Hum Genet 16:1526–1534
373. Ciarleglio CM, Ryckman KK, Servick SV, Hida A, Robbins S, Wells N, Hicks J, Larson SA, Wiedermann JP, Carver K, Hamilton N, Kidd KK, Kidd JR, Smith JR, Friedlaender J, McMahon DG, Williams SM, Summar ML, Johnson CH (2008) Genetic differences in human circadian clock genes among worldwide populations. J Biol Rhythms 23:330–340
374. Hawkins GA, Meyers DA, Bleecker ER, Pack AI (2008) Identification of coding polymorphisms in human circadian rhythm genes PER1, PER2, PER3, CLOCK, ARNTL, CRY1, CRY2 and TIMELESS in a multi-ethnic screening panel. DNA Seq 19:44–49
375. Pedrazzoli M, Louzada FM, Pereira DS, Benedito-Silva AA, Lopez AR, Martynhak BJ, Korczak AL, Koike Bdel V, Barbosa AA, D'Almeida V, Tufik S (2007) Clock polymorphisms and circadian rhythms phenotypes in a sample of the Brazilian population. Chronobiol Int 24:1–8
376. Carpen JD, von Schantz M, Smits M, Skene DJ, Archer SN (2006) A silent polymorphism in the PER1 gene associates with extreme diurnal preference in humans. J Hum Genet 51:1122–1125
377. Ebisawa T, Uchiyama M, Kajimura N, Mishima K, Kamei Y, Katoh M, Watanabe T, Sekimoto M, Shibui K, Kim K, Kudo Y, Ozeki Y, Sugishita M, Toyoshima R, Inoue Y, Yamada N, Nagase T, Ozaki N, Ohara O, Ishida N, Okawa M, Takahashi K, Yamauchi T (2001) Association of structural polymorphisms in the human period3 gene with delayed sleep phase syndrome. EMBO Rep 2:342–346
378. Archer SN, Robilliard DL, Skene DJ, Smits M, Williams A, Arendt J, von Schantz M (2003) A length polymorphism in the circadian clock gene Per3 is linked to delayed sleep phase syndrome and extreme diurnal preference. Sleep 26:413–415
379. Viola AU, Archer SN, James LM, Groeger JA, Lo JC, Skene DJ, von Schantz M, Dijk DJ (2007) PER3 polymorphism predicts sleep structure and waking performance. Curr Biol 17:613–618
380. Benedetti F, Radaelli D, Bernasconi A, Dallaspezia S, Falini A, Scotti G, Lorenzi C, Colombo C, Smeraldi E (2008) Clock genes beyond the clock: CLOCK genotype biases neural correlates of moral valence decision in depressed patients. Genes Brain Behav 7:20–25
381. Sookoian S, Gemma C, Gianotti TF, Burgueno A, Castano G, Pirola CJ (2008) Genetic variants of Clock transcription factor are associated with individual susceptibility to obesity. Am J Clin Nutr 87:1606–1615

382. Johansson C, Willeit M, Smedh C, Ekholm J, Paunio T, Kieseppa T, Lichtermann D, Praschak-Rieder N, Neumeister A, Nilsson LG, Kasper S, Peltonen L, Adolfsson R, Schalling M, Partonen T (2003) Circadian clock-related polymorphisms in seasonal affective disorder and their relevance to diurnal preference. Neuropsychopharmacology 28:734–739
383. Woon PY, Kaisaki PJ, Braganca J, Bihoreau MT, Levy JC, Farrall M, Gauguier D (2007) Aryl hydrocarbon receptor nuclear translocator-like (BMAL1) is associated with susceptibility to hypertension and type 2 diabetes. Proc Natl Acad Sci U S A 104:14412–14417
384. Peng ZW, Chen XG, Wei Z (2007) Cryptochrome1 maybe a candidate gene of schizophrenia. Med Hypotheses 69:849–851
385. Nievergelt CM, Kripke DF, Remick RA, Sadovnick AD, McElroy SL, Keck PE Jr, Kelsoe JR (2005) Examination of the clock gene Cryptochrome 1 in bipolar disorder: mutational analysis and absence of evidence for linkage or association. Psychiatr Genet 15:45–52
386. Simerly R (2006) Feeding signals and drugs meet in the midbrain. Nat Med 12:1244–1246
387. Van Someren EJ (2000) Circadian rhythms and sleep in human aging. Chronobiol Int 17:233–243
388. Dijk DJ, Duffy JF, Czeisler CA (2000) Contribution of circadian physiology and sleep homeostasis to age-related changes in human sleep. Chronobiol Int 17:285–311
389. Xu Y, Toh KL, Jones CR, Shin JY, Fu YH, Ptacek LJ (2007) Modeling of a human circadian mutation yields insights into clock regulation by PER2. Cell 128:59–70
390. McClung CA (2007) Circadian genes, rhythms and the biology of mood disorders. Pharmacol Ther 114:222–232
391. Ferreira JJ, Desboeuf K, Galitzky M, Thalamas C, Brefel-Courbon C, Fabre N, Senard JM, Montastruc JL, Sampaio C, Rascol O (2006) Sleep disruption, daytime somnolence and "sleep attacks" in Parkinson's disease: a clinical survey in PD patients and age-matched healthy volunteeers. Eur J Neurol 13:209–214
392. Fleming SM, Chesselet MF (2006) Behavioral phenotypes and pharmacology in genetic mouse models of Parkinsonism. Behav Pharmacol 17:383–391
393. Barnard AR, Nolan PM (2008) When clocks go bad: neurobehavioural consequences of disrupted circadian timing. PLoS Genet 4:e1000040

Chapter 11
Systems Biology and Modeling of Circadian Rhythms

Thomas d'Eysmond and Felix Naef

The mathematics of circadian rhythms has been an intensive and productive field in mathematical biology, starting way before the molecular era of modern biology. These earlier approaches illustrate very nicely how abstract reasoning can lead to great insight into the structure and function of molecular clocks. For example, phase singularities, which must exist based on general properties of oscillating systems (limit-cycles), are at the heart of circadian biology, leading to strong predictions on measurable quantities such as phase shifting curves. These concepts are presented in the excellent reference book by Arthur T. Winfree [1] bearing the eloquent title "The geometry of biological time". The reason why biomolecular cycles are very attractive for modelers is that they represent one of the simplest dynamical instabilities found in many areas of biology. Cycles bring with them the definition of the *period*, which plays a central role in circadian biology as it is linked with the periodicity of our earthly habitat. In recent years, thanks to the great wealth of genetic and biochemical data available from studies model organisms, modeling circadian oscillators has evolved toward including more and more molecular details using the language of chemical kinetics. This has enabled to gain a deeper insight into the function of the intricate genetic networks controlling circadian oscillators, and has even allowed making explicit quantitative predictions about the role of individual biochemical processes in controlling different aspects of the cycles. For example, a large amount of work has been dedicated to the sensitivity of the period to changes in model parameters, or to understand the mechanism behind a defining property of circadian clocks known as temperature compensation, whereby the period is relatively stable in function of the temperature. Another layer of dynamical richness is introduced by the coupling of autonomous oscillators to external Zeitgebers such as light or temperature signals, which leads to the phenomenon of entrainment or mode locking. This interaction with the environment is fundamental as circadian rhythms must realign constantly with the environment in order to cope with noise, seasonal variations in photoperiod length,

F. Naef(✉)
School of life sciences, Ecole Polytechnique Fédérale de Lausanne (EPFL), CH-1015, Lausanne, Switzerland, Swiss Institute of Bioinformatics (SIB), Lausanne, Switzerland

U. Albrecht (ed.), *The Circadian Clock*, Protein Reviews 12,
DOI 10.1007/978-1-4419-1262-6_11,

or shifts as experienced in jetlag. Finally, these questions must be discussed in the light of the coupling between autonomous oscillators from electrical or chemical signals, which leads to rich phenomenology such as collective synchronization.

This short chapter is concerned with the application of concepts from nonlinear oscillators to circadian biology, and with systems biology approaches that attempt the development of detailed and predictive mathematical models for circadian clocks. Other systems biology approaches such as the study of these cycles, using comprehensive functional genomics datasets will not be discussed here. The field of modeling circadian oscillators is rooted in solid landmark studies and is now rapidly expanding. Many excellent reviews have been written on the topic [2, 3], and therefore we will present only a selection of some of the most important concepts and related findings.

11.1 Limit Cycle Oscillators and Phase Models

Even before the identification of the molecular basis for circadian clocks, strong evidence suggested that circadian rhythms are the manifestation of an underlying stable limit-cycle attractor. Such dynamical behavior represents a closed orbit (trajectory) in a nonlinear dynamical system (a system of ordinary differential equations or ODEs), with the additional property that the nearby trajectories, or deviations from perturbations, are attracted back to the cycle (Fig 11.1a). These properties allow us to parameterize the phase space around the cycle with a phase variable (an angle). The surfaces of constant phase are called isochrons and intersect in the phase singularity surface (Fig. 11.1b). Even if the biophysical events and trajectories live in a high-dimensional (the number of species) space, we are mostly accustomed to think of limit cycles in two dimensions. To remind the general case, we emphasize in the figure legend the dimensionality of the various surfaces. Perturbations that kick the trajectory near the phase singularity surface have the effect of resetting the phase of the oscillation to essentially random values (after the trajectory has returned to the attractor). In a population of oscillators, this leads to desynchronyzation, and two recent experiments in single cells have shown this property [4, 5].

11.1.1 Phase Response Curves

Perhaps the strongest evidence that circadian cycles are limit-cycle oscillators is provided by phase response or phase shift experiments. These have been studied extensively in a variety of models (http://www.cas.vanderbilt.edu/johnsonlab/prcatlas/index.html), but we give here just a few references [1, 6, 7]. In these experiments, the system is perturbed off its attractor, for example, by a light or temperature pulse. If the perturbation is short, the net effect is that the system almost instantaneously jumps from one isochron to another one. The relationship between the final (postperturbation) and initial (preperturbation) defines the phase response curve (PRC, Fig. 11.1c–d).

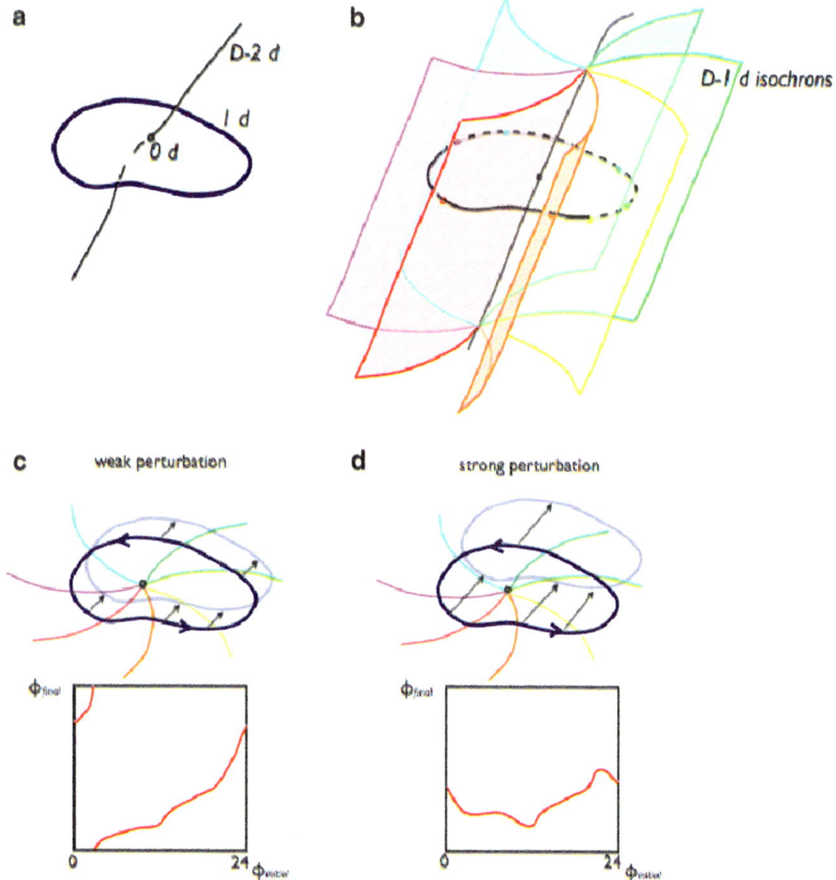

Fig. 11.1 Limit-cycle, isochrons, and phase-response curves. (**a**) Limit-cycle trajectory (*blue*, 1-dimensional) in a model with D=3 dimensions. The (unstable fixed point (zero-dimensional) and phase singularity surface (D-2=1-dimensional) are shown. (**b**) The isochrones (D-1 dimensional surfaces), shown in different colors, intersect at the phase singularity surface (a line in D=3). (**c**, **d**) Phase responses; weak perturbation or type 1 (**c**), and strong or type 0 (**d**). The final phase φ_f is plotted against the initial phase φ_i (before the perturbation). The limit-cycle (*blue*, here in D=2) is shown together with the perturbed initial conditions (*light blue*) that determine the final phase φ_f

We distinguish two qualitatively different regimes: that of weak perturbations, in which all final phases are accessible (also termed type 1 PRC, Fig. 11.1c), and a situation of strong perturbation, in which only a subinterval of phases is reached (Fig. 11.1d). The difference can be conveniently visualized geometrically: in one case, the set of perturbed conditions still encloses the phase singularity, while in the other it does not (Figs. 11.1c–d). Phase response curves have practical relevance: its knowledge allows to optimally design countermeasures to align a shifted circadian pacemaker to a new schedule, for example, after trans-meridian travels.

11.1.2 Coupled oscillators and synchronization

It has been shown that coupling in the central pacemaker, located in the suprachiasmatic nucleus (SCN) in the brain, is important for maintaining accurate rhythms. This coupling among neurons depends on synaptic transmission [8–10] and can thus be blocked chemically [11]. The benefit of coupled cells, when it comes to timing accuracy is quite intuitive. Coupled oscillators have attracted huge attention from modelers, and one of the generic emerging properties is the phenomenon of collective synchronization, by which strong enough coupling can overcome heterogeneity (or noise) in a population. The conceptual framework for synchronization uses simple phase oscillators, as in the Kuramoto models [12] and variations thereof [13]. In these, it is possible to derive the strength of the interaction that is required to synchronize a population of cells as a function of other parameters, such as the variability in the cell frequencies. Recently such models have been used to estimate coupling strength in prokaryotic [14, 15] and eukaryotic oscillators [16].

11.2 Connecting Experiments and Models in Circadian Biology

The circadian clock has become a prime example of a system in which model predictions can be tested against quantitative data, in particular, period and phase data. We give here a brief and incomplete overview of the types of experiments that provide useful data in conjunction with models. Most of these data are designed to probe the cycle and its responses to perturbations by tracking cycle amplitude, period, and phase. We discuss which types of experiments are most useful for model building and calibration, as well as model testing or validation.

11.2.1 Period Phenotypes and Sensitivities

The most common type of data is from genetic perturbation and their measured effect on period. In rare cases, mutations can lead to arrythmicity, but in most cases particularly in higher organisms where most circadian genes have paralogs with at least partially redundant function, involvement in clock function is detected as a period phenotype, either short or long. In mammals, only one gene (Bmal1) is currently known to cause arrythmicity. The first identified mammalian clock mutation was the short-period tau mutant [17], which was later found to encode a casein kinase CK1 [18] that is involved in the FASPS syndrome in humans [19, 20]. Period phenotypes have been exploited most extensively in plants to calibrate detailed kinetic ordinary differential equation models [21]. Recently, perturbation experiments using siRNA [22, 23] or libraries of chemical compounds [24] have allowed more precise measurements of period phenotypes and revealed new kinases involved in period control in human cells. Global perturbation can also give insight into clock mechanisms. An interesting experiment studied oscillator period under a decreased transcription rate, using drugs like α-amanitin, which resulted in an unexpected shortening of the period [25]. Period phenotypes are powerful in conjunction with models, for example, for discriminating between candidate models. Several theoretical studies have systematically investigated the dependency of period on network structure and model parameters [26–28].

11.2.2 Time Series Data

Another complementary strategy is to calibrate models directly from time series data. In mammals, mRNA and protein time course data were used to calibrate a high-dimensional model [2], but lack of precise data, particularly for proteins, remains an issue. While with current time resolution, typically 4 h, this boiled down to essentially period and phase information, it is possible that increased sampling will allow in the future to discriminate models, based on the waveforms of the profiles.

11.2.3 Single Cell Properties

The first single-cell measurements were done using GFP [7] and luminescence reporter techniques [29] and opened the door to a detailed real-time analysis of cell autonomous oscillators and their fluctuation profiles in the cellular milieu. One theoretically robust method to compare the data with models is then the use of auto-correlation analysis and their decay properties [16]. Since, more extensive studies of individual oscillations have revealed the difference between mutants and tissues [24]. An important tool consisting in adding melanopsin to fibroblast cells was recently developed [4, 5], making fibroblast cells entrainable by short light pulses. These studies have allowed to measure accurate PRCs and shown how perturbations by pulses at the appropriate phases can efficiently scatter phases and lead to desynchronyzation. Moreover, transitions from type 1 to type 0 PRCs in function of the strength of the stimulus were observed as predicted (Fig. 11.1) [5].

11.3 Explicit Kinetic Models and Systems Biology

11.3.1 From Conceptual Models to Detailed Kinetic Models

In the pioneering years of chronobiology, few details were known experimentally on the circadian clock mechanism. Hence, the level of detail in mathematical models was restricted to the essential features of circadian oscillations. Typically, this consisted in modeling the negative feedback exerted by a protein on the expression of its own gene [30, 31]. Consequently, their use is to describe circadian oscillations

Table 11.1 Models discussed in this section.

	Leloup &Goldbeter	Forger &Peskin	Becker-Weimann
Year of Publication	2003	2003	2004
# of variables	16 or 19 (extended)	74	7
# of parameters	53 or 64 (extended)	36	24
Fit of parameters to experimental data	No	Yes	No

regardless of the species in consideration by renaming the variables and eventually rescaling the parameters. Usually, parameters are chosen such that the system admits limit-cycle solutions. Model parameters can have known constraints, but are often tuned in order to achieve an agreement on observables such as the period length or amplitude. Recent years have seen the emergence of more detailed and high-dimensional models that explicit biomolecular findings such as novel clock components or phosphorylation states in a number of model organisms (reviewed in). In essence, many of these models can still be viewed as delayed feedback oscillators, but their properties can be compared theoretically in terms fulfilling certain optimality principles [32]. We will focus here on the discussion of a few models for the mammalian circadian oscillator, summarized in Table 11.1.

11.3.2 Explicit Kinetic and Higher Dimensional Models

The richness of mammalian timing is that it consists in a hierarchical clock network, with a central SCN pacemaker subordinating local timing to slave oscillators in peripheral tissues and organs. Often a problem that modelers face is that while parts of the regulatory network were characterized at the suited level of detail, some parts and interactions remain unresolved. At this point, the model can sometimes also help to discriminate between possibilities. Here, we will discuss three detailed mammalian models, and for each, summarize their original purpose, level of detail, and their degree of agreement with the current experimental knowledge.

Before we discuss the specifics, we mention a recent experiment that helped to prioritize certain models over others [25]. In the latter, bioluminescence signal of Bmal1-luc NIH3T3 mouse fibroblasts was measured under conditions of reduced transcription rates (up to about a third of the transcription rates in the untreated cells) in wild-type and Per1 knockout cells. Upon reduction of the transcription rates, it was observed that sustained oscillations are maintained, but that period shortens and amplitude decreases. Furthermore, the period shortening was absent in Per1 knockout cells. This experiment was accompanied with a simple theoretical analysis of a few models, in which transcription rates were decreased *in silico*. It was found that the observed period shortening was consistently reproduced in some models but not in others; models implementing delayed negative feedbacks "a la

Goodwin" were in better agreement [25]. This suggested that while increasing the modeling detail level helps to more specifically address certain precise experimental questions, it can at the same time lead to a loss of predictive power for other aspects. Each model has a limited range of applicability, depending on the questions that the modelers tried to address with it.

11.3.3 The Leloup and Goldbeter Model

This model consisted in a first try toward a more detailed model including the main players in the mammalian clock networks and some of their main modifications, e.g., phosphorylations [33]. It comes in a reduced and extended version and explicitly states the interactions between the genes Per, Cry Bmal1, and Clock in the reduced, and additionally, Rev-Erbα in the extended version. Even if several paralogs of the Per and Cry genes are known to be involved in the circadian clock mechanism, only a single copy was included for simplicity. In both versions, the authors chose model parameters within a physiological range that led to self-sustained oscillations with a period of about 24 hours. They speculate that these are adapted to describe circadian oscillations from neurons in the SCN, the *master clock*, while damped oscillations can be obtained in a range of parameters below the Hopf bifurcation and should be used to describe circadian rhythms in peripheral tissues, the slave clocks. However, it is now known that peripheral oscillators are mostly self-sustained [7]. An interesting prediction of this model is the contribution of redundant loops to circadian oscillations in the same clock. Indeed when the indirect autoregulation of Per and Cry expression is suppressed, simulations show that circadian oscillations are still possible through the negative autoregulation of Bmal1 expression through Rev-Erbα. From an evolutionary point of view, this suggests that the more complex mammalian circadian clock mechanism has evolved toward a clock with several intertwined regulatory loops in order to increase the robustness of circadian rhythms to mutations or even physiological perturbations. The model also can incorporate a light-induced expression of the Per gene to simulate entrainment by a light-dark cycle. The reduced model exhibits period shortening under reduced transcription rate reduction, in agreement with recent findings [25], despite its narrow oscillatory regime for some of the proposed parameter sets. The extended model, however has the opposite behavior, namely period lengthening upon transcription rate reduction. This indicates that this trend is sensitive to parameter modifications and that the somewhat arbitrary choice of parameters with almost no constraints can lead to a variety of outcomes.

11.3.4 The Forger and Peskin Model

This model is currently one of the most detailed mammalian models. It was built to take advantage of the vast quantity of experimental knowledge on the mammalian circadian clock. In particular, clock gene duplications, such as for Cry and Per, were taken into account. For instance, this permits to easily simulate knockout experiments. Another originality of this model is that model parameters were found from experimental data with an optimization procedure called 'coordinate search method'. The result of the fit can be further validated by experimental observations, for instance, the protein CRY1 degrades more quickly than CRY2 and the initial phosphorylation rate of PER1 is higher than PER2, the degradation rates of phosphorylated PER proteins are higher than of unphosphorylated ones. The success of the model is to accurately predict the phase of entrainment, amplitude of oscillation, and shape of time profiles of clock mRNAs and proteins. It also exhibits robustness upon parameter variations and mutations.

The model also agrees on particular differences observed experimentally between included duplicated clock genes. For instance, more PER2 is found in the nucleus than PER1, because PER2 is expressed at higher concentrations and requires less phosphorylation to enter the nucleus. However, a significant difference is found between the fitted and measured kinase concentrations. Compared with the experiment [25], the period tends to increase upon reduction of the transcription rates and abruptly drops at very low transcription rates. The calibration of parameters found from experiments suggests that the structure of the model itself, despite its high complexity of over seventy parameters, does not include the details of this transcription rate compensation mechanism.

11.3.5 The Becker-Weimann Model

The objective was to use the minimal number of variables to reflect essential qualitative features of mammalian circadian oscillators and thus characterize the differential roles of negative and positive feedback loops. The genes included in the model correspond to those that are crucial for circadian oscillations. The available results from knockout experiments are used to identify these genes: Per2, Cry (one variable stands for both known paralogs), and Bmal1. The latter is the key gene responsible for the negative feedback mechanism. Bmal1 alone is described by four of the seven model variables, standing for the level of mRNA, and levels of the protein present in the cytosolic and nuclear cellular compartments and in the transcriptionally activated state. The gene Rev-Erbα has an implicit presence through the network structure.

In particular, the authors show that when reducing the positive feedback strength, circadian oscillations can be rescued if negative feedback strength is also reduced. This fact has been observed experimentally [34] and can be validated by the model by varying several parameters gating the positive and negative feedback loops. Interestingly, upon reduction of the transcription rate to a third of the nominal rate, the model has a slight period shortening of several hours in agreement with the experimental result discussed earlier [25]. Furthermore its relative low dimensionality is an advantage for *in silico* studies of intercellular coupling when the circadian gene expression activity has to be simulated for huge numbers of interacting cells. It has, for instance, been used to describe damped SCN neurons [35] and to show that a sufficiently large number of such oscillators with a varying period arranged in a structure mimicking the one in the SCN could spontaneously synchronize and oscillate in a self-sustained manner.

11.3.6 Conclusion

The attractiveness of oscillatory phenomena for theorists (essentially due to their simplicity) has attracted modelers to circadian biology since its early days. Thus, the field has a long history of pioneering works that laid the conceptual basis and the structure of the circadian problem. The current challenge for circadian biology in the postgenomic and systems biology era is to fully characterize the molecular basis and networks that implement these functions. This will ultimately reveal how a key component of biological timing is implemented and how it constitutes and is an integral part of the cellular machinery. We feel that the time for making significant progress in that direction has come, as illustrated by some of the recent steps toward this goal. In our opinion, one key issue for circadian modeling will be to keep the right balance between conceptual and detailed modeling. While it is tempting to seek answers in the latter, experience has already shown the limitations of overly detailed and underdetermined models. We believe that conceptual models, obtained through the right simplifications, will continue to play a prominent role. Not only do these tend to be more interesting and rewarding for the modeler, but they also provide in the end a better tool for making useful abstractions that will eventually lead the field of systems chronobiology forward.

References

1. Winfree AT (2001) The geometry of biological time. Springer
2. Forger DB, Dean DA II, Gurdziel K, Leloup JC, Lee C, Von Gall C, Etchegaray JP, Kronauer RE, Goldbeter A, Peskin CS et al (2003) Development and validation of computational models for mammalian circadian oscillators. OMICS 7:387–400
3. Roenneberg T, Chua EJ, Bernardo R, Mendoza E (2008) Modelling biological rhythms. Curr Biol 18:R826–R835
4. Pulivarthy SR, Tanaka N, Welsh DK, De Haro L, Verma IM, Panda S (2007) Reciprocity between phase shifts and amplitude changes in the mammalian circadian clock. Proc Natl Acad Sci USA 104:20356–20361
5. Ukai H, Kobayashi TJ, Nagano M, Masumoto KH, Sujino M, Kondo T, Yagita K, Shigeyoshi Y, Ueda HR (2007) Melanopsin-dependent photo-perturbation reveals desynchronization underlying the singularity of mammalian circadian clocks. Nat Cell Biol 9:1327–1334
6. Myers MP, Wager-Smith K, Rothenfluh-Hilfiker A, Young MW (1996) Light-induced degradation of TIMELESS and entrainment of the Drosophila circadian clock. Science 271:1736–1740
7. Nagoshi E, Saini C, Bauer C, Laroche T, Naef F, Schibler U (2004) Circadian gene expression in individual fibroblasts: cell-autonomous and self-sustained oscillators pass time to daughter cells. Cell 119:693–705

8. Liu C, Reppert SM (2000) GABA synchronizes clock cells within the suprachiasmatic circadian clock. Neuron 25:123–128
9. Maywood ES, Reddy AB, Wong GK, O'Neill JS, O'Brien JA, McMahon DG, Harmar AJ, Okamura H, Hastings MH (2006) Synchronization and maintenance of timekeeping in suprachiasmatic circadian clock cells by neuropeptidergic signaling. Curr Biol 16:599–605
10. Ohta H, Yamazaki S, McMahon DG (2005) Constant light desynchronizes mammalian clock neurons. Nat Neurosci 8:267–269
11. Yamaguchi S, Isejima H, Matsuo T, Okura R, Yagita K, Kobayashi M, Okamura H (2003) Synchronization of cellular clocks in the suprachiasmatic nucleus. Science 302:1408–1412
12. Kuramoto Y (1984) Chemical Oscilations, Waves and Turbulence. Springer-Verlag, Berlin 164
13. Rougemont J, Naef F (2006) Collective synchronization in populations of globally coupled phase oscillators with drifting frequencies. Phys Rev E 73:1–5
14. Mihalcescu I, Hsing W, Leibler S (2004) Resilient circadian oscillator revealed in individual cyanobacteria. Nature 430:81–85
15. Amdaoud M, Vallade M, Weiss-Schaber C, Mihalcescu I (2007) Cyanobacterial clock, a stable phase oscillator with negligible intercellular coupling. Proc Natl Acad Sci USA 104:7051–7056
16. Rougemont J, Naef F (2007) Dynamical signatures of cellular fluctuations and oscillator stability in peripheral circadian clocks. Mol Syst Biol 3:93
17. Ralph MR, Menaker M (1988) A mutation of the circadian system in golden hamsters. Science 241:1225–1227
18. Lowrey PL, Shimomura K, Antoch MP, Yamazaki S, Zemenides PD, Ralph MR, Menaker M, Takahashi JS (2000) Positional syntenic cloning and functional characterization of the mammalian circadian mutation tau. Science 288:483–492
19. Toh KL, Jones CR, He Y, Eide EJ, Hinz WA, Virshup DM, Ptácek LJ, Fu YH (2001) An hPer2 phosphorylation site mutation in familial advanced sleep phase syndrome. Science 291:1040–1043
20. Xu Y, Padiath QS, Shapiro RE, Jones CR, Wu SC, Saigoh N, Saigoh K, Ptácek LJ, Fu YH (2005) Functional consequences of a CKIdelta mutation causing familial advanced sleep phase syndrome. Nature 434:640–644
21. Locke JC, Southern MM, Kozma-Bognár L, Hibberd V, Brown PE, Turner MS, and Millar AJ (2005) Extension of a genetic network model by iterative experimentation and mathematical analysis. Mol Syst Biol 1, 2005.0013
22. Maier B, Wendt S, Vanselow JT, Wallach T, Reischl S, Oehmke S, Schlosser A, Kramer A (2009) A large-scale functional RNAi screen reveals a role for CK2 in the mammalian circadian clock. Genes Dev 23:708–718
23. Baggs JE, Price TS, DiTacchio L, Panda S, Fitzgerald GA, Hogenesch JB (2009) Network features of the mammalian circadian clock. PLoS Biol 7:e52
24. Hirota T, Lewis WG, Liu AC, Lee JW, Schultz PG, Kay SA (2008) A chemical biology approach reveals period shortening of the mammalian circadian clock by specific inhibition of GSK-3beta. Proc Natl Acad Sci USA 105:20746–20751
25. Dibner C, Sage D, Unser M, Bauer C, d'Eysmond T, Naef F, Schibler U (2009) Circadian gene expression is resilient to large fluctuations in overall transcription rates. EMBO J 28:123–134
26. Stelling J, Gilles ED, Doyle FJ (2004) Robustness properties of circadian clock architectures. Proc Natl Acad Sci USA 101:13210–13215
27. Bagheri N, Stelling J, Doyle FJ (2007) Quantitative performance metrics for robustness in circadian rhythms. Bioinformatics 23:358–364
28. Kurosawa G, Iwasa Y (2005) Temperature compensation in circadian clock models. J Theor Biol 233:453–468

29. Welsh DK, Yoo SH, Liu AC, Takahashi JS, Kay SA (2004) Bioluminescence imaging of individual fibroblasts reveals persistent, independently phased circadian rhythms of clock gene expression. Curr Biol 14:2289–2295
30. Goldbeter A (1995) A model for circadian oscillations in the Drosophila period protein (PER). Proc Biol Sci 261:319–324
31. Goodwin BC (1965) Oscillatory behavior in enzymatic control processes. Adv Enzyme Regul 3:425–438
32. Rand DA, Shulgin BV, Salazar D, Millar AJ (2004) Design principles underlying circadian clocks. J R Soc Interface 1:119–130
33. Leloup JC, Goldbeter A (2003) Toward a detailed computational model for the mammalian circadian clock. Proc Natl Acad Sci USA 100:7051–7056
34. Oster H, Yasui A, van der Horst GT, Albrecht U (2002) Disruption of mCry2 restores circadian rhythmicity in mPer2 mutant mice. Genes Dev 16:2633–2638
35. Bernard S, Gonze D, Cajavec B, Herzel H, Kramer A (2007) Synchronization-induced rhythmicity of circadian oscillators in the suprachiasmatic nucleus. PLoS Comput Biol 3:e68

Index

The manufacturer's authorised representative in the EU is Springer Nature Customer Service Centre GmbH, Europaplatz 3, 69115 Heidelberg, Germany. If you have any concerns regarding our products, please contact ProductSafety@springernature.com

Printed and bound by CPI Group (UK) Ltd, Croydon, CR0 4YY
15/07/2026
02167634-0005